CERAMIC MATERIALS IN ARCHAEOLOGY

ISABELLE C. DRUC
BRUCE VELDE

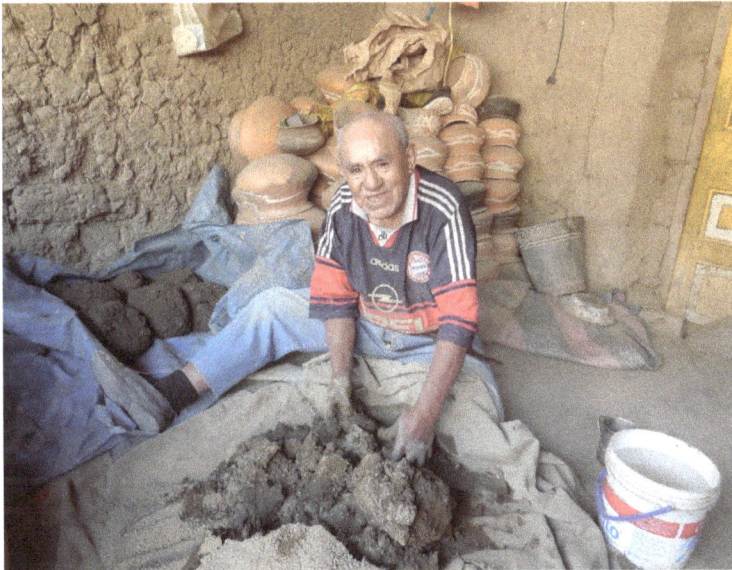

DEEP
EDUCATION PRESS

Blue Mounds, Wisconsin

Ceramic Materials in Archaeology
ISBN 978-1-939755-49-0
Copyright © 2021 by *Deep Education Press*
A subsidiary of Deep University Inc., Wisconsin, USA
Member of Independent Book Publishers Association (IBPA)

The printed version of this book is available in print at most online retailers.
The e-book version is available at https://deepeducationpress.org/index.html

Library of Congress Cataloging-in-Publication Data

Keywords: 1. Archaeology. 2. Ceramic analysis. 3. Applied Sciences handbook. 4. Ceramics. 5. Laboratory manual.

Target audience: students, professors and researchers in archaeology, ethnography, ceramic, geology and ceramic analysis.

Cover back photo: Nicanor Saavedra, Taricá, Ancash, Peru. 2017. Druc ©

To contact the Press: **publisher@deepeducationpress.org**

Content

	Foreword	9
	Acknowledgements	**10**
1	**Introduction**	**11**
1.1	Objectives	11
1.2	Methods	13
1.3	Vocabulary	14
1.4	Structure of the Book	17
2	**Rocks and Minerals**	**20**
2.1	Chemical Elements and Their Chemical Affinities	20
2.2	Major Rock Types	24
2.3	Minerals	26
2.3.1	Mineral Formulae	26
2.3.2	Major Mineral Families	27
2.3.2.1	Silicates	27
2.3.2.1.1	The Silica Minerals and Quartz	28
2.3.2.1.2	Feldspars	29
2.3.2.1.3	Pyroxenes and Amphiboles	30
2.3.2.1.4	Olivine	30
2.3.2.1.5	Micas and Chlorite	31
2.3.2.2	Carbonates	32
2.3.2.3	Oxides	33
2.3.3	Mineral Grain Shapes	33
2.4	Minerals in Rocks	35
2.4.1	Sedimentary Rocks	35
2.3.2	Igneous Rocks	35
2.4.3	Metamorphic Rocks	37
2.4.3.1	Metamorphic Pelites	37
2.4.3.2	Metamorphosed Carbonates	38
2.4.3.3	Metamorphosed Igneous Rocks	38
3	**Clay Minerals and Their Properties**	**41**
3.1	General Concepts	41
3.2	Chemical Constitution of Clay Minerals and Clay Mineral Families	44
3.2.1	Mica-Like Clays (Illite, Celadonite and Glauconite)	45
3.2.2	Smectites	46
3.2.3	Kaolinite Minerals	49
3.2.4	Chlorites and Related Minerals	50
3.2.5	General Chemical Identity of the Clays	51
3.3	Physical Properties of Clay Minerals	51
3.3.1	Clay-Water Mixtures	51
3.3.2	Clay Shapes	53

3.4	Thermal Stability of Clays and Clay-Water Mixtures	55
3.5	Kinetics	58
3.5.1	Grain Size	59
3.5.2	Heating Rate	59
3.6	Summary	60
4	**Origin of Clay Resources**	**63**
4.1	Segregation of the Elements by Weathering	63
4.2	Weathering of Minerals	65
4.3	From Rocks to Soils to Sediments	66
4.3.1	Weathering Profiles	66
4.3.2	Transportation by Water Flow, Grain-Size Sorting	69
4.3.3	Transport and Deposition of Clays	71
4.3.4	Wind Transport	73
4.3.5	Burial of Sediments	74
4.4	Hydrothermal Alteration	74
4.5	Sources of Materials Suitable for Ceramics	75
4.5.1	Clays	75
4.5.2	Non-Clay Grains	76
4.6	Factors Favorable for Clay Formation and Concentration	78
4.7	Sediment Types of Use for Ceramic Making and Their Mineral Characteristics	79
5	**Physical and Chemical Processes of Making Ceramics**	**86**
5.1	Plasticity	86
5.2	Mineral Temper Grains	88
5.2.1	Natural Mineral Grains	88
5.2.2	Decantation and Separation of Natural Mineral Temper Grains	89
5.2.3	Tempering by Mixtures of Source Materials	92
5.2.3.1	Mixtures of Clays and Non-Clay Grains from Different Sources	92
5.2.3.2	Mixtures of Clay Sources	97
5.3	Decorations and Surfaces	99
5.3.1	Surface Smoothing and Polishing	99
5.3.2	Slip	102
5.3.3	Paint	104
5.3.4	Glazes	106
6	**Physical and Chemical Processes in Firing Ceramics**	**110**
6.1	Firing and Material Transformations	110
6.1.1	Variables of Transformation to Make a Ceramic	110
6.1.1.1	The Firing Process: Time and Temperature . .	111
6.1.1.2	Stages of Transformation in Time-Temperature Coordinates	112
6.1.1.3	Paste Composition and Fusing Agents	116
6.1.1.4	Reduction of Iron	118
6.1.1.5	Firing Programs and Surface Color	118
6.1.1.6	Particle Size	120

6.2	Firing Practices	121
6.2.1	Firing on the Ground (Bonfire)	121
6.2.2	Pit Firing	123
6.2.3	Kilns	123
6.3	Summary of Factors in the Formation of a Ceramic Body During Firing	124
6.4	Structure, Porosity and Density of Ceramics: Non-Plastics, Clays and Pores	124
6.4.1	Pores	125
6.4.1.1	Primary Pores	125
6.4.1.2	Secondary Pores	126
6.4.1.3	Microporosity	128
6.4.2	Temper, Material and Firing	128
6.4.3	Thermal Properties of Oriented Clays	130
6.4.4	Hardness	131
6.5	Oxidation-Reduction Effects	132
6.5.1	Oxidation and Reduction Effects on Raw and Fired Materials	133
6.6	Oxidation-Reduction Cycles	136
6.7	Mineral Reactions During Firing	136
6.8	Families of Ceramic Products	139
6.8.1	Earthenware	140
6.8.2	Pottery, Terra Cotta, and Faience	141
6.8.3	Stoneware	142
6.8.4	Porcelain	142
6.9	Summary	143
7	**The Making of Pots**	**147**
7.1	Temper and Tempering	148
7.2	Raw Materials	149
7.2.1	Clay Material	149
7.2.2	Tempering Materials and Methods of Tempering	152
7.2.3	Tempering and Temper Identification	157
7.3	Making a Pot: Physical and Chemical Reactions	161
7.3.1	Needs as a Function of the Object	161
7.3.1.1	Plasticity and the Role of Temper and Non-Plastics	162
7.3.1.2	Drying and Shrinkage	163
7.3.1.3	Material Expansion	164
7.3.1.4	Grain Angularity	165
7.3.2	Paste as Related to Function, Form and Manufacturing Requirements	165
7.3.3	Needs as a Function of Use of the Object	166
7.3.3.1	Durability and Breakage Resistance; Strength and Hardness	167

7.3.3.2	Porosity, Density, Permeability, Impermeability	168
7.3.3.3	Thermal Stress Resistance and Thermal Conductivity	168
7.4	Material Preparation	169
7.5	Forming Techniques	172
7.6	Surface Coatings as Related to the Function of the Ware	177
7.7	Firing and Furnaces	180
7.7.1	Open Fires	181
7.7.2	Pit-Kilns, Semiclosed Structures, Open Kilns	182
7.7.3	Closed Kilns	183
7.7.4	Paste Type and Surface Color Related to the Type of Firing	185
7.8	Summary	187
8	**Optical Observation of Ceramics**	**189**
8.1	Methods: How Can One See a Ceramic Sherd?	189
8.1.1	Computer Scanner	189
8.1.2	Binocular Microscope	190
8.1.3	Portable Digital USB Microscope	190
8.1.4	Petrographic Microscope and Thin Section Preparation	194
8.1.4.1	To Count or Not to Count: Granulometric and Modal Analysis	196
8.1.5	Reflected Light Microscopy	198
8.1.6	Scanning Electron Microscope	199
8.2	Characteristics Observed: What Can One See in a Ceramic Sherd	200
8.2.1	Slip, Glaze or Paint	200
8.2.2	Temper Grains and Clays	202
8.2.3	Temper Grains and Size Distribution	203
8.2.4	Grain Shapes	208
8.2.4.1	Crystal Shapes	208
8.2.4.2	Angularity	209
8.2.4.3	Size Distribution of Temper Grains	212
8.3	Identification of Different Techniques in Paste Preparation	214
8.4	Paste Texture	214
8.5	Summary	217
9	**Ceramics and Archaeology: Case Studies**	**222**
9.1	Yellow Garnets and Trafficking Wine	223
9.2	Iron Age Pottery in Southwestern England and its Geological Sources.	226
9.3	Whole-Sample Compositions of Some Sigillate Ware Produced in France	229
9.3.1	Lezoux Samples	230
9.3.2	Identifying Production Areas	231
9.3.3	Lezoux Coarse or Common Ware	236
9.3.4	Specific Problems of Archaeological Interest Using Sigillate Ware Data	239
9.3.4.1	Arezzo Moulds	239

9.3.4.2	Atevis Workshop	240
9.4	Prehistoric Peru.	242
9.4.1	Petrographic Analysis	243
9.4.2	Modal Analysis	247
9.4.3	Chemical Analysis - XRF and INAA. .	250
9.5	Modern Ceramic Production in the Andes	253
9.5.1	Production Setting	254
9.5.2	Petrographic Analysis	256
9.5.2.1	The Unprepared Black Clay	256
9.5.2.2	The Unprepared Yellow Temper	256
9.5.2.3	The Clay-Temper Mix	258
9.5.3	The Fired Pot Fragment	259
9.5.4	Image Analysis	260
9.6	Clay Characterization by SEM	262
9.7	Determination of Firing Temperature	265
9.8	Mössbauer Spectroscopy	267
10	**Some Current Analysis Methods**	**272**
10.1	Ceramic Analysis: What For and How?	272
10.1.1	Classification	272
10.1.2	The Study of Pottery Technology	273
10.1.3	Provenance Studies	273
10.1.4	Quantitative Studies	274
10.1.5	Use of Qualitative and Quantitative Studies	274
10.1.6	Sample Size, Qualitative and Quantitative Studies vs. Time and Cost Invested	276
10.1.7	Use of Comparative Studies: Sampling Geological Comparative Materials: Compositional Profiles and Petrofacies Maps	278
10.2	Physical and Chemical Analysis Methods	279
10.3	Visual Methods	281
10.3.1.	Binocular Microscope	281
10.3.2	Portable Digital Microscope	282
10.3.3	Petrographic Microscope	282
10.3.4	Computer Scanner and Video Systems	283
10.3.5	Electron Microscopes	284
10.3.5.1	Scanning Electron Microscope, SEM-EDXRF, Scanning Micro-XRF	284
10.3.5.2	Transmission Electron Microscopes (TEM)	285
10.3.5.3	High-Resolution Transmission Electron Microscope (HRTEWM)	285
10.4	Mineral Identification by Non-Optical Methods	286
10.4.1	X-Ray Diffraction (XRD)	286
10.4.2	QEMSCAN	287
10.4.3	Thermogravimetric Analysis (TGA)	287
10.4.4	Differential Thermal Analysis (DTA)	288
10.4.5	Infrared Spectral Analysis (IR) and FTIR	288
10.4.6	Electron Microprobe (EMP)	289

10.4.7	Raman Spectroscopy	290
10.5	Identification of Chemical Elements	292
10.5.1	Laser Ablation Inductively Coupled Mass Spectroscopy (LA-ICP-MS)	292
10.5.2	Whole Sample Analysis	294
10.5.2.1	X-Ray Fluorescence (XRF)	295
10.5.2.2	Portable X-Ray Fluorescence (pXRF) Analysis	295
10.5.2.3	Proton Activated X-Radiation (PIXE)	296
10.5.2.4	Neutron Activation Analysis (NAA)	297
10.5.2.5	Mössbauer Analysis	297
10.6	Provenance based on Zr age and U-Pb isotopes	298
10.7	Age Determinations by Thermoluminescence (TL)	298
10.8	Density, Porosity, and Hardness	299
10.9	Magnetic Analysis	301
11	**How to Acquire the Knowledge to Do the Job**	**306**
11.1	Courses in Geology, Chemistry and Physics	306
11.2	Some Journals, Books, and Laboratories Active in the Field of Interest	309
11.3	Video Documentaries Related to Topics Discussed in This Book	311
	Biographies	**312**

Foreword

This is an updated, revised and extended version of the 1999 edition of the book *Archaeological Ceramic Materials*. There has been many new studies conducted in ceramic analysis since the original publication of this work and advances in the technologies used to gather chemical and mineral information about archaeological ceramics. We felt important to update our book and to offer a color, digital version easily accessible to all. There are also aspects we wanted to develop more, in particular the types of clays and sediments available to potters, their mineralogy, and if possible, how to distinguish them in the ceramic paste, reinforcing even more the geological aspect of the text. Many wonderful and very useful books are now published and much information is available on the Internet, including image data banks for petrographic analysis, class material, discussions, articles, and tutorials for mineral identification. Consequently, this new version of our text is meant to offer one way to use all this material and complement what already exists.

Already present since the ceramic ecology mouvement in the 70s, but now much more in the forefront of data interpretation is the need to consider the potter, her/his work as part of a community of practice, the study of the chaîne opératoire, and the impact of the potter's choice on every step of the production. However, our initial focus on 'materials' remains, simply because there are already a great many books and articles offering this perspective. The understanding of the materials and their transformation during production is just one part of ceramic studies, a step, which does not prevale over the complexity of human impact. Data cannot be interpreted without this in mind.

New techniques, smaller and portable devices, recent developments combining different analytical instruments, are now available for the analysis of archaeological materials, ceramics included. Many investigators are adapting instruments to their needs and we see an explosion of ingenious projects and applications. We attempted to portray this in our discussion on analysis methods, but this is only a glimpse of what is out there.

Our biais, in terms of content and case studies reflect our expertise. We also kept our initial focus on ceramic vessels, rather than the enlarged understanding of the word, which is now more often presented and includes all objects made of hardened clay or clay-related materials. We hope nevertheless that this text will be of use.

In appreciation for the work we have been able to conduct, for the artisans past and present, for the beauty of ceramics and for the people involved in their study.

Isabelle Druc and Bruce Velde
December 2020

Acknowledgements

We are indebted to many people, colleagues, researchers, ceramists, artisans, who helped us shape this book, one way or another. In particular, we would like to thank the colleagues and potters who asked us or allowed us to study their material, from which many examples and illustrations derive. Notably, Kinya Inokuchi, Saitama University, kindly allowed us to use examples taken from our collaborative analysis of ceramics from the ceremonial site of Kuntur Wasi, Peru; similarly, we thank Michelle Young, National Museum of American Indian, for comparative material from the area of Atalla, Peruvian Andes; Milosz Giersz, Warsaw University, for samples from the coastal area of Huarmey, Peru; Alicia Espinosa, Université Paris 1, for samples from the north coast of Peru. The images of excavated ceramics from the Longshan period (ca. 2500-1900 B.C.) site of Liangchengzhen and clays from southeastern Shandong, China were made possible by the ongoing research collaboration with Anne Underhill (Yale University) and Wang Fen and Luan Fengshi (Shandong University). One example of glazed ware comes from the work of Mary Ownby, University of Arizona, who graciously provided us the photographs. Jim Stoltman, a great figure in ceramic petrography who passed away in 2019, allowed us to use a few examples of his work. Our thanks also go to Qingyu Lu, Shandong University, who sent us a great picture illustrating a transmitted light microscope. All other cases illustrated come from personal research with ancient and traditional ceramic productions.

Photos credits

Unless otherwise stated, all graphs, photographs and microphotographs are by the authors.

Abbreviations

μm   micron (10-6 meter or 0.001mm)
ppm   parts per million
ppl   plain polarized light
xpl   crossed polarized light
   (terms referring to the way light passes through a petrographic microscope)

Note

When referring to potters, the use of gender articles (he/she, his/her) is impartial, and may alternate. This does not imply that the trade is only conducted by one or the other.

1 Introduction

1.1
Objectives

The aim of this book is to introduce students in archaeology and perhaps others to the materials that form ancient ceramics. It is by studying the ceramic materials, their nature and function in making the object that their use by potters through the ages can be explained, and this will lead to a better understanding of the potter's behaviour and the influences on his/her ceramic production. This path leads to a use of chemical and petrographic analysis of these natural materials. Many such analyses are currently used and new ones are becoming better known in archaeological study. This book, then, is conceived as an introduction to the origin of the materials which form ceramics in an archaeological context, their selection and use by potters and the eventual detailed, modern analysis of the most abundant material found in archaeology, which is ceramics.

The great abundance of broken pots unearthed in archaeological digs of all ages and their fragmentation has, in turn, led to a large number of carefully labeled inventories. However, stylistic and aesthetic criteria of evaluation and identification (shape, designs on the surface, color) of a potsherd often do not lead to satisfactory identification in the cases of fragmentary materials which are often very similar, without diagnostic criteria. Such "classical" studies often do not answer the usual archaeological questions of: what was it used for, how did it get here, who made it, what techniques were used to produce it? A study of the material composing the ceramics and their physical state can, in many cases, lead to the answers to many of these questions, and the systematic investigation of ceramic materials is now a common part of archaeology as an aid to furthering the understanding of ancient civilizations.

However, the texts relating geology to ceramics and archaeology, designed for non-geologist students are few. This book attempts to explain the *origin* of the components of ceramic materials, which involves the *choice* of these materials by potters as a function of their physical properties, tradition, and use, i.e. the effect of *firing* on ceramic materials. Further, it is necessary to know the means which can be used to *analyze* the ceramics in a post-use context. This book gives elements of the knowledge in geology necessary to understand the materials of ceramics and the information one can obtain by looking at the ceramic material. The relation of geology to ceramics is very important. It is the link to understanding much of the archaeological ceramic material. We wish to give a useful overview of the ceramic arts as it can be used to establish origins, techniques and use of ceramics. These are the goals of much archaeological research. The goals should be related to concrete archaeological problems and examples.

With the current trends of widespread adoption of materials science methods in archaeological research, it has become more frequent to find archaeologists, geologists, physicists and chemists specializing in different, specific analysis methods (and working together). This has led to the creation of the field of investigation called archaeometry. Archaeometry is the study of the chemical and physical properties of archaeological materials. Its goal is to measure, numerically, the objects of archaeological interest. Thus, for those not engaged in ceramic analysis per se, it is more and more important to be able to follow the discourses of these specialists. It is time to open this field to the non-specialist. To move from studies of stylistic trends to those of provenance of materials and production techniques, it is necessary to break down some of the barriers set up by specialists in different fields of analysis to open the general study of ceramic materials to a larger number of interested workers. Specialists will always remain specialists, but it is hoped that a useful level of understanding of current ceramic analysis techniques can be developed among a larger number of archaeologists, so that they can understand which problems can be solved by what machine in the laboratory.

The problems which confront a neophyte are legions in the modern world. In the not-so-distant past, one needed only a sharp pencil and a good memory to master the basic knowledge of ceramic study. Now there are electron microprobes, differential thermal analyses, statistical significance studies and so forth. One wishes to know what tools can be used to solve which problems. It is evident that an archaeologist cannot reasonably be asked to master the panoply of tools with which the physicists and chemists play. And a chemist or physicist may not fully understand the archaeologist's concerns: hence the archaeometricians, who specialize in the use of one or another technique in the material sciences. As with most things, there are some parts which are more important than others. We wish to introduce an interested person into the realm of the possible. This is an understanding of the basis for the existence of raw materials, their combination to make an object and the treatment which they have received are variables which are mastered with reasonable ease. This is our thesis.

The origin of ceramic material, in archaeological contexts, is geology. Clays, sands and many other ingredients in ceramics are of geologic origin, i.e. coming from rocks and their interaction with the elements (rain, wind and so forth). Clays are the major constituents when making a pot. Clays come from the destruction of rocks. Therefore it is important to know how the different elements make clays, what are the leftovers in the clay-making process and what are the chemical traces which can give one a clue to the geographic origin of the clay and temper materials found in the finished product.

Today, clay constituents, fusing agents, oxides and glazes can come from sites thousands of miles from the production site of a ceramic. However, in the past, transportation costs were greater than ceramic resources and products compared to present-day criteria. Clays usually did not travel more than several miles, tens at most if water transport did not intervene, sands and so forth even less. Thus, the potter was more tied to the local resources for his production than pottery factories today. This situation, then, influenced pottery production and distribution. Given this, the potter would have adapted local resources to answer specific needs, functional or aesthetic. The type of resources and how they were used are very important to deciphering the provenance of a ceramic object, and attention should be paid to the production possibilities of at a given site.

1.2
Methods

What, then, do we propose? First, a general idea of the origins of clays and other materials commonly found in ceramic manufacture. Geology and geological processes are important elements of such an analysis. Next, one should have an idea of the steps necessary to form a wet, clay-rich mass into the object of archaeological study, a pot. What are the ingredients that aid in making a wheel-turned bowl, what are those that do not? The next step in the sequence is the firing and decoration process. What happens when a ceramic paste is fired? In understanding this, one can understand the needs and use of a potter of his basic raw materials. This leads to an understanding of the cultural and socio economical contexts of the object and its production.

The last problem is how to analyze the different components of a ceramic object in order to establish the ingredients, their evolution during the ceramic-making process, and the importance this has in identifying the site and period of a ceramic production. We hope to give some insight into these problems and thus promote an interest in the investigation of ceramics from their "inside". By a knowledge of the constituents and the processes of their origin, one can better know the object itself and hence its maker.

One problem of crucial importance to a use of materials science and geological science in the study of archaeological materials is vocabulary. It is of greatest necessity that everyone uses the same words to express the same ideas or concepts. It is even more important to use the same words to describe the same objects. If this is not done, no true cooperation between the disciplines can be made. In general, the latter part of the 20th century has been shown to be one of interdisciplinary sciences combining different knowledge and approaches from different specific fields of research in order to solve a given problem. It is necessary to continue this trend to its greatest efficiency. One discipline can always aid another to solve its problems. The starting point is *vocabulary.*

However, as is the case in most disciplines, there is disagreement within the field of archaeology itself as to how one can express observations and ideas. The problem can be resumed, most likely, as follows: certain terms should be used to describe known or easily knowable facts. These will be observations. They should be based to the greatest extent on previous usage. Those facts, which one must deduce, not being directly accessible, should have, if possible, another expression. However, this is not always easy in the context of available words and names. Another maxim is not to create unnecessary vocabulary. Hence, common names should retain old uses, and new combinations, with qualifiers, can be used in the case of uncertainty.

An example: clays and non-clay particles. Clays and clay minerals have a definite meaning in geology. Engineering science also has a specific use for these terms and their physical meaning. A long period of scientific investigation has produced a reasonable understanding of what a clay particle is and what happens when it reacts with water. However, in ceramics, not only archaeological ceramics, clay means something much more general. It is what is formed into a pot. Here, there is a need to unify or at least define what one means when one says "clay".

In this text, we do not wish to impose a viewpoint on anyone. Our objective is to use the different sciences of geology and ceramics engineering to describe the physical and chemical effects of mineral and physical variability on the production of ceramics. These tools can help an archaeologist to understand how a ceramic object was formed and to find methods to unravel its geographic and technical existence. Hence, we attempt to use terms in a way that is consistent with geological and ceramic engineering as well as archaeological experience.

1.3
Vocabulary

In order to keep things straight, and to allow the reader to follow our arguments, we have provided a list of key words and concepts which is designed to avoid confusion. These are used in this text as a suggestion of general use.

Ceramic: a general term for all objects made from a dominantly silicate material which have been transformed in physical state by heat (firing). In potter's terminology, ceramics refers to a material and not to a specific ware type. It can be tiles, sewer pipes, airplane components or cast decorative pieces as well as pots made on a potter's wheel.

Clay (potter's): the basic ingredient in ceramic manufacture, composed of plastic particles (clay) and natural non-plastic grains. The proportions can vary according to circumstances.

Clay: a natural material composed of about 80% clay minerals and other fine-grained material of different mineral species. This material has a natural grain size of less than 2 μm (micrometers) in diameter.

Clay mineral: particles smaller than 2 μm in diameter which have a specific mineral structure, called a layer or sheet silicate. Their grain size and particle structure give them special physical properties. Clay mineral particles form the plastic phase in most materials used to make ceramics.

Clay resource: the natural occurrence of a clay-rich material (mineral and grain size) from which one can extract a pure clay mineral fraction. Such a material can suit the needs of a potter in its initial state or in a state modified either by extraction or enrichment of a clay mineral fraction or in its combination with another clay resource.

Decantation: the action of separating granulometric fractions (clay- silt-sand) by letting the sediment settle in water. The coarser fraction will be found at the bottom, the finer on top. The clay minerals settle out the last.

Earthenware: ceramics fired at low firing conditions (time and temperature), allowing it to still be porous (the body is not vitrified). The major part of the initial material is not transformed physically by the destruction of mineral structures and their melting. The color of its mass gives it an earthy appearance. This color is due more to the choice of clay resources than to other factors. White earthenware is akin to a faience and displays a white, porous but strong body.

Faience: a term which represents a specific type of ceramic typified by a white porous body coated with a transparent or white (tin-opacified) glaze layer. This method of ceramic decoration was developed in the later Middle Ages in the Islamic areas of the Mediterranean area and then in Europe, especially Italy (Majolica). The equivalent in England was Delftware, while Faience was the term used in France. The term 'white earthenware' is sometimes used to refer to a faience with a stronger glaze and very white body developed later (18th c). Currently, potters use this term to designate a type of well-fired material developed at medium to high states of transformation of the paste.

Glaze: vitrified material (glass substance) on the surface of a ceramic. The material can become glassy upon firing or be applied as a glass powder and remelted upon firing. Glaze materials can be composed of oxides, natural minerals or, in some cases, of clays which melt at low temperatures.

Grit: mineral grains of sand size or coarser granulometry found in the paste of a pot. This term is used to define a temper agent but also a mineral fraction in the paste. Finer grain-sized material in pastes is usually of natural origin, i.e. associated with the clay minerals.

Grog: fired clay or crushed ceramic fragments used as tempering material.

Inclusions: non-plastic materials in the ceramic paste, which can be of diverse origins.

Levigation: action of separating granulometric fractions by driving the sediment through successive inclined pans with running water. The coarser fractions will stay behind, while the finer, clay-rich sediment will be carried further away. The result is a sorted material that the potter can readily use.

Lithics, lithic fragment: a piece of a rock, a multigrain or multimineral fragment found in a ceramic paste.

Non-plastics: non-clay grains (as opposed to the plastic clay mineral material), which include all inclusions larger than the clay particles (> 2 μm). Non-plastic grains can be of mineral or organic nature. They have a tempering function of reducing the plasticity of the clay material.

Paste: the mixture of materials which is used to form the core or body of a ceramic.

Porcelain: a highly transformed ceramic material which has undergone a profound transformation of the materials in the initial paste from prolonged firing at high temperatures (usually above 1300° C). The Asian use of the term does not correspond to that current in Europe. European porcelain is produced by the fusion of the paste to form a new mineral, mullite, which binds the remnant glass into a translucid, durable material. Chinese terminology indicates the fusion of the clay minerals in the paste as well as feldspar grains leaving quartz and a glass structure.

Pottery: in potter's terminology, pottery is a general term for all vessels made of clay resources. They are generally utilitarian and not formed by casting. In other disciplines the term is used at times for an earthen- ware.

Sand: granulometric fraction: all mineral grains between 50 μm to 2 mm in diameter. Most of what are called grits are of sand size.

Silt: granulometric fraction: all grains between 2 and 50 μm in diameter.

Slip: fine clay-rich suspension (in water) applied on the surface of a ceramic, as a surface finish to give a smooth aspect or as a decoration, in the fired or unfired state. A slip can be colored with oxides. A self-slip is a made out of the clay mineral fraction of the paste base used by the potter to make his pot. This clay fraction is brought to the surface by smoothing the clay paste under pressure.

Stoneware: is opposed to faience in potter's terminology, carrying a specific clay quality, color and firing properties. It is a material that turns brown or dark when fired and withstands higher firing conditions than faience, but less than porcelain. It is also recognized by ceramic analysts as a type of ware which is more resistant and of better quality than earthenware. Generally, a significant portion of the temper or grits in the paste is melted in stoneware. It is hard like a stone.

Temper grains: refers to any non-plastic grains in the ceramic paste or in the raw materials used to make a paste. They usually have a structural function besides reducing the plasticity of the clay.

Temper: (archaeological literature) refers to the material added by the potter to modify the plasticity of his clay base, to give it workability, or some specific properties. It can be any material, including clay, sand or other mineral, vegetal, organic, or human-transformed material (crushed bones, shells, animal dung, cotton balls, grog, etc.). This is the definition currently used, even in material sciences. However, considering the physical properties of particles in the fine-grained size fractions, temper refers to any non-plastic, non-clay-sized (and-shaped) grains present naturally in or added to the clay material. The non-clay material is less plastic than clays. Temper decreases plasticity. For potters, traditional and modern, the word temper is rarely used and has no meaning. They use different materials to make their paste, but there is no such distinction as clay and temper.

Tempering material: (1) Any material used to temper a clay (plastic and non-plastic). (2) Natural or added non-plastics in a raw material or ceramic paste reducing the plasticity of the clay base.

Tempering: this word, used alone, refers to the voluntary act of adding material to the clay base. It describes a conscious action done by the potter.

1.4
Structure of the Book

Figure 1.1 shows the object of our study, a fragment of a pot. This particular segment has been sliced across its body to show the internal structure of the sherd. The clay-rich paste of the core of the ceramic contains grits of various sizes, sand to silt. They are roughly equally distributed in the mass of the once plastic material.

The surface of the sherd is asymmetric, the inside shows only the paste material while the outside is layered. This may or may not be glaze or slip, and in our case there are traces of paint on the surface. The surface treatments can then contain materials not found in the core of the ceramic.

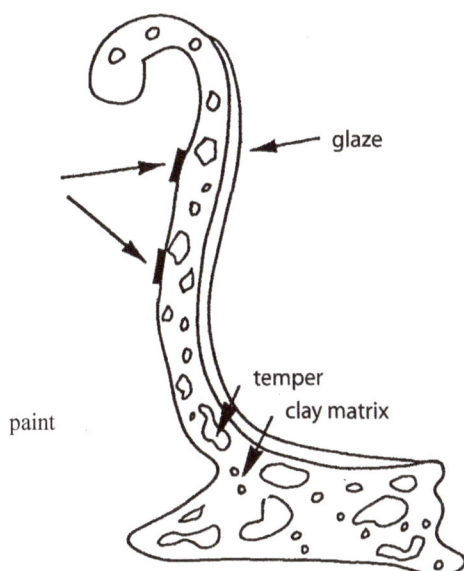

Fig. 1.1. Schematic representation of a ceramic sherd. The internal material is a clay matrix (very fine-grained materials) and larger, non-clay materials called temper. These grains temper the plasticity of the clay materials which are plastic during the forming and preparation of the ceramic. Surface treatments can be paints and slips or glaze in different combinations. These different elements in the ceramic can be analyzed and identified in order to understand the manufacturing process, the provenance of the materials used in the confection of the ceramic and hence the possible geographic area and sometimes epoch of manufacture of the object.

The portions of the sherd we can analyze are (1) the clay material, or the fine-grained part of the paste, (2) the sand and silt grains of the paste and (3) the surface treatment materials such as slip, glaze and paint. Each of these structural and surface elements has a specific origin which can be natural or due to a selection made among natural materials by the potter. In most cases, in the study of archaeological material the origin of the different elements in the structure of the sherd will be determined by geologic phenomena. Rocks and minerals are the basis of the ceramic materials.

Therefore, it is important to have an idea of the geological phenomena which form and distribute the components of ceramics which can be used by a potter.

Once the elements are assembled, it is important to know how they react under the conditions of the ceramic producing cycle. This understanding can help to understand why a given ceramic production was produced with its component materials. Which clays and tempers are useful for which pots? Given the resources, which materials will be used in a given type of production? What are the limits of production, given a certain type of furnace and firing technique? The questions are endless, the responses can mostly be found through a knowledge of the geological aspects of the materials used.

The structure of this book follows several questions:

- What are the rock and minerals common on the surface of the earth? These are the mineral grains usually found in the paste.
- What is a clay mineral? What are the properties which clay gives to a mixture of clay and water which allows the potter to work?
- Where can one find the clay resources necessary to make a pot? What are the places likely to be a clay resource?
- Once the clay resource and temper materials are gathered, what are the steps which lead to a ceramic? The properties of the different components used to make a ceramic determine its behaviour under drying and firing and ultimately the physical properties which are sought for the use of the ceramic object.
- What methods are used to make a surface impermeable, pretty, etc.? What paints can be used?

All of these elements of ceramic production are as many keys to product production, distribution, provenance, craft organization in a given community. The use and distribution of ceramics give an enormous amount of information which leads to an understanding of the cultural, social, and economic functions of a society.

A series of case studies of the analysis and use of these analyses of the different components of ceramics is given in order to show how one can use the knowledge indicated in the initial chapters of the book. Finally, a number of current analytical methods are indicated as they can be used to identify and quantify the components of the different parts of a ceramic material. We hope that this series of descriptions and explanations will lead a student of archaeology to a better understanding of the potential of the study of the internal part of ceramic materials.

Some Basic References and Potters-oriented works of interest for ceramic studies

Albero Santacreu, D (2015) *Materiality, techniques and society in pottery production: The technological study of archaeological ceramics through paste analysis*, De Gruyter, Warsaw/Berlin, pp 337

Balfet, H, Fauvet-Berthelot, M-F, and Monzon, S (1983) *Pour la normalisation de la description des poteries*. Editions du CNRS, Paris, France, pp 135

Beauvoit, E, Amaraa, AB, Cantin, N, Lemasson, Q, Sireix, Ch, Marache, V, and Chapoulie, R. (2020) Technological investigation on ceramic bodies of 19th century French white earthenware from the Bordeaux region. *J of Arch Sci Reports* 31, online. https://doi.org/10.1016/j.jasrep.2020.102314

Bloomfield, L (2019) *Science for potters*. The American Ceramic Society, Ohio, pp 148

Bronitsky, G (ed) (1989) *Pottery Technoloy: ideas and approaches*. Westview Special Studies in Archaeological Research, Westview Press, Colorado, pp 165

Chappell, J (1991) *The potter's complete book of clay and glazes*. Watson-Guphill Publications, NY, pp 416

Chavarria, J (1993) *The big book of ceramics: a guide to the history, materials, equipment and techniques of hand-building, throwing, molding, kiln firing and glazing pottery and other ceramic objects*. Watson-Guphill Publications, NY, pp 192

Echallier, J-C (1984) *Elements de technologie céramique et d'analyse des terres cuites archéologiques*. Documents d'archéologie méridionale, Méthodes et Techniques no 3, Salon, France.

Hamer, F (1975) *The potter's dictionary of materials and techniques*. Pitman, Longon, Watson-Guptill, New York, pp 349

Nelson, G (1966) *Ceramics: a potter's handbook*. Holt, Rinehart and Winston Jr, NY, pp 348

Sentance B (2004) *Ceramics. A World Guide to Traditional Techniques*. Thames and Hudson, London, pp 216

Shepard, A (1965) *Ceramics for the archaeologist*. Carnegie Institution of Washington, Washington, DC, pp 414

Rice, PM (2015) *Pottery analysis, A source book second edition*. The University of Chicago Press, Chicago, pp 592

Rye, OW (1981) *Pottery technology: principles and reconstruction*. Taraxacum, Washington.

Vander Leeuw, SE, Prichard, AC (eds) (1984) *The many dimensions of pottery: Ceramics in archaeology and anthropology*. University of Amsterdam, Amsterdam.

2 Rocks and Minerals

In order to understand the origin of the components or the materials which make up pottery, it is useful to make a slight excursion into chemistry and geology. Chemistry is important because it is a major tool in understanding and interpreting the inner characteristics of ceramic materials. These are the parts which are to a certain extent hidden from the naked eye. These are the components which can give much information to the archaeologist which was not available in the recent past. There are two parts to such studies, the chemical identity of the different components and the grouping of the chemical elements into their mineral forms. These relations must be explained in their natural context, that of geology. The chemical identity and geological-mineralogical identities of the different components will be discussed as a first step.

2.1
Chemical Elements and Their Chemical Affinities

There are many chemical elements found in nature (some 100) but few are encountered often in any significant abundance in nature settings. In order to generalize distinctions of relative abundance for the different elements, they have been given the classification of major elements, minor elements and trace elements by geologists, those people who study the chemistry of the earth, among other things. Oxygen is common in air (about 23%) but it is more abundant in rocks (about 50% of the atoms present). Because of this abundance in common materials, chemists have traditionally isolated elements as oxygen compounds. The chemical identity of materials in the natural world is generally given as oxide weight percent of an element. Oxygen is the common anion (negative charge in a compound) and most of the elements analyzed in a rock or ceramic material are the cations (positive charge) of oxide compounds. In geology, the relative abundance of the different cation elements analyzed is given in three categories. Generally speaking, *major elements* are present in a rock sample above several percent oxide weight. *Minor elements* are abundant in the 1-0.05% levels and *trace elements* are those of less than 0.05% abundance. Major elements are counted in percent, minor elements in fractions of a percent and trace elements in parts per million (ppm). An example is gold.

Of course some normally minor elements are occasionally found in significant concentrations as are, less occasionally, trace elements. In the case of metals, these unusual occurrences are usually called ores, because they can be extracted in economically viable quantities.

Major Elements. Major elements given as oxides generally compose 90% of a rock or mineral analysis. In ceramic archaeology, one deals mostly with the major and minor elements, and occasionally the trace elements. Trace elements can be used in certain cases to identify differences in origin of materials which are not obvious when using more simple methods of investigation, such as optical identification. At times, trace or minor elements can be found in concentrations on a ceramic to form a useful effect, such as paints or glaze. For example, gold and silver were at times used in decorations, in pure or diluted form, on ceramic productions. A common glaze is lead (Pb), a very uncommon element in the paste materials of a ceramic.

Most minerals found in rocks, soils and sediments, which are the source materials used by potters, are silicates, and to a lesser extent, carbonates. The cation elements found in these materials are limited in number. Each natural mineral which is used in a ceramic has a specific range of chemical composition. Minerals are entities of more or less fixed composition which form crystalline grains in a rock or sediment. A coherent aggregate of mineral grains is called a rock. A rock, then, is an assemblage of many, many crystal grains. Usually, your hand holding a rock holds millions of crystals. Some rocks have more and some fewer crystals in them, which is a function of their individual sizes.

The most common cation elements are carbon (C), sodium (Na), magnesium (Mg), aluminium (Al), potassium (K), calcium (Ca) and iron (Fe). These elements occur in different combinations and ratios to form different minerals. Some minerals concentrate iron (Fe) others aluminium (Al), for example. Thus, in the same rock, one can find some minerals that concentrate two or three specific elements and other minerals which concentrate other elements.

A rock can be defined either as a given assemblage of minerals or as an assemblage of different chemical elements which form a whole. If one assumes a given composition of a rock, there will be a certain type of mineral assemblage formed and the relative abundance of each mineral species present will depend upon the total composition of the rock. Minerals express the composition of a rock by their type (chemistry) and their relative abundance. Figure 2.1 attempts to illustrate these relationships.

Let us assume that an initial rock is composed of two chemical parts, element A and element B. Possible minerals can be pure element A, as is mineral X, or one half A and the other B, as is mineral Y, or pure element B, mineral Z. In order to make up the rock in the upper part of Fig. 2.1a, which has three parts A and one part element B, one can combine different amounts of mineral X, Y and Z. We will need 62% of mineral X, 25% of mineral Y and 13% of mineral Z. This will give the requisite proportions of the elemental components which make up the rock made from the minerals of a fixed composition. The mineral compositions cannot vary, but their proportions in the rock can. Thus, one sees that there is a quantitative relation between the chemical components present in a rock and their pro- portions of minerals found in the rocks.

ROCK COMPOSITION

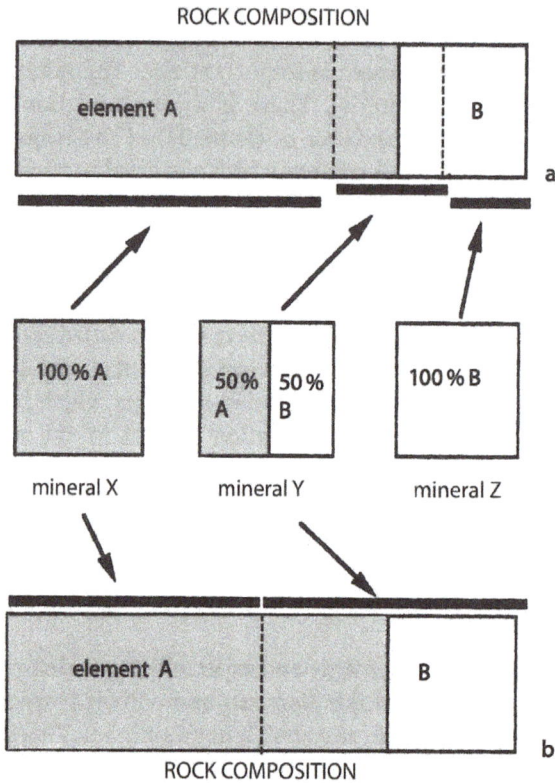

Fig. 2.1 a, b. Schematic diagrams of the relations of rock chemistry and mineral chemistry. **a** A rock made up of two components, with three minerals. One mineral (X) is composed of pure element A the second (*Y)* of elements A and B and the third (Z) of pure B. The rock has more A than B (3:1 in proportions) so that there will be more minerals of type X than Z. Mineral Y will be of intermediate abundance. The abundances are shown by the *length of the heavy line* under the rock composition. **b** Rock composition is the same, but only two minerals are present, *X* and *Y,* whose abundance is the same. Relative abundances are shown by the heavy bar under the rock composition.

Alternatively, one can form the rock composition made of element A and B from just two minerals. This is shown in the lower portion of Fig. 2.1b. Half of the rock is made of mineral X, pure element A and half of mineral Y, made of 50% A and 50% B. Thus, two different combinations of minerals can be used to make up the same rock composition.

Hence, the composition of a rock is not the only variable necessary to describe it; the mineral content is important also. The same is true when considering a ceramic material. There is a direct relation between the chemistry (the relative abundance of elements) of the major elements in the ceramic and the mineral content which was used to produce the paste of the ceramic.

Minor and Trace Elements. Minor elements and trace elements can behave in two ways in their incorporation in minerals. Either they fit easily in a dominant phase made of major elements in what is called solid solution, or they can form specific phases on their own. Elements in solid solution find their place in minerals along with major elements which have a similar chemical affinity. An interesting relation is that in the same mineral species, different minor elements can be present in solid solution depending upon the origin of the rock in which the mineral was formed. Hence, minor elements can at times be used as tracers as to the origin of a given mineral species. High-temperature feldspars from magmatic rocks do not have the same minor elements as those coming from rocks of lower temperature origin.

However, some minor elements and especially trace elements will often not fit well in the place of major elements and will form specific minerals in which they are greatly concentrated, if not dominant. There will be a few of these mineral grains present in a rock. For example, sulphur concentrates in iron sulphide (pyrite, FeS) though the sulphur content of a rock is normally in the range of minor elements ($< 1\%$). Phosphate can be concentrated in apatite (Ca_5PO_4), (Cl, F) and so forth. Even more striking elements such as gold (Au), zirconium (Zr) or platinum (Pt) which are present in rocks as trace elements, most often form minerals of their "own", being only very slightly soluble in other minerals of greater abundance. Hence, only one or two grains of these minerals will be found in a hand-sized rock specimen but each grain contains 100% of the "minor" or "trace" element. If the grains are separated from the rock matrix, concentration will change the abundance of the element in question. As example of this effect is gold panning from sands in mountain streams. Large amounts of sterile sand are sifted to produce a few small grains of gold. Thus, minor and trace elements can be found in the major mineral components of a rock or in just a few mineral grains in a rock.

Therefore, the use of the terms major, minor and trace element concentration can mean different things in minerals or in rock descriptions. The same is true, of course, for ceramics. These arguments may seem abstract and rather useless in a first analysis, but they are, in fact, very important. If one makes a chemical analysis of a ceramic sample for major, minor and trace element abundance it is very important to remember the differences. Major elements will give an idea of the type of mineral material which makes up the ceramic. For example, if there is about as much aluminium as there is silicon (in oxide percent), the ceramic sample was made of a very clay rich paste. Two to four percent potassium (oxide) indicates the presence of a micaceous mineral in great abundance. If magnesium is above 6 %, the sample has been derived from basic rocks in one way or another, etc. These are the guides that can be used for the major elements as they are expressed by the minerals used to make a ceramic.

Minor element abundance will indicate the variations which can occur from one source of material to another. For example, in ceramic sherds from the same region, if the iron content (Fe) is twice that in one sample than in another, the sources of materials (clays, coloring agent, temper) or production methods were different. Minor element abundance can give information on variants in a group. Most often, minor elements express major mineral abundance, which varies according to its geologic source.

Trace element analysis is much more difficult to interpret. If the trace element is present in trace abundance in most of the different minerals (solid solution) which formed the ceramic, that trace element can be used to determine provenance of the mineral components in the paste or distinguish different ceramic groups. However, if the trace element is present in only one mineral, i.e. in high local concentration, the ceramic sample must be of such a size as to contain this grain in a representative amount. If the sample is too small, it could contain two grains by chance, or none at all. In one case there will be a "high" trace element concentration and in the other there will be none at all. Trace element abundance can be highly variable from one sample to another.

The obvious problem is to know which elements are highly concentrated in minerals of very low abundance and when they are nicely dispersed in the several minerals present in abundance in the ceramic paste. There are no rules, but in general, the heavier the element (the higher its atomic number, the further down the atomic table it lies, the higher the probability that it will occur in a specific mineral of high concentration and low abundance.

2.2
Major Rock Types

Geologists have developed a certain vocabulary in the description of rocks found at the surface of the earth. It is useful for an archaeologist to know the meaning of these terms so that she or he can use the accumulated knowledge of geology to determine the parameters of ceramic origin. There are three large groups of rock types according to their origin.

Sedimentary rocks are derived from sediments, as one would expect. Sediments are surface materials deposited usually in water, sometimes by wind transport. Sedimentary rocks are sediments which have been buried at some depth. A sediment buried under other sediments, say at the bottom of a sedimentary basin on the edge of a continent, will be little transformed from its initial state until it reaches a depth of 1km or so. At less than 1km depth it is still soft and has little difference in mineral content compared to its state upon its deposition. However, as this material is buried by successive layers of sediment weighing 2.5 times the weight of water, the sediment becomes hard, losing much of its absorbed water and taking on the appearance of a rock. All sedimentary rocks contain minerals much like those of the sediments from which they were derived. This means that the material coming from erosion of continents, clay and sand for the most part, is little changed.

Some carbonate minerals are likely to be found in the sediments. They are the product of animal shell life for the most part. These carbonates can become dominant and form almost pure calcite rocks ($CaCO_3$). More rarely, plants can accumulate to form coal beds made of carbonaceous concentrations. Thus, sedimentary rocks for the most part have two origins, one physical and the other animal. In some cases, precipitation from solution (chemical in origin) can form rock layers (salt beds for example).

Typically, sedimentary rocks have parallel banded structures which represent the regular layers formed as their sediments settled onto a sub- stratum. Bedding or banding is the typical macroscopic feature of sedimentary rocks. The names of sedimentary rocks are based upon their mineral content and the proportions of the minerals present.

Metamorphic rocks are formed at depth or under conditions of local heating. Sediments become metamorphosed by the action of higher temperature. The term metamorphic means that the initial minerals have been recrystallized to give other mineral phases due to the action of heat. These rocks are denser than the sedimentary rocks and they are very hard, forming in tightly banded and foliated materials. At times, the new minerals are dominant and can be seen easily with the naked eye. High concentrations of different minerals give rise to specific names of metamorphic rocks. For example, clay-rich sedimentary rocks give shales, carbonate (shell)-rich rocks give marble, sand gives quartzite. In metamorphic rocks one can find many different species of minerals, usually in minor quantities.

The names of metamorphic rocks are based upon the intensity of heating or transformation and the minerals present in the rock.

Igneous rocks are those that have been melted. The melt (magma), dominated by silica, is generally lighter than the rocks which surround it at the great depth at which rocks melt, at tens to a hundred kilometers, and it is lighter, especially if it contains some gas. The difference in density gives igneous rocks buoyancy. Magmas tend to move upward in the earth's crust. As one would suspect, it is necessary to go very deep in the earth's crust to realise the temperatures necessary to melt a rock, many kilometers. The melted rock then moves upward. If it reaches the surface of the earth it is called *volcanic* when it cools to form a rock. Volcanoes are made this way. If it fails to come to the surface it is called an *intrusive* rock, when it cools and hardens. The intrusion of melted rock (magmas) into a sedimentary rock can metamorphose it into a metamorphic rock.

The magmatic rocks then have characteristics which indicate their high-temperature history, a special set of minerals and textures that are very dense, and highly colored. Basic lavas from volcanoes are black and lustrous showing glassy edges when broken. Most often these rocks are basalts. Intrusive rocks often grow large crystals which are clearly visible to the naked eye. The most common intrusive rock on continents is granite formed in this way. The chemical composition of magmatic rocks varies rather regularly. In a simple manner, when silica increases, there is more sodium (Na) and potassium (K) in the magma. When silica is lower, magnesium (Mg) and iron (Fe) are more abundant. The silica-rich magmatic or igneous rocks are called *acidic* and those containing more magnesium and iron are called *basic.* Hence there are families of elements which "go" together in magmatic rocks. There are, of course, intermediate types of rocks, between acidic and basic. However, the great mass of magmatic rocks is in these two categories.

For the two types of geologic manifestation of igneous rocks, intrusive in the earth's crust or extrusive as volcanoes, both acidic and basic rock are present. However, if one considers the total mass of material, there is a definite preference for the different types to occur. Basic rocks are largely extrusive, volcanoes, and acidic rocks are largely intrusive.

The nomenclature of igneous rocks is based upon their chemical composition, mineral types and grain sizes. The surface of the earth itself is divided topographically into two great zones, ocean and continent, land and water. The ocean bottoms are dominated by basalts, the most common extrusive basic lavas which form basalt rock. The central portions of continents are dominated by acidic igneous rocks, intrusive granites and some extrusive forms. The continents are also covered by layers of sediments which have become sedimentary rocks. At the edges of continents one finds thick layers of sediments.

2.3
Minerals

2.3.1
Mineral Formulae

In science there are always conventions of expression, that is, there is a way of saying or writing things. This is especially important in chemistry, where the order of the expression is very strict. In mineralogy a major criterion is a description of a mineral by its chemistry. The key to this scheme is the number of anions present. Ions in chemistry are divided into two types, those having a negative charge (anions) and those having a positive charge (cations). In nature there is a strong tendency for negative ions to find positive ones with which they can cohabit. The tendency is to form an electrostatically neutral unit, in one way or another. Nature abhors imbalance. Thus, in solution (water for instance), ions can exist as negative (anions) or positive (cations) species but the total of ions will always balance, i.e. the plus equals the minus. In solids, minerals, the same is true. Cations balance anions. However, one counts the number of cation charges by the number of anions assumed to be present.

In most rocks, oxygen provides the great majority of the anions. Since anions balance cations, the number of oxygens represents about half the ions in minerals found at the earth's surface. Oxygen ions (anions) have a charge of minus two, O^{2-}. For example, two oxygen anions are associated with one silicon to form quartz; this gives $SiO2$. In this formulation the subscript indicates the number of the ions preceding the symbol of the element present in the mineral unit formula. In $SiO2$ there is one silicon cation and two oxygen anions. The positive charge on the silicon is then by definition equal to that of the oxygens (two times two). Silicon ions are indicated as $Si4+$. Hence, two oxygens give a negative charge of minus four and the silicon has by consequence a charge of plus four since there is a 1:2 ratio.

In some cases, a single element can have two or more different forms or totals of positive charge. Iron, for example, can be either divalent ($Fe2+$) or trivalent ($Fe3+$) having a positive charge of two or three. Iron can have then several mineral formulae, oxides, such as FeO, Fe2O3. In the first case, iron is divalent (a doubly charged cation, $Fe2+$) and it is associated as a consequence with but one oxygen anion of two negative charges. If iron has a charge (positive) of three it cannot be associated with an even number of oxygens. One has a combination of two iron trivalent ions with three oxygen ions. This is expressed as Fe2O3 where the total of iron-positive charges is two times three (six) and the total of negative charges supplied by the oxygens is six (three times two). In the first instance, one divalent iron ion is associated with one divalent oxygen ion. The subscript one is not used in the chemical mineral formula. In the second instance, one trivalent iron ion is associated with one and a half oxygen ions but the number of trivalent ($^{3+}$) iron ions is doubled to make an integral number of associated oxygen ions, three. Thus, two trivalent iron ions having a total of six positive charges are associated with three divalent oxygen ions, with, of course, a total of six negative charges.

These are the conventions of writing mineral formulae, the total of cation (positive) charges equals the total of anion (negative) charges.

2.3.2
Major Mineral Families

The next step now is to know the type of minerals most commonly found in rocks. These minerals can often identify the rocks themselves. The identification of minerals is, of course, not an easy task; there are many minerals and many rock types, at least according to mineralogists and geologists. However, the majority of minerals and rock types in the three categories described above, sedimentary, metamorphic and igneous, are rather simple and occur in great abundance. For this reason, one can give a rather brief summary of rocks and minerals, which should represent 80-90% of mineral types which will be found in a ceramic. However, the eye is used to finding the exception. If one looks at a distribution of grains, say 100, it will pick out the exceptions, say 5%, and ignore the other grains. Thus, a geologist will look at a rock thin section under the microscope and say, "Oh, look at the staurolite!", but he sees only 5-8 grains among the 5000 present. His eye is trained to find the exception and is not a good numerical gage. In many instances, however the exception is more important than the rule, and geological training will help to find subtle clues as to rock and source identity. All this is to say, if not trained, look at the forest and not the trees, even less their leaves. Leave leaves to an expert.

2.3.2.1
Silicates

We will consider the major groups of minerals, based upon their structure and chemistry. The key to understanding the minerals found at the surface of the earth is the fact that most minerals, a very great majority, are silicates. This means that silicon (Si) is the major cationic (positively charged) element. It is intimately and very strongly combined with oxygen. All other elements present in silicates are combined with oxygen also but most of them less strongly. Only aluminium ions can take the place of silicon in silicates and they are bound with the same force.

Nearly half of the atoms in complex silicates are oxygen. It is so pervasive that it is rarely analyzed for its abundance. Most modern chemistry assumes that a given number of oxygen anions are associated with cations (those of positive charge).

The classification of silicates is based upon how many silicon atoms are present with respect to other cations, i.e. the relative abundance of Si atoms. Aluminium (AI) is important also as it proxies for silicon in silicate structures. This situation gives an Si-Al-O complex of very tightly bonded cation-anion molecules.

2.3.2.1.1
The Silica Minerals and Quartz

Quartz is the most silica-rich mineral and the most common mineral on the earth's continental (land) surface. It is a major component in soils, sands and rocks formed near the surface. It is the most abundant mineral in granites. Quartz has a very simple chemical formula, $SiO2$. It is the silicate par excellence. It resists chemical attack and it remains present through many episodes of geologic activity, such as weathering, erosion, water transport and deposition, and further through sedimentation and the formation of sedimentary rocks. The origin of quartz grains can occasionally be determined, but rarely. At times they show evidence of metamorphism and deformation, several greyed areas can be seen under the petrographic microscope (Sec. 7.2.3) in the same grain under analyzed light. However, for the most part quartz grains reveal little of their origin.

Special conditions of temperature, either low or high, can create other crystalline forms of silica, i.e. different silica minerals. However, these minerals are rare and difficult to identify. They are called cristobalite, trydimite, coesite, stishovite. These forms are, in fact, more for specialists than for practicians of mineralogy in an archaeologicai context. Hence, we will basically ignore them.

However, a common, poorly crystallized form of silica which has different crystalline forms is called chert. This is a mineral formed in the sedimentary environment. By its hardness and resistance to weathering and its fine crystal size, it is difficult to break. When broken by percussion, it flakes. With clever blows or firmly applied pressure, one can make hard, durable tools with a cutting edge. Either a scraper, arrow-head or perhaps a spear point can be fashioned from chert. Chert is a fairly common form of silica present in carbonate sedimentary rocks.

High concentrations of quartz give the name of sandstone to a sedimentary rock and quartzite to a metamorphic rock.

2.3.2.1.2
Feldspars

This is, in fact, a family of minerals which has two parts. One is made of a potassic end member (K). Potassium feldspar can have such names as orthoclase, microcline, sanidine; all minerals have very nearly the same chemical composition. The differences are in the organization of the atoms in the crystalline structure. This crystalline structure reveals the type of rock from which the potassic feldspar came, igneous and metamorphic of slow cooling (intrusive) or volcanic of rapid cooling origin. Most potassium feldspar found in ceramics comes from intrusive, slowly cooled rocks, i.e. granites.

The chemical formula of these potassic feldspar minerals is close to

$KAlSi_3O_8$ potassium feldspar.

One sees that silicon (Si) is the dominant cation in the formulation, three out of five cations. When one assimilates aluminium (Al) with silicon (Si), the silicate part is yet more dominant. The importance of silicon and aluminium (Al) in the minerals gives them a high stability at the earth's surface.

The second feldspars' family are the plagioclase series. In these minerals there are two poles, one sodic (Na) and the other calcic (Ca). Their chemical formulae have all proportions between the poles

$NaAlSi_3O_8$ and $CaAl_2Si_2O_8$ plagioclases.

The sodic form is called albite, the calcic end-member anorthite. Plagio- lases with intermediate compositions have different names.

Again, the silicon (Si) plus aluminium (Al) cations are dominant and the minerals are relatively stable. In fact, those plagioclase minerals with high sodium (Na) content are much more stable in soils than those which are highly calcic (Ca). Plagioclase feldspar minerals are often twinned where crystalline units are repeated adjacent to one another to form a composite crystal. Under the petrographic microscope they have stripes visible in their structure seen under analyzed light. The more sodic plagioclases are found in granites and metamorphic rocks while calcic plagioclase is found in lavas and basic rocks. Plagioclase minerals are then found in a variety of igneous rocks, and also in metamorphic rocks. Plagioclase is a rather common mineral.

One often finds the term alkali feldspar in the current literature. It designates a feldspar dominated by either potassium or sodium. The potassium feldspar is one type whereas the sodium feldspar is the pole of the plagioclase series. They are not in the same structural-compositional group but they have similar characteristics in that they both melt at relatively low temperatures compared to quartz or calcic feldspar and hence are sought to lower the melting temperature and the time-temperature relations needed to transform the clays into a non-clay state and to attain a transformed paste.

2.3.2.1.3
Pyroxenes and Amphiboles

These are two groups of minerals closely related, by their structure and composition. The major difference between them is that the amphiboles contain crystalline water (hydrogen ions) and the pyroxenes do not. The importance to an archaeologist is that the amphiboles will be rapidly attacked by heating and in a ceramic body. They will be oxidized, turning brownish red in color. Pyroxenes are almost not affected at all by normal firing temperatures, remaining greenish in color. Both mineral groups are of a chain structure. This is translated physically by a long crystal form, like a thick stick. The ends of the stick in pyroxenes fracture at right angles and the amphiboles at angles of 120°, a diamond form.

Both mineral groups contain, most often, calcium (Ca), iron (Fe) silicon (Si) aluminium (Al) and magnesium (Mg). Varying amounts of sodium (Na) and small amounts of potassium (K) and manganese (Mn) can be present. The high quantities of iron and magnesium indicate that they will be found in basic magmatic rocks, those whose compositions are dominated by these elements.

General chemical formulae are

$(Ca,Na,Mg,Fe,Al)\ 2(Si,Al)205$ pyroxene.
$(Ca,Na,K)(Mg,Fe,Al)(Si,Al)O_{20}\ (OH)_2$ amphibole.

Amphibole is most often found in metamorphic rocks, more commonly in basic (Fe, and Mg) metamorphics, those metamorphic rocks which were derived by the transformation of old basic igneous rocks. Pyroxenes are found in magmatic rocks and more rarely in metamorphic rocks. They are rare in sedimentary rocks.

2.3.2.1.4
Olivine

Olivine is a mineral almost entirely restricted to basalt lavas and other rocks coming from very deep areas in the earth's crust. The lavas are basic in composition, because olivine contains a lot of magnesium (Mg). Olivine is very unstable under conditions of weathering and when found as an unaltered mineral it should be present at or near the outcrop area or the source rock. For example, black sands at the edges of volcanic islands in the Pacific Ocean are often largely composed of olivine crystals. The presence of olivine and pyroxene give a black color to the sands of these islands. For the most part olivine is

Mg_2SiO_4 olivine.

Olivine contains less silica than other ions, by a factor of 2. A small amount of iron (Fe) is present instead of magnesium (Mg) in olivines. It is indeed a very "basic" mineral, very magnesium- and iron-rich, being found as a major component of basic igneous rocks. Olivine is very rare in metamorphic rocks and almost absent from sedimentary rocks.

2.3.2.1.5
Micas and Chlorite

These minerals are often very obvious in the paste of a ceramic object because they tend to reflect the light. A glinting mineral will often indicate the presence of mica. They have the structure of clay minerals only they are bigger. They have a sheet-like structure (see Chap. 3 on clay minerals).

Micas contain hydrogen ions, as do amphiboles, which contribute to their instability upon firing n the ceramic making process. Micas are almost always potassic (K) and form two groups, one dominated by aluminium (Al) and the other by iron (Fe) and/or magnesium (Mg). The aluminous micas are called muscovite and the iron-magnesium types biotite. The thermal stability of micas is largely determined by their iron (Fe) content, high iron content promoting early destabilization in the firing process. As with amphiboles, iron-rich micas become brown in a well-fired ceramic. The compositions are

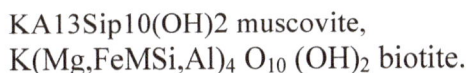

$KA13Sip10(OH)2$ muscovite,
$K(Mg,FeMSi,Al)_4 O_{10} (OH)_2$ biotite.

Micas are very common and often form the major part of metamorphic rocks. They are frequent but are of lower abundance in intrusive igneous rocks. Micas are typical of rocks found in mountain ranges on continents. They resist weathering to a great extent and can be found in many sand fractions on beaches or in stream beds. Hence, they will be found in sedi- ments and sedimentary rocks. They are, of course, very common in soils.

Chlorite is a sheet silicate, similar to those of clays, but is rarely encoun- tered in the clay fraction. Chlorites are normally of larger grain size. Chlorites are different from micas in that they do not contain potassium. They are essentially composed of iron (Fe), magnesium (Mg), aluminium (Al) and silicon (Si). Of course, they contain oxygen and water in the form of hydroxyl ions (OH). The chlorite formula is

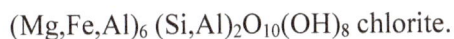

$(Mg,Fe,Al)_6 (Si,Al)_2O_{10}(OH)_8$ chlorite.

Chlorites are less stable during weathering than micas due to the fact that they have no alkali ions and that they contain significant amounts of iron. They are not stable upon firing in the ceramic-making process either, becoming brown.

High concentrations of micas in sedimentary rocks give the name shales, and in metamorphic rocks abundant micas and chlorite determine the general name of schist or gneiss terms used to designate rocks of increasing higher temperature of formation.

2.3.2.2
Carbonates

In sediments and sedimentary rocks, one finds concentrations of quartz and clays which determine two categories of sedimentary rocks, sand- stones and shales. A third large category of sediments and sedimentary rocks is that of the carbonate facies. The major minerals of carbonates are, for the most part of two sorts,

$CaCO_3$ calcite,
$CaMg(CO_3)_2$ dolomite.

These minerals contain the carbonate anionic group CO_3 as is evident in the formulae. The cations are calcium (Ca) and magnesium (Mg). Carbonates are a combination of air (the carbonate part which is carbon dioxide, $CO2$ found in the air) and calcium or magnesium which are elements highly soluble in the weathering process, and hence found in seawater. Carbonates are then a combination of air and (sea)water. The work of forming the carbonate minerals is done for the most part by living organisms which use the combination as a protective coating, a shell.

Carbonates are shell deposits formed in shallow seas or deep ones. Often, these deposits are almost pure shell, and thus they have a special composition, one rich in carbonate, which is, in turn, very easily attacked by slightly acidified rainwater, the aggressive elements of weathering.

The important property of carbonates is that they are unstable upon heating. High temperatures, above 800 °C, break the structure, making a gas, CO_2, and leaving behind a highly reactive oxide, CaO. In a ceramic this residual oxide combines with silicates, especially clays, to form a melt. The destruction of carbonates helps in forming a solid and impermeable ceramic. However, when the carbonate is present, as it is destroyed by heating, it liberates gas which can cause damage to the ceramic. Thus, carbonates in a ceramic paste are handled with caution.

Carbonates in sedimentary rocks give the name limestone (calcite-dominant) or dolomite (dolomite-dominant) and in metamorphic rocks the term marble is used. This name has a much more restricted use in geology than in archaeology, where marble can be used for many other types of metamorphic or occasionally igneous rocks.

Another form of carbonate is aragonite. This mineral has exactly the same composition as calcite but a different crystal structure. In natural rocks, calcite is the stable phase. Aragonite is a very high-pressure mineral. However, shell-building animals tend to make their protective structures in the form of aragonite crystals. The additional energy of biologic activity can create this high-pressure form of calcium carbonate.

2.3.2.3
Oxides

A minor (in quantity) but very important group of minerals are the oxides. The greatest part of oxides are those of iron (Fe). The grain size of iron oxides is usually so small that they cannot be identified by optical microscopic means. In most clays and clay bodies they are of such low abundance that they cannot be identified by means of X-ray diffraction either. However, they have one very distinctive feature, they give a strong color to the matter in which they are found.

Iron (Fe) Oxides. The iron oxides have various colors which follow their oxidation state and hydration state. The most prevalent iron oxide is hematite or Fe_2O_3. This is a form of trivalent iron (Fe3+) which is red in color. Most pottery whose clay source is a mixture of materials in soils has some iron in it, which becomes oxidized during firing. This gives a red to pink color to the paste. Another natural oxide is goethite. This mineral is yellow, containing some water molecules. It is easily destabilized upon firing to become red hematite. Lepidocrocite is also an oxide of iron which contains some water molecules. It is orange in color. A more reduced form of iron oxide (Fe^{2+}) is maghaemite, found in swampy areas which are associated with acidic waters. It is black in color. These minerals are found in soils, mixed with other minerals. In some cases, iron oxides are concentrated by either surface alteration mechanisms (ochre is a good example. It is a hematite-rich deposit of iron oxide and clay) or hydro- thermal vein alteration.

Manganese (Mn) Oxide. Manganese forms complex oxides which are generally black or dark brown in color. They do not change dramatically in color upon heating, but the depth of their color changes to a lighter form of grey. They are used in paints on ceramics which are not subjected to a high temperature or a long heating time during the pottery-making process. Most manganese concentrations are of metamorphic or hydrothermal origin. Some small quantities are found forming dendrites (veined ramifications) on surfaces in rocks. They are pretty but not useful as a source of manganese oxide.

Very few rocks contain enough oxides to give them a rock name. They are typically accessory minerals. However, the coloring capacity of oxides (reds, purple, brown) makes them very evident.

2.3.3
Mineral Grain Shapes

Each mineral family, the micas, feldspars, amphiboles, olivines and pyroxenes, has a particular crystal shape, commonly distinct in the different rocks in which they are found. These shapes help to identify each mineral type. Also, the shape can explain their evolution under the conditions of weathering and water transport (Chap. 4). Figure 2.2 indicates in a very general way the general shapes of some mineral crystals.

The quartz and feldspar minerals, which normally have a blocky shape, tend to occur in more or less equant forms upon weathering and transportation. They have a fairly equant, blocky shape to begin with. The crystals break at their edges and not internally. The amphiboles and pyroxenes to a lesser extent tend to break along their long cleavage axis, which is a plane of weakness. These minerals tend to remain long and stick-shaped during different weathering processes. The micas have the same characteristic of splitting or being cleaved when transported. They cleave or break along the sheet-like planes of their structure and remain sheets during their geologic history. Therefore, in general, rounded grains are **in** the quartz-feldspar families. The flat grains are micas and the rod- shaped forms will be amphiboles and pyroxenes.

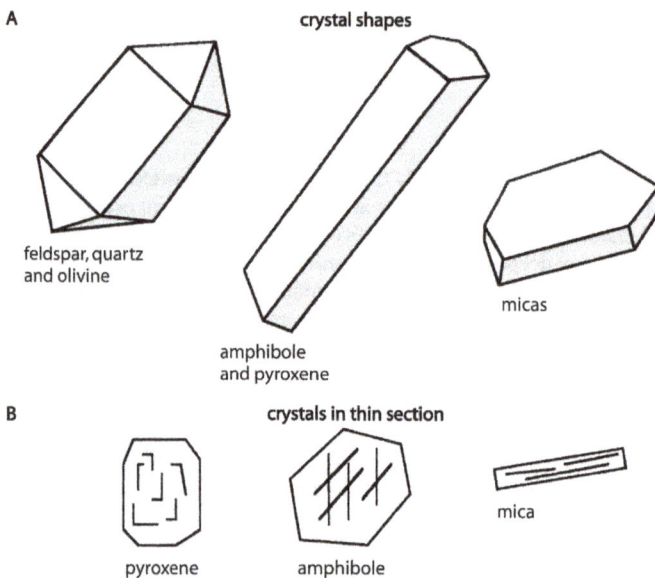

Fig. 2.2 A, B. Representation of mineral crystal shapes. A Olivine, feldspar and quartz crystals are equant in form, without strong cleavage planes. Their shapes are squarish. Amphiboles are prismatic in shape, longer and thinner. Pyroxenes slightly less so. Micas are flat and plate-shaped. **B** Crystals, when cut show different shapes and cleavage (internal cracks due to crystallographic structures). The amphibole crystals show cleavage cracks at 60- 120° angles while pyroxenes show cracks at near right angles when cut in cross-section. The micas show cracks parallel to their long shape

These general notions will help in most instances but they are not infallible. There are many more minerals present in the sediments and soils of the surface. However, the greatest part of the crystals encountered will be found to have come from the above groups.

2.4
Minerals in Rocks

A rock is most often made up of several types of mineral. It is rare to find a rock of one mineral species. The sedimentary rocks which are almost monomineralic, sandstone (quartz) and limestone (calcite), are easy to define. However, the metamorphic and igneous rocks are another problem.

2.4.1
Sedimentary Rocks

Sediments are surface materials transported and deposited by means of water (rivers, lakes and oceans) or air, or the product of an accumulation of shell material grown in large bodies of water. They become rocks as they are buried more deeply beneath other sediments. The initial effect of burial is the expulsion of the water of sedimentation. This loss of fluid gives a certain substance to the sediments which are initially about 80% water when deposited. Upon loss of water they harden into rocks. Thus sedimentary rocks resemble sediments very much because little mineralogical transformation has occurred in their formation.

Sediments are composed for the most part of three mineral groups; carbonate, quartz and clays. The last type can for the moment be assimilated to very fine-grained micas. Chapter 3 deals with clay minerals in more detail. Sediments are then grouped into three general categories: clay-rich, sand-rich and carbonate-rich. These are the poles of sedimentary rock types. Different rocks can have different proportions of these components or poles, some clay, some sand and some carbonate. Sedimentary rocks can be at times made of pure sand, carbonate or clay, but this is rare. With compaction, the sand becomes sandstone, the clay becomes shale and the carbonate material becomes carbonate rock, most often limestone. The nomenclature of sedimentary rocks is simple.

Sands and sandstones are siliceous (SiO_2) containing quartz with minor feldspar. Carbonate rocks are made of calcite as a major component and commonly called limestone. Lime is the degassed calcite (CaO). At times, the mixed carbonate, dolomite, is found in carbonate rocks and its presence gives the name of dolomite rock. Minor amounts of clay and sand (quartz) are often present in carbonates. When clay content is near 30 and more the common name for this type of carbonate is marl.

High clay content can make a sedimentary rock useful in ceramics.

2.4.2
Igneous Rocks

Igneous rocks are of basic and acidic composition with many varieties in between, which are of minor abundance. The major minerals in igneous rocks are quartz, feldspars, micas, pyroxene and olivine. One can make a graphic representation of the mineral abundances as a function of the acidic or basic character of the igneous rock, as shown in Fig. 2.3.

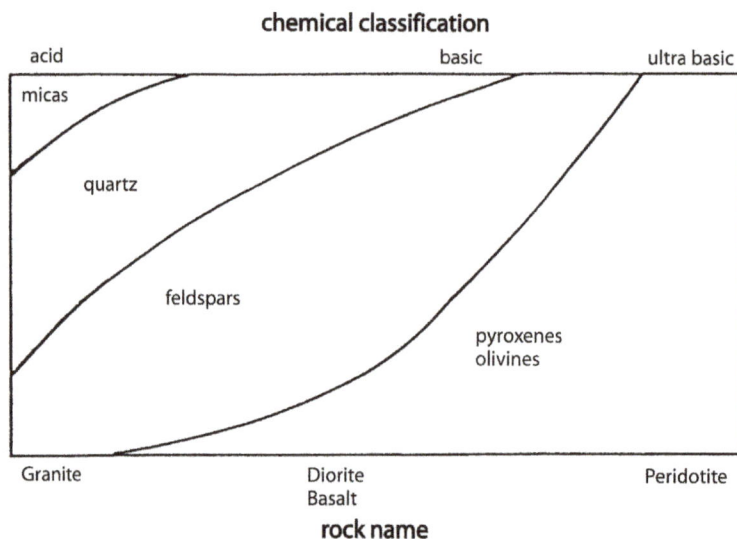

chemical classification

acid   basic   ultra basic

micas

quartz

feldspars

pyroxenes
olivines

Granite   Diorite   Peridotite
                     Basalt

rock name

Fig. 2.3. Relations between rock type by composition and the minerals present. Granites are made up of micas, feldspars and quartz. The extrusive lava expression is called rhyolite. Basalts (extrusive lavas) and basic rocks (gabbros) contain some feldspar but also pyroxenes and olivine. The diagram shows the effect of bulk chemical composition *(left to right* granite to basalt, and relative mineral abundance *from top to bottom. A line* on the granite composition *(vertical)* has a high proportion of feldspars, less quartz and a little mica.

If one reads the diagram in Figure 2.3 vertically, the relative proportions of the minerals are indicated by the width of the band given them. The further one moves from acidic to basic, the greater the field and abundance of plagioclase, the presence of pyroxene is noted and finally in the basic and highly basic rocks olivine is dominant. As one moves from acidic to basic rocks, the proportion of aluminum diminishes in favour of iron and magnesium. Potassium gives way to sodium and calcium.

Geologists have divided magmatic rocks into two types, those with a dominance of small crystals (millimeters) and glass and those with a dominance of larger crystals and no glassy material. The rocks which contain glass indicate an eruptive (extrusion of magma onto the earth's surface) origin and intrusive rocks with coarse grains were formed in the crust of the earth. Rocks of the same chemical composition can have different origins and hence different names. For example, granite is an intrusive rock and rhyolite is the extrusive equivalent. Gabbro is an intrusive basic rock (olivine and pyroxene) while basalt with the same minerals but containing glass is an extrusive rock, a lava. These names are indicated in Figure 2.3 along with the mineral contents.

If one finds a mineral assemblage (several grains in the same fragment) in a pottery sherd of quartz and feldspar, one is dealing with an acidic rock; if one finds olivine and pyroxene with plagioclase, one has a basic rock. Quartz is almost unknown in basic rocks, olivine is almost unknown in acidic rocks.

On weathering quartz and potassium feldspars are relatively resistant, hence they persist in sands. On weathering, olivine and pyroxene are easily attacked, and plagioclase is even more vulnerable. They will be more rare in sands. Bits and pieces of acidic rocks will be found in sediments and sedimentary rocks while basic rocks tend to be lost, forming clays or being dissolved as ions in ocean water (magnesium, calcium, and sodium). When basic magmatic minerals occur in sands, they will indicate proximity to the rocks of this type, i.e. little transport and weathering.

2.4.3
Metamorphic Rocks

The major minerals (speaking from a volumetric point of view) in a very large part of metamorphic rocks are micas, feldspars, carbonates and quartz. This reflects the most common origin of metamorphic rocks, which is sediments. Some metamorphic rocks are formed from magmatic materials, granites, basalts and so forth. These rocks are less important, in volume, than the metamorphic rocks formed from sedimentary materials. They contain a great variety of minerals.

There are, then, many rock names for metamorphic types because they come from varied sources. The rock names are based upon the minerals present and very often upon minor minerals found in them which are special, representing special conditions of formation, most often temperature and pressures.

2.4.3.1
Metamorphic Pelites

The clay-rich sediments (pelites) form shales and metamorphic pelitic rocks upon metamorphism. This origin from clay-rich materials leads to the formation of phyllosilicates, micas, in the large grain category. The clays become micas, changing slightly in their composition but retaining their sheet-like shape. The grains are bigger than in soils or sediments, but are much the same. In metamorphic rocks one can identify the micas by optical methods. They are flat, and sheet-shaped. In thin section (optical observation under the microscope) they most often form long lath shapes because they are cut across their thin dimension. In a hand specimen, a metamorphic pelite is easy to recognize by the number of shiny flat faces of the mica crystals which reflect the light. These micas are often accompanied by quartz grains and also by plagioclase feldspars. The composition of the plagioclase can be often used to determine the temperature maxi- mum reached by the rock during its metamorphism.

Metamorphic rocks can contain a great variety of minerals present in rather small quantity. Often the accessory minerals constitute 15-20 % of the rock. However, these minerals are very important as a diagnostic of the conditions of metamorphism. Thus, in metamorphic rocks one looks for the exception more than for the rule. A small number of garnets or staurolite crystals will change the name of a rock and indicate an origin different from one containing only mica and chlorite for example.

However, given the great number and complexity of identification of the many metamorphic mineral species, we will not go into their identification here. This is a matter best left to an expert because there is too much chance of error in trying to identify metamorphic minerals in pelites.

2.4.3.2
Metamorphosed Carbonates

Little mineralogical change is seen as carbonates are metamorphosed. The crystal size is bigger. The minerals calcite and dolomite are preserved for the most part, at least under most metamorphic conditions. The impurities in carbonate sediments such as sand and clays do not generally combine with the carbonate minerals until very high temperatures of metamorphism are reached, above 700 °C or so. Such rocks are relatively rare, but, of course, interesting. The general name given to metamorphic carbonates is marble; but marble is also used to describe a rock used for decorative purposes. The term can be used to identify any hard, easily cut rock that is not an igneous one. One finds references made to granite and marble. However, the marble is frequently not strictly a carbonate-rich rock of metamorphic origin.

Thus, one can expect that most metamorphic carbonates will contain calcite, dolomite, some quartz and feldspars and micas. The micas in metamorphic rocks are the metamorphic equivalent of clays, the same mineral structures but of a larger grain size.

2.4.3.3
Metamorphosed Igneous Rocks

The biggest changes which occur in metamorphism are those affecting basic igneous rocks. Most of the initial minerals in a basic igneous rock (plagioclase, pyroxene, olivine) will be changed in their composition and mineral species. The pyroxenes and olivines become amphiboles (hydrated iron-magnesium silicate minerals) and the high temperature, calcic plagioclase becomes a more sodic species. Basic metamorphic rocks are then composed very often of sodic plagioclase, and amphibole. The general name of such rocks is amphibolite.

Acidic igneous rocks change little in their mineralogy (species present) upon metamorphism, but the crystallographic form of the minerals changes with temperature and the proportion of the minerals can change. In general, metamorphic rocks of acidic origin contain potassic feldspar, some plagioclase and micas plus, in general, much quartz. The metamorphism is most remarked by the change in rock structure, the micas tend to be segregated together to form bands or layers in the rocks. These are called gneisses and schists. Minor minerals (in abundance) can give clues to the temperature history of the rocks and hence such minerals can be used to identify the metamorphic rock.

Acidic metamorphic rocks are then dominated by quartz, potassic feldspar, some plagioclase and micas, especially muscovite. Individual mineral assemblages will be difficult to distinguish from those of granites.

The following mineral groups can be used to distinguish some of the abundant general rock-type categories as shown in Table 2.1.

Table 2.1. Major minerals found in different rocks

Minerals	Original Rocks
Quartz (high abundance)	Sandstone or metamorphic sandstone (quartzite)
Micas + quartz	Metamorphic and clay-rich sedimentary rocks (pelites)
Quartz + potassium Feldspar + mica	Granite and high-temperature metamorphic rocks formed from clay-rich sediments or metamorphic clay-rich sediments
Amphiboles+ plagioclase	Metamorphosed basic magmatic rocks, now amphibolites
Pyroxene + olivine	Basalts (lavas) and basic magmatic rocks
Calcite + dolomite	Carbonate rocks, either sedimentary or metamorphic

Bibliography: Mineral and Rock Identification

Adams, AE, MacKenzie, WS, and Guilford, C (1984) *Atlas of sedimentary rocks under the microscope*. Longmans, London, pp 104

Best, MG (1982) *Igneous and metamorphic petrology*. Freeman, San Francisco, pp 630

Blatt, HC (1992) *Sedimentary petrology*. Freeman, N.Y., pp 514

Carozzi, AV (1993) *Sedimentary petrography*. Prentice Hall, New Jersey, pp 263

Clark, A (1993) *Hey's mineralogical index: mineral species, varieties and synonyms*. Chapman and Hall, pp 848

Deer, WA, Howie, RA, and Zussman, J (1966) *An introduction to rock-forming minerals*. Longmans, London, pp 582

Delvigne, JE (1998) *Atlas of micromorphology of mineral alteration and weathering*. Mineralogical Association of Canada, Ottawa and Orstom, Paris, pp 516

Dietrich, RV, and Skinner, BJ (1990) *Gems, granites and gravels: knowing and using rocks and minerals*. Cambridge Univ. Press, Cambridge, pp 173

Frye, K (1993) *Mineral science; an introductory survey*. Macmillan, N.Y., pp 360

Greensmith, JT (1989) *Petrology of sedimentary rocks*. Unwin Hyman, Boston, pp 262

Gribble, CD, and Hall, AJ (1992) *Optical mineralogy: principles and practice*. UCL Press, London, pp 303

Hall, A (1996) *Igneous petrology*. Longmans, London, pp 551

Hyndman, DW (1985) *Petrology of igneous and metamorphic rocks*. McGraw Hill, N.Y., pp 786

Klein, C (1989) *Minerals and rocks: exercise in crystallography, mineralogy and hand specimen petrology*. John Wiley, N.Y., pp 402

Krumbein, CW (1988) *Manual of sedimentary petrology*. Society Economic Geologists, Tulsa, OK, pp 549

MacKenzie, WS, and Guilford, C (1980) *Atlas of rock-forming minerals in thin section*. Longmans, London, pp 98

MacKenzie, WS, Donaldson, CH, and Guilford, C (1991) *Atlas of igneous rocks and their textures*. Longman Scientific and Technical, England and New York, pp 148

Mason, R (1991) *Petrology of metamorphic rocks* (2nd ed). Springer, Netherlands, pp 230

Miyashiro, A (1994) *Metamorphic petrology*. UCL Press, London, pp 404

Nesse, WD (1991) *Introduction to optical mineralogy*. Oxford Univ. Press, pp 335

Perkins, D (2002) *Mineralogy*. Prentice Hall, NJ, pp 483

Perkins, D, and Henke KR (2000) *Minerals in Thin Section*. Prentice Hall, NJ, pp 125

Rapp, G (2009, 2nd ed) *Archaeomineralogy*. Springer Verlag, Berlin, Heidelberg, pp 348

Scholle PA, and Ulmer-Scholle, DS (2003) *A color guide to the petrography of carbonate rocks*. AAPG Memoir 77, The American Association of Petroleum Geologists, Tulsa, OK, pp 474

Tucker, ME (1991) *Sedimentary petrology: an introduction to the origin of sedimentary rocks*. Blackwell, Oxford, pp 260

Welton, JE (1984) *SEM petrology atlas*. American Association of Petroleum Geologists, Tulsa, OK, pp 237

Yardley, BW, MacKenzie, WS, and Guilford, C (1990) *Atlas of metamorphic rocks and their textures*. Longman Scientific and Technical, England and New York, pp 120

Ulmer-Scholle, DS, Scholle PA, Schieber, J, Raine, RJ (2014*) A color guide to the petrography of sandstones, siltstones, shales and associated rocks*. AAPG Memoir 109, The American Association of Petroleum Geologists, Tulsa, Oklahoma, pp.526

Sources on the Internet

Many websites can now be found on the Internet helping with mineral and rock identification, with extremely useful image databases. Even so geology (and geologists) oriented, these sources are valuable reference materials, including for petrography and comparative thin section photomicrographs. Also recommended is the excellent and very well illustrated book about mineral weathering and alteration written by Jean Delvigne (1998). This allows one to understand (and see) the changes occuring at the mineral and chemical level, affecting rocks and minerals, with 'morphism' of one mineral into another one. When analyzing the composition of a ceramic paste, and considering the raw materials involved, it is good to keep this in mind. The picture-perfect crystal characteristics and shapes are not that clear in the real world! Archaeological materials have also spent much time in the ground and been subjected to alterations (in particular chemical ones) that can affect the interpretation of the analysis data.

3 Clay Minerals and Their Properties

3.1
General Concepts

Due to their great importance in the field of ceramics, clay minerals are treated here as a separate chapter. They find their geologic place as sediments and sedimentary rocks, as discussed in Chapter 2.

Clay minerals are the plastic part of ceramics, and it is because of this plasticity that ceramics can be made. As one might expect, not all clay minerals are the same. They form different mineral groups which are defined in geology and soil science by their chemical composition. Again, as one would expect, changes in chemistry will result in different physical changes also. To a certain extent, the chemistry of clays can control their interaction with water, and their plastic properties. Also differences in chemical composition will determine their behaviour under conditions of firing in the ceramic process. Some clays are destroyed or for all purposes become a glass or amorphous at different temperatures, depending upon their chemistry. Therefore it is important to know a little about clay miner- al composition in order to understand why a potter has chosen a given clay type to make his pots. Also the clay type will determine the need to use more or less tempering material.

What is a *clay mineral?* The general definition of clay minerals is that of particle size. In most disciplines, particles of less than 2 μm (micrometers or 0.002 mm) diameter are considered to be clays. This is, of course, not a definition of a mineral type or group. Minerals are defined as a function of their chemistry and crystal structures. Then why do we call particles of less than 2 μm clay minerals? The reason is that most of the grains in nature of this size have a common structure and certain chemical characteristics which can be defined according to the criteria of mineralogy which are chemical and spatial arrangement of the constituent atoms. Of course, clay-sized material is not totally made up of clay minerals, but most of it is. Some other minerals are reduced to this small grain size, by natural or anthropogenic means, and they are found along with clay minerals in the clay size fraction.

Hence there are two terms, one, *clays,* which are all minerals in the less than 2 μm fraction and the other, *clay minerals,* which are minerals in this size fraction with a specific mineral structure.

The structure common to most clay minerals is that of a thin, sheet-like form. Clay particles are somewhat like the pages of paper in a book. As is the case with minerals, the external structure reflects the arrangement of the atoms within the crystal. Clay structures are large two-dimensional linked networks which have strong chemical bonds in two directions and weaker ones in the third direction. Within the crystal, cations are linked to oxygen ions, and in fact cross-linked, i.e. an oxygen ion is shared between two or more cations which creates the tightly bound, *linked network* (Fig. 3.1A). Linked networks can be formed from two or three layers of cations. Always one outer cation layer is made up dominantly of silicon ions. This layer is called the *tetrahedral ion layer* because the silicon ions are surrounded by four oxygen ions which form a tetrahedron figure. The next cross-linked layer of cations is surrounded by six oxygen anions. These ions form an octahedron around the cations present. These are *octahedral ion sites* in the structure. The oxygens between the cation layers are cross-linked, that is, they are shared by electronic attraction to more than one cation (oxygen ion). They can also form part of both the tetrahedral and octahedral configuration (Fig. 3.1 labeled Band C, respectively).

The number of cross-linked cation layers can be either two or three. In the language of clay mineralogists the first is a 1:1 layer structure where two cation layers are present and the second is a 2:1 layer structure where three cation layers are present, one tetrahedral and the other octahedrally coordinated. The first number gives the silicon, tetrahedrally coordinated layer count and the second that of the octahedrally coordinated ion layers.

These cross-linked cation and anion networks (2:1 or 1:1) are stacked one on the other to form a clay crystal (Fig. 3.2a). The structural arrangement of atoms in layers gives the name of layer silicate to clay minerals. The internal arrangement of the atoms in the layer silicates is expressed in the exterior shape of clay minerals. As the crystals grow, they follow their internal structure, forming tabular crystallites. Clay minerals are also called *sheet silicates* because of their sheet-like crystal shapes.

As can be seen in the schematic diagram of Figure 3.1, there are about as many oxygens as cations in clay structures. This is true of most silicates. In the Figure the oxygen ions are linked by electronic covalent bonding to cations on the inside of the unit layers, the surface oxygens have little or no electronic charge on them. They are most often without an electrostatic charge or neutral charge.

A very important feature of clays, of course, is that they contain *hydrogen ions.* This is the chemical reason for weathering and hydrothermal alteration, the substitution of hydrogen ions for other cations as rock-forming minerals are transformed into clays. The hydrogen ions are linked to oxygens, most often in the inner parts of the unit layers. They are associated with some of the oxygens of the octahedrally coordinated cations. A very important fact in this hydrogen content is that hydrogen tends to escape from crystals and other substances as heat is applied. Heating clays produces water. Part of the water comes from the interior of the clay structure (hydrogen ions plus oxygens). As the water or hydrogen ions leave the structure, the clay mineral becomes unstable and it is destroyed to become, most often, a glassy substance. This transformation leaves the clay materi- al in another, more rigid and non-hydratable state. The clay destroyed by heating in ceramic manufacture gives rigidity and physical stability to the object. The hydrogen ions form what is called *crystalline water,* which is expelled as H_2O when the minerals are destroyed by heating.

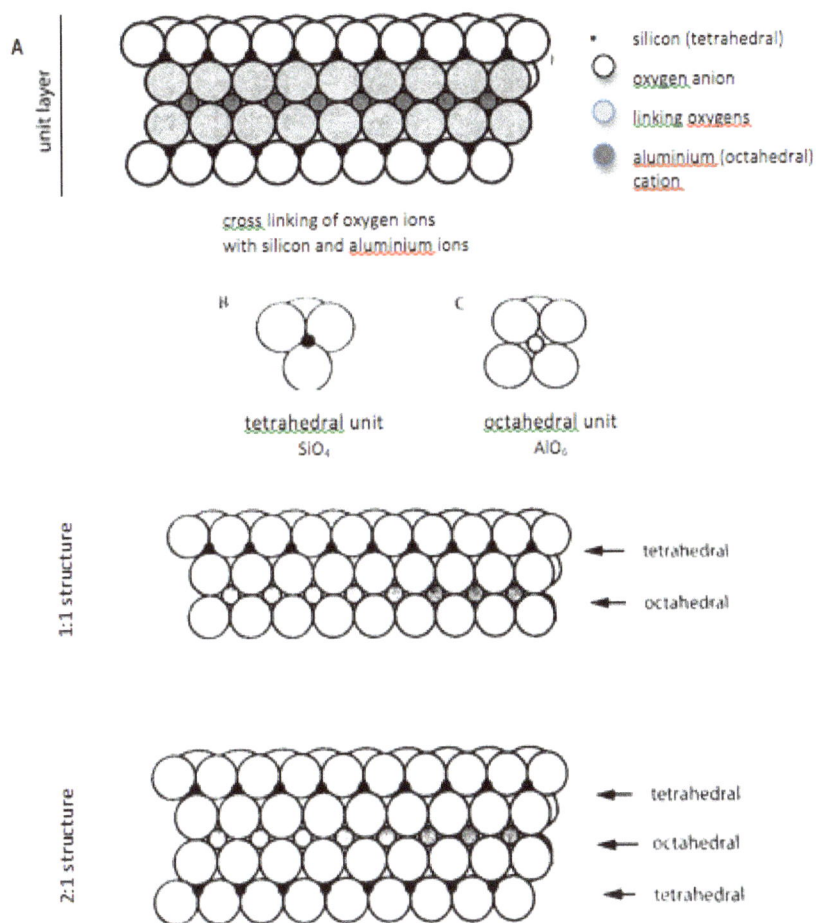

Fig. 3.1. A Diagram indicating the types of atomic stacking and linkage in a clay mineral structure. The largest atoms or ions are oxygen, followed by aluminium and then silicon. Aluminium and silicon ions fit into the holes left when oxygen ions are stacked together. **B** Silicon (Si) forms a coordination unit with four oxygens into a tetrahedral form. **C** Aluminium (Al) coordinates with oxygens to form a unit with six ions in an octahedral form. The *lower figures* indicate the two basic types of cross-linked structures found in clay minerals, 1:1 structure where there is a single layer of tetrahedral ions and a layer of octahedrally linked ions and the 2:1 structure where there are two layers of tetrahedrally linked ions on either side of an octahedrally linked layer.

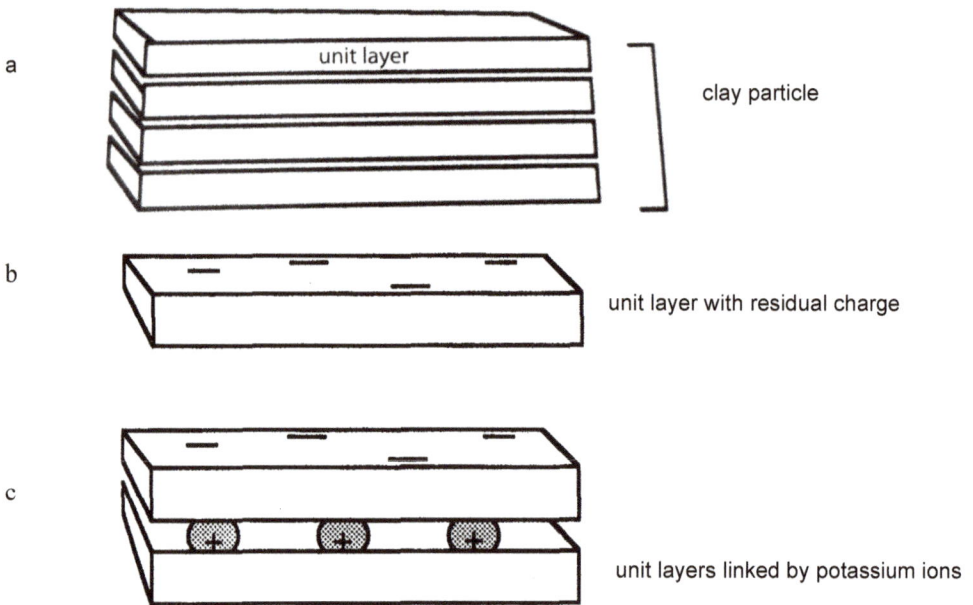

Fig. 3.2. a-c. The cross-linked oxygen cations and coordinated cations form sheet- shaped blocks in clay structures, which are repeated one on top of the other. They are stacked to form a crystal or clay particle (a). The fundamental structural units (1:1 or 2:1 layers as shown in Fig. 3.1) in the clay particle can be electronically neutral or charged through ionic substitution inside the structure (b). These charged unit layers can incorporate other ions in between the layers, creating a complex structure (c).

3.2
Chemical Constitution of Clay Minerals and Clay Mineral Families

We will consider first the composition of the major mineral groups of clays, then the interaction of clays with water and finally the effect of heating on the different clay minerals. This will lay the basis for understanding the use of clay minerals in ceramic manufacture.

Basically, clay minerals can be divided into four large categories, for our purposes. They are mica-like minerals (illite, glauconite and detrital micas), smectites, kaolinite minerals and chlorite and chlorite-related minerals. There are many more mineral names in the geological and mineralogical literature. However, they belong, as defined by mineralogists, to these four groups or mineral types. The clay mineral type is based upon major, fundamental chemical and structural characteristics. These chemi- cal and structural differences lead to differences in behaviour during the ceramic-making process (plasticity, change in volume and resistance to firing) as well as their occurrence in nature or in their geologic origin.

In the discussion which follows, mineral formulae are given which segregate the constituent ions by the type of coordination or site in the structure, interlayer, octahedrally coordinated layer and tetrahedrally coordinated layer.

3.2.1
Mica-Like Clays (Illite, Celadonite and Glauconite)

Mica minerals are those which have the greatest coherence between the unit layers in the clay crystallites. This is due to the fact that there is a residual charge on the unit layers because cations of low charge are substituted for cations of higher charge. For example, when Al^+ takes the place of Si^{4+} in a mica structure, there is a negative residual charge on the unit layer. This charge is compensated by one potassium in micas and slightly less than one potassium ion, K^+, per unit formula in mica-like, clay minerals. This potassium is rather loosely bonded to the surface of the silicate unit layer (Fig. 3.2b) but it holds them in place and allows them to form a three- dimensional crystal. Since the adjacent silicate unit layers have the same residual charge, the potassium ions bond the layers together to form a single, relatively robust crystal (Fig. 3.2c). These crystals can form large units, beyond the clay size definition of 2 p.m. In this size range they become micas in mineralogical terminology. The bonding of potassium between two clay sheet unit layers is not as strong as the silicon Si-O or aluminium Al-O covalent linking and sharing bonds. The site in the crystal structure where potassium is found is the *interlayer* site, indicating that it is inserted between the clay unit structures. If a mica crystal is subjected to differential pressure, it will break along the potassium ion bonding planes or interlayer sites and not across the sheet structures. This is called mica cleaving.

The bonding or interlayer ions in clay micas are almost exclusively potassium. There are two main types of clay micas, those which are essentially silicoaluminous (Si-Al) called *illites* and those which contain large amounts of iron and silicon (Fe-Si) called *glauconites*. Celadonites contain magnesium (Mg) and iron (Fe). These minerals have the following simplified chemical mineral formulae

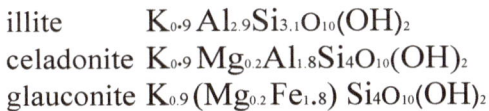

illite $K_{0.9} Al_{2.9} Si_{3.1} O_{10}(OH)_2$
celadonite $K_{0.9} Mg_{0.2} Al_{1.8} Si_4 O_{10}(OH)_2$
glauconite $K_{0.9} (Mg_{0.2} Fe_{1.8}) Si_4 O_{10}(OH)_2$

The cations are traditionally shown on the left of the formula and the oxygen and oxygen-hydroxyl (hydrogen and oxygen) groups on the right. Subscripts show the relative numbers of ions of each sort preceded by the subscript. For example, $Mg_{0.2}$ indicates that there only two tenths as many magnesium ions as potassium in the glauconite formula. There is nine times as much iron (Fe) as magnesium (Mg), and so forth. In fact, the potassium content of metamorphic minerals is 1.0 while that of sedimentary and soil micas (mica-like) is 0.9 ions, which distinguishes the origin of these minerals, aside from their grain size.

Illite is a very common, major component of many soils and clay deposits which originate through weathering. Some illite is, in fact, the result of weathering of larger micas in rocks. Glauconite is much more rare, being found in deposits which form on shallow ocean floors at water depth of less than 200 m. Most glauconite is found in sedimentary rocks (see Sec. 2.3). The glauconite grains are green to green-brown in their natural state, and very easy to identify. It is especially important because it contains a high quantity of iron. When glauconite weathers it becomes oxidized, the iron it contains leaves the silicate clay mineral to form an iron oxide based upon the formula Fe_2O_3. Glauconites and celadonites (bluish green in color) which are more stable under weathering conditions, have been used as a coloring agents in paints from the time of the Palaeolithic or older cave paintings, and as a decorative color on walls in pre-Roman times in Europe, as well as a paint pigment in its oxidized state for interior and exterior buildings into the 19th century in France and Italy among other areas under the name of ochre.

Glauconites and perhaps celadonite have probably also been used, in their natural state, as a pigment and glazing material in Europe in Greek and Roman potteries. If fired in an open, oxidizing atmosphere, glauconites will become red. Under reducing conditions they become black.

3.2.2
Smectites

Smectites are very special clays because they have the property of incorporating water within their structure in a reversible manner. In the presence of water or water vapour, smectites take up 70% of their volume as molecular water. In dry conditions, they lose this water and shrink. They are *swelling clays.* The water molecules are taken into the structure be- tween the sheet structural units of the clay mineral. This is the site where the potassium ions are found in micas.

The reason that smectites take up water in between the structural layers in the interlayer sites is that although they have ionic substitutions in their unit layer structure creating a charge on the surface of the layer as do micas, in smectites these substitutions are "incomplete". Micas have a layer charge of one ion per unit formula (as in the glauconite or illite formulae above). Smectites have lower charges, usually between 0.25 to 0.4 ions per unit formula. Instead of fixing a potassium ion solidly between the unit layers of the crystal as do micas, smectites fix hydrated cations. The reason for the accompaniment of water with the cation is that cations in aqueous solution are surrounded by water molecules. Since the charge between the clay layers is not strong (less than a mica) the layers cannot be ionically bonded in a strong manner and there is "room" for a charged cation-water complex. The cation-water-complexes are incorporated into the interlayer sites of the smectite crystals.

There are three *hydration states* possible for these cations as they are incorporated between the smectite layers. One state is where two layers of water molecules are present, another where only one water layer is present associated with the cations and the last where no water layers are present. Figure 3.3 shows the structural relations of the hydrated layers of cations between the unit layers of the clay.

The hydration state of smectites, the number of water layers associated with cations held between the smectite layers, is a function of the amount of water present in the environment of the clay. Taken out of liquid water the stability of the clay-water complex will change as temperature and humidity change. Water content of the cation-water complex held in smectite clays decreases as temperature increases and humidity decreases. This relation is shown schematically in Figure 3.4. In terms of ceramic processes, smectite clays will contain most water during the plastic phase when clay and water are mixed in the initial stages of pottery forming. Drying in air will release some of the interlayer water, heating will release even more.

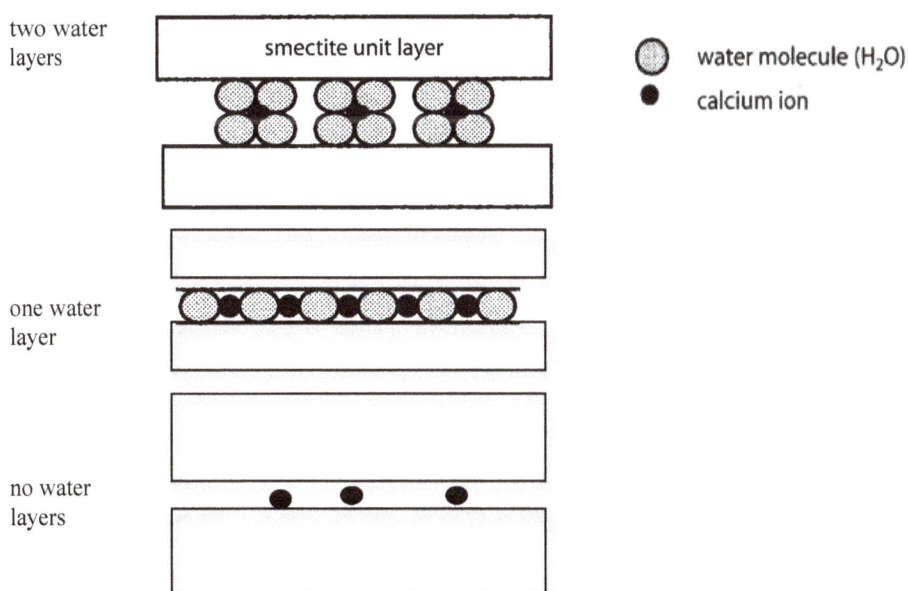

Fig. 3.3. Layer units of low charge (smectites) in a clay crystal can incorporate charged ions in between the layers, which are associated with water molecules in aqueous solution, as indicated in the figure. When the clays are extracted from the water, the charged cations between the clay layers will be associated with more or fewer water molecules, depending upon the conditions of humidity or temperature. The ions can be associated with two layers of water molecules, one molecule, or none.

The cations incorporated between the smectite layers are loosely held. This being the case, one ionic type can be exchanged for another easily when the clays are in an aqueous solution where various cations (hydrated, of course) are present. For example, if a calcic (Ca-substituted) smectite is introduced into an aqueous solution containing sodium (Na), the calcium is exchanged for sodium. This is due to the effects of the law of mass action; the more asymmetry between the ionic types in the clay and in the solution, the greater the tendency for interdiffusion which will equalize (at least partially) the distribution of ions in the clay and in solution.

Fig. 3.4. Relations between humidity and temperature and water retention of charged smectite clay particles. Low humidity and high temperature leave ions without water whereas increasing the humidity or lowering the temperature favor incorporation of water molecules between the layers of charged clay particles.

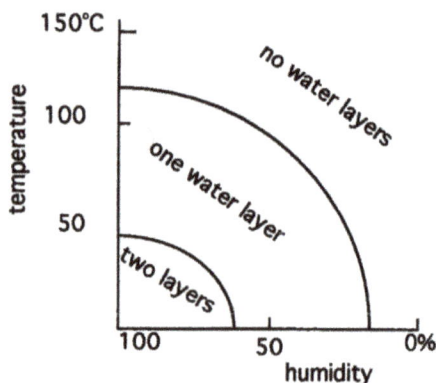

The normal tendency in aqueous chemistry is to make things equal, to iron out segregations of matter. This effect is indicated in Figure 3.5, where hydrated calcium ions are exchanged from the clay interlayer site for sodium ions, which are more abundant in the aqueous solution.

Smectite clays can attract charged organic as well as charged inorganic ions into and on their structure. This capacity to incorporate organic matter into the clays has been used at times to color the clays and to use this mixture as a paint.

The general mineralogical formulations for smectite mineral compositions is

$$M+(Al,Mg,Fe)_2(Si,Al)_4O_{10}(OH)_2 \, nH_2O$$
$$M+(Mg,Fe,Al)_3(Si, Al)_4O_{10}(OH)_2 \, nH_2O.$$

Two formulae are given. This is because two major types of smectite clays can be found in nature. The most common is the first where there are two ions in the octahedral site (see Fig. 3.1), given by the first series of ions between parentheses. These two ionic sites (di-octahedral) can be dominated by aluminium and the minerals are called *aluminous smectite* or the ions in the octahedral site can be dominantly trivalent iron (Fe_3+) and the mineral is called *nontronite*. When three, or nearly three ions are present (the tri-octahedral structure), the site is usually dominated by magnesium and the smectite mineral is called *saponite.*

The interlayer ions, in the interlayer site where potassium is found in micas, is designated *M* because the ionic type can be exchanged from one geologic or laboratory environment to another and thus it is not characteristic of the clay itself. When a smectite has a sodic interlayer-hydrated ion content it is called a sodic smectite, as one would expect. When calcium is present it is calcic, etc.

The nH_2O portion of the formulation, at the end, indicates the presence a variable number (n) of water ions, associated with the exchangeable cations. In fact, the number of ions, n, is variable following the ambient conditions of temperature and relative humidity.

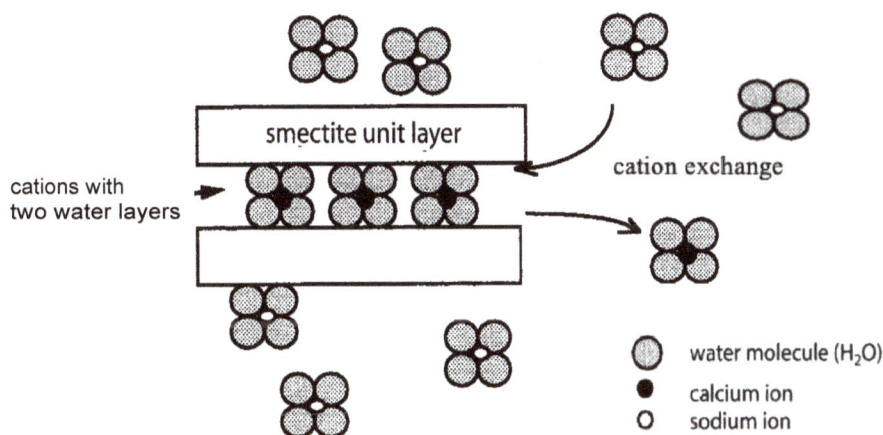

Fig. 3.5. Cation exchange. Charged, smectite clay particles and interlayer ions with their associated water molecules. In aqueous solution, the ions are hydrated to a maximum state. The principle of ionic exchange is shown where the clay particle is initially calcium (Ca)-rich and the solution sodic (Na). The ions in solution are more numerous than those between the clay layers and hence an equilibrating exchange takes place to equalize the ionic concentrations between water and clay particle. Calcium ions leave the sites between the smectite layers and sodium is substituted in their stead.

3.2.3
Kaolinite Minerals

Kaolinite clays are neutral structures, i.e. there is no substitution of ions of different charge within the unit structures of the clay. In fact, kaolinites are of a very restricted composition. Their formula is

$$Al_2Si_2O_5(OH)_4$$

Silicon and aluminium ions are present in the same proportions, occasionally some trivalent iron (Fe_3+) is present. The structure is more hydrated, by crystalline water, i.e. there are more hydroxyl or hydrogen ions present in the unit layers than in the micas or smectites. Micas and smectites have two (OH) groups per ten oyxgens while kaolinites have an equivalent of eight hydroxyls for the same number of oxygens. In the commonly used mineralogical formulation, micas have ten oxygens, while in that used for kaolinite there are only five oxygens. Because there are one octahedrally coordinated layer of Al ions and one tetrahedrally coordinated layer of Si cations, the structure is designated by clay mineralogists a 1:1 structure or mineral.

The small amount of iron or the lack of it in kaolinites is important for ceramists. Pure kaolinite will never change color with heating. It is initially white and will always be white.

3.2.4
Chlorites and Related Minerals

These minerals can be found in clay deposits but are more frequently encountered in rocks that have formed at higher temperatures than those of clay mineral stability. They are most common in low-temperature metamorphic rocks. Nevertheless, chlorites can be found in clay deposits, in soils and sedimentary rocks. They can form in these environments, rarely, but are most often relicts of material formed under other conditions. Some hydrothermal deposits are chlorite-rich. Their concentration makes them attractive in certain cases, therefore one must consider chlorites in ceramic archaeology.

The chlorite structure is based upon the 2:1 model, two silicon-dominated (Si) tetrahedral layers on either side of an octahedrally coordinated cation layer. The octahedral ions are usually mostly magnesium (Mg), containing some aluminium (Al) and iron (Fe). There are three ions in this layer, per mineralogical unit, and hence they are called tri-octahedral minerals. The special characteristic of chlorites is that they have a supplementary layer of ions between the 2:1 units. This is similar to micas, where potassium is found between the unit layers but not quite the same. The chlorite unit layers do not present charge imbalance calling for a cationic substitution between the unit layers. The interlayer unit is composed of hydroxyl-coordinated cations, magnesium, aluminium and iron. Instead of having simply oxygen anions linked to the cations, there are OH units in these sites. The structure can be called a 2:1+1 structure, indicating at the same time its relation to the mica-smectite minerals and its difference due to the hydroxyl interlayer ion structure. The mineral formula is

$$(Mg,Al,Fe)_6 (Si,Al)p_{10}(OH)_8$$

In this mineral the silicon atoms are proportionally less abundant than in the other 2:1 minerals, the micas and smectites. The hydrogen ions are more abundant. However, these ions are tightly held in the structure and, as a result, chlorites are usually stable to rather high temperatures. The structures can be visualized as in Figure 3.6.

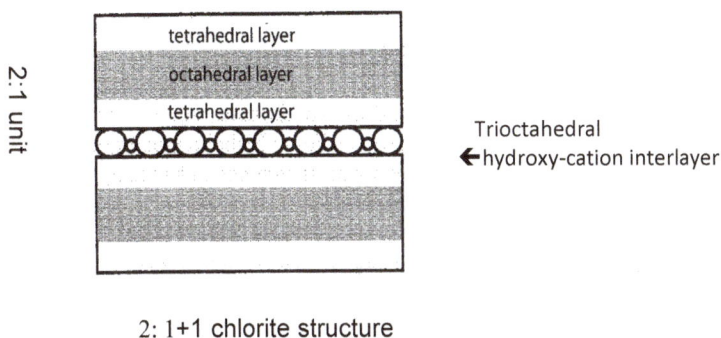

2: 1+1 chlorite structure

Fig. 3.6. Diagram illustrating a hydroxy-cation layer in a clay particle between two 2:1 (tetrahedral-octahedral-tetrahedral layers). These are 2:1+1 structures or chlorites.

3.2.5
General Chemical Identity of Clays

From the above brief description of clays, one can deduce that the clay minerals are made of silica and alumina for the most part. Most of the minerals will be identified by these elements. Hence, it is difficult to distinguish between one or another species of clay using a bulk chemical analysis. Some elements do indicate the presence of a specific clay. For example, the mica- like minerals illite and glauconite are the only minerals to contain potassium (K). Illite has little iron in it and hence high potassium content and little iron would indicate the presence of illite. A high potassium content and high iron content could indicate glauconite; but beware of mixtures! Clays mixed with iron oxides could produce a confusing compositon. Illite plus iron oxide would resemble glauconite. High iron content without potassium indicates a mix of kaolinite or smectite and iron oxide. Chlorites are the only clay minerals which contain notable amounts of magnesium (Mg) and hence the presence of notable amounts of Mg would indicate the presence of chlorite. White clays with roughly equal quantities of aluminium oxide and silicon oxide ($Al_2O_3 = SiO_2$ in weight) indicate that the paste was obtained with almost pure kaolinite.

3.3
Physical Properties of Clay Minerals

3.3.1
Clay-Water Mixtures

The fact that clays are of small grain size lends them to use in ceramic production. The smallest grain size materials will give the most plastic response when mixed with water. Why so? All grains have small, but important charges on their surfaces. This is especially true at the ends, where the structure is incomplete, or unfinished. In micas these ends are on the edges of the sheets or layers of the crystal.

The second factor is that water molecules are very special. The H_2O structure is electronically equilibrated but has a polar configuration. This is easily seen when one considers that two hydrogen ions are attached to an oxygen ion. The oxygen is much larger than the hydrogen ions, and, as a result, they are fixed on a small surface of the oxygen ion. The situation where two positively charged ions must be fixed on the surface of a negatively charged ion makes it evident that one part of the oxygen will not have compensating ions and will have a slightly residual negative charge at its surface. Further, the two hydrogen ions are situated at 120 °C from one another on the surface of the oxygen ion, presenting an aggregate positive portion of the three-ion assemblage. Hence, the water molecules are polar, or have a positive and a negative side to them. This being the case, when water molecules surround a clay particle, they tend to form several attached layers around the silicate crystal surface, compensating the residual charges on the mineral surface. If the mineral has positive charges, the negative end of the water molecule is attracted to the surface and the other water molecules line up, positive to negative, and vice versa out from the clay particle. These water molecules are not firmly attached and can be removed by heating to some 110 °C or so, but all the same they are persistently present at lower temperatures.

Clays have a high surface-to-volume ratio due to their sheet-like form. Figure 3.7 indicates this concept, where three grain shapes are represented in two dimensions, a sphere (by a circle), a cube and a sheet silicate. The different shapes have the same volumes: the circle has the smallest surface, followed by a square and the sheet. Clays, with a sheet structure, will have a greater surface area than other grains of the same volume. They will attract more water to their surfaces than other grain shapes and mineral species. It takes more water to wet them and more energy to dry them out. Adding water to a clay powder gives first a sticky, coherent mass. The clays absorb the water onto their surfaces and bind it to them. Since there is a limited amount of water, the clays are stuck together by sharing a water molecule. The several water layer "skin" developed on the surface of the clay acts as a lubricant so that the clay particles can glide one over the other under the pressure of a potter's hands. A higher water content of the clay-water mixture is plastic, i.e. easily deformable but retaining the shape given by the deformation.

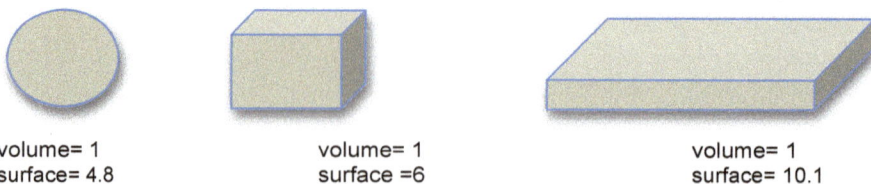

volume= 1
surface= 4.8

volume= 1
surface =6

volume= 1
surface= 10.1

Fig. 3.7. Illustration of the relations between surfaces and volumes of different shaped particles. Each has the same volume but the surface is much greater for the sheet-shaped particle (a clay shape) than for a sphere.

Adding much water to clays destroys the plastic properties, making a slurry suspension which is a liquid, not a plastic. However, the clay particles remain in suspension, they "float" due to their small grain size, which allows larger, or more dense particles to settle to the bottom of the suspension. If the suspension is not too dilute, the clay materials can be applied to the surface of a pottery piece as a slip or paint. The mixture is liquid and hence easy to apply in a thin coat. Drying, or water loss, leaves the solids on the pottery piece (see Sect. 5.3.2).

Adding still larger quantities of water is a method of cleaning natural clay deposits of their impurities. Only the very fine grains remain in suspension and the larger particles fall to the bottom of the aqueous mixture. In dilute mixtures clays can attract other molecules to their surface which can be coloring agents or have other affects. For example, the black paints on classical era ceramic objects and the shiny black glazes on Campanian (pre-Roman ceramics from southern Italy) have been obtained by mixing organic matter with the clays and firing them. The interaction of organic combustion during the transformation of the clays promotes the crystal- lization of reduced iron oxide giving an indelible black color which does not alter during the subsequent firing cycle. The same effect can be seen on some pre-Columbian ceramics from South America.

3.3.2
Clay Shapes

The internal crystal structure of clays gives them an outward physical expression of a sheet-like particle. This shape confers specific properties in itself. If one takes a sheet of paper, or a leaf, and lets it fall through the air, it automatically orients itself against the forces that make it move. Tree leaves float in a zigzag pattern as they resist the air in their fall to the ground. Clay particles in water suspension orient themselves against the forces of gravity and resist their fall to the bottom of the beaker. In their resistance they orient themselves in a parallel fashion, creating a dense, thin layer of material. This is indicated in Figure 3.8. Larger clay particles will settle in a parallel structure forming mud cakes or shiny surfaces at the bottom of puddles. When a clay-water mixture is painted on a ceramic surface or applied by dipping the ceramic in the viscous mixture, the clays will orient on the surface of the ceramic as the water is evaporated by drying. The clay deposit is oriented into a homogeneous, dense mass.

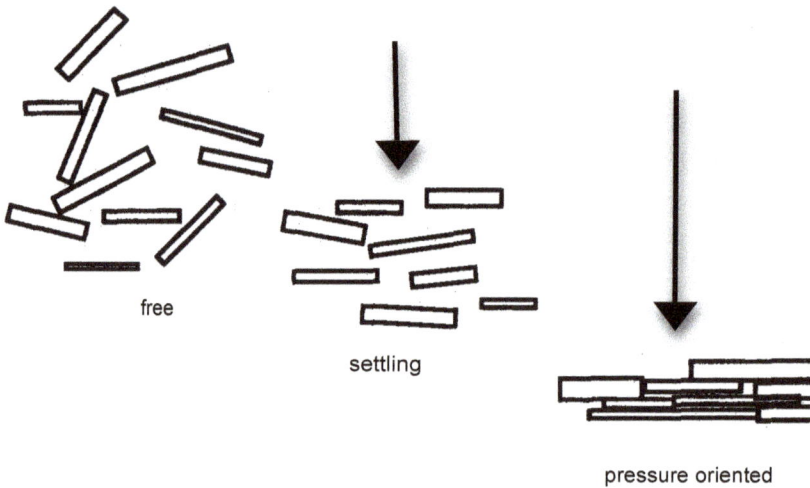

Fig. 3.8. Illustration of clay particle orientation against an applied force. Irregular, open associations of clay particles are typical of a situation where no force is applied. Orientation begins with force, such as particles settling under the force of gravity. Total orientation occurs with a directed effort, and the particles show a maximum orientation, perpendicular to the force. Pressure orientation is common in clay pastes formed by a potter.

Many times in the past, potters wished to give a lustrous surface to a ceramic. In many cases, they could simply use a stone or another hard, smooth object to "polish" the surface by pressing it on the object as it rotated on the potter's wheel, orienting the clay particles at the surface and forcing temper grains below the surface of the object. The polish by pressure is effected by orienting the clay grains perpendicular to the force applied. Hence, the structure of clay particles lends itself to physical orientation of the clays and a certain number of effects which can be used for aesthetic reasons or for technical ones.

When the structure of a ceramic is modified by the work forces applied to it in forming the clay paste into a desired object, the clays, voids, and even the sand grains in the interior of the ceramic may show an oriented, parallel arrangement indicating the direction of forces applied to the clay-temper mixture in its plastic state. They are oriented in a direction perpendicular to the major force for the most part, as illustrated in Figure 3.9. The example chosen is not one of clays, they are too small to be seen conveniently under a microscope (less than 2 μm), but the general aspect, orientation of voids and elongated minerals (such as micas) are visible, and many are part of the clay matrix. The pressure of the potter's hands shaped the clay mass strongly and the flat particles aligned perpendicular to the pressure (and parallel to the ceramic walls).

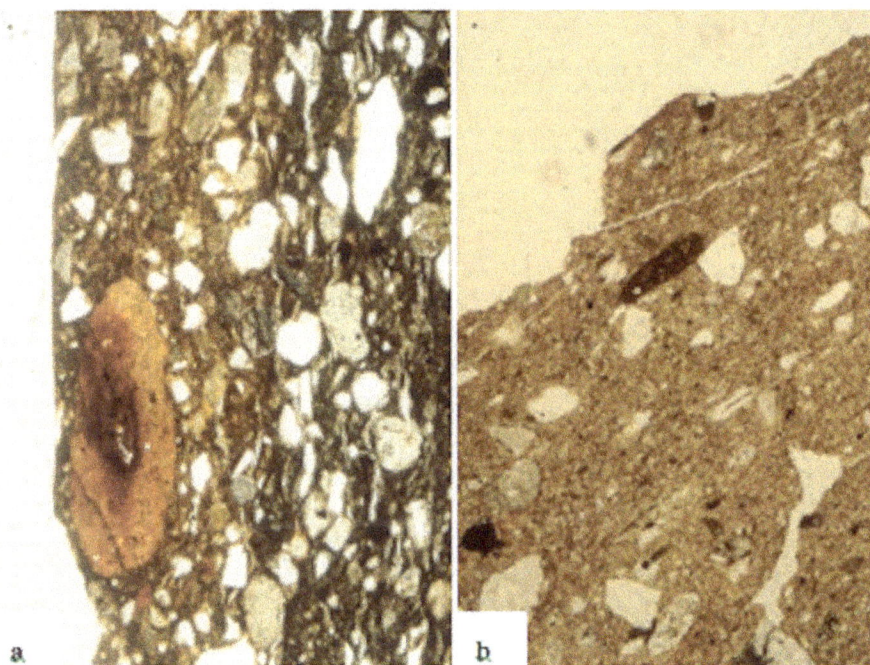

Fig. 3.9 Most of the elongated sand grains and voids in these two photomicrographs are oriented in the same direction, parallel to the ceramic wall. This is also seen in (**b**) with a general orientation of the finer fraction in the clay matrix parallel to the surface. In both cases, this feature results from pressure applied perpendicular to the walls, giving the strong particle orientation seen here. The fine diagonal fracture in (b) is post-depositional and not part of the manufacture.. (**a**) KW41, Peru, 1st mill B.C. coarse Fe-rich mudstone is 1mm long, 4x, ppl (**b**). PU146 4x ppl.

In summary, clays have a strong effect on the physical state of a clay-water mixture, ranging from a plastic mass to a thick slurry useful for covering an object entirely (slip), Thus, they make up the basic matter of a ceramic, as well as the coating of the surface. The shape of clay particles gives them the property of orientation against the forces of pottery making, which can help in determining the method used to form and finish the ceramic product.

3.4
Thermal Stability of Clays and Clay-Water Mixtures

In this section we will talk about the stability of certain organisations of materials. Stability has a sense of the absolute. It is an ultimate term of the existence of an object in a given form. This means that if the physical conditions are maintained for an infinite period of time, things will be always the same. Time is the key word in considerations of stability of materials under heating conditions. Stability assumes that time does not count, that it is infinitely long. In the world of a potter, this cannot be. One heats a furnace, it costs money or effort to keep it hot, and the pots must be fired. The amount of time and the temperature attained are strictly controlled by economics. If it takes 100 h (t) for a clay body to attain its thermal stability at a given temperature x, this is too long. Therefore, one can create a temperature of x + 1and obtain the same result as one of stability in t-1 time. In the processes of mineral stability one can use time and temperature as complementary factors. Both time and temperature are factors in the reactions of clay and water. However, initially we will consider the cases of laboratory work, which does not have an infinite amount of time to operate in.

Clays in water tend to attract a significant amount of this water on to their surfaces. This water helps to lubrify them, one against another, and gives them plasticity. However, during the process of transforming a plastic mass into a solid and rigid one, forming a ceramic, this water must be taken away. There are other types of water found in the clays and it must also be expelled before one can make a rigid and stable ceramic vessel. The loss of the different water types comes about in four steps. The first is the loss of free water which is not chemically attracted or fixed on the clays but present as a general lubricant. The second water is surface water. This is water which is chemically attracted to the surface of the clay particles. The third water is that bound chemically to the clay through interlayer or other hydrated cations. The fourth water is that found within the structure, crystalline water which is the hydroxyl function in the formulae of clay minerals shown in Section 3.2.

The dehydration of clays is measured in the laboratory by heating the clay-water mass at a given rate, a certain number of degrees per minute, usually 10. One observes the loss of weight as each temperature is attained. Figure 3.10 gives these analytical results for several common clay types. Typically, the laboratory analyses do not account for the excess water used by a potter to create a highly plastic material. The laboratory analysis starts from a clay hydrated only by water in the atmosphere, the ambient humidity. Thus, in the Figure 3.10 one does not see the importance of losing the free and abundant water of plasticity. This water can, in fact, be largely lost in a dry atmosphere over a period of several days or less. The amount present is variable, more so than indicted in Figure 3.10. If a potter has a highly plastic material, it can contain up to 20-30% H_2O. This water must be eliminated slowly so as not to deform the structure of the object by cracking or slaking of the paste. Initially, potters use a drying stage in a dry, well-aired space. The free water can be eliminated at temperatures well below 100°C.

The second operation consists of eliminating the weakly held water which is associated with the surfaces of the clay particles. This water is loosely held but should be lost at temperatures near 100 °C. In the laboratory example given, the clays kaolinite and illite have only two types of water that absorbed on their surface and crystalline water. The absorbed water is lost at temperatures near 100 °C. This is true for all clay minerals and, in fact, for non-clay minerals. However, the case of smectite gives the greatest amount of weight loss. Most other clays have only several percent of water attached to their surfaces in the air-dried state.

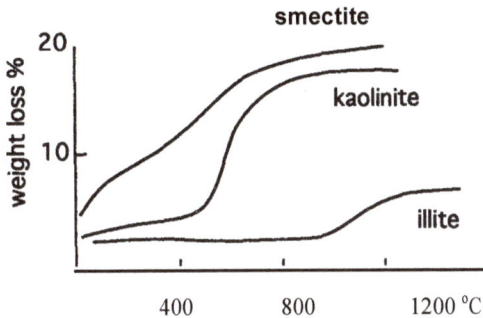

Fig. 3.10. Diagram showing the weight loss (essentially water) as a function of temperature for firing at 1h intervals for different clay mineral species. Smectite loses water before kaolinite, before illite. In general, the more water a clay contains, in weight percent, the lower the temperature at which this water is lost.

The third type of water in clay-water assemblages is that associated with cations that are attracted to the clay surfaces and specifically to the internal sites of smectites (Sect. 3.2.2). This water is lost in the temperature range of 120-200 °C. The curve of weight loss in Figure 3.13 shows a plateau or constant weight for smectite when this interlayer of chemically bound water is lost. In these first three stages the quantity of water lost is variable but can be important, several percent to tens of percent. If the loss is abrupt, or if the critical temperature is attained in a short period of time, the structure of the ceramic is likely to be lost. Thus, these stages of water loss are undertaken with caution and slowly.

The loss of crystalline water, the hydroxyl units (OH), occurs at different temperatures for different minerals. Smectite and kaolinite lose this water near 550 °C while illite or muscovite lose their crystalline water at 1000 °C at a heating rate of 10 °C/min.

The correlative of weight loss is loss in volume of the clay mineral. Figure 3.11 indicates the loss in volume for the three minerals considered in Figure 3.10. The volume change is, in fact, the important factor in ceramics. This is the shrinkage factor which must govern the shape of the initial clay body so that it can find its desired dimensions upon firing as a finished product. A rapid loss of volume cannot be accommodated in a thick-bodied object. There is not enough time for the inside to adjust to the change in the shape of the outside. Thus, temperature programs for thick jars will show a slower increase in temperature than could be accomplished in a thin-walled vessel.

In general, the initial loss of free water can be effected by air drying. The loss of loosely bound water is accomplished during the initial stages of heating in a firing cycle. In the case of smectites, the bound water must be lost slowly in this initial stage also. The temperature range of 100-200 °C is critical here. The next step is the loss of crystalline water, where the clay is transformed forever into another mineral or an amorphous form. This occurs in the higher temperature range 500-1000 °C. The change in weight and volume is less important here and the firing program takes this stage less into account.

Finally, at still higher temperatures, the clays lose their crystalline structure and eventually melt. Here, the volume change is great and after some time at high temperature, new minerals form. For the most part, the temperatures, after long heating periods above 1000 °C, effect melting and re-crystallization. Figure 3.12 indicates, in a general way, the different reactions of clays to heating.

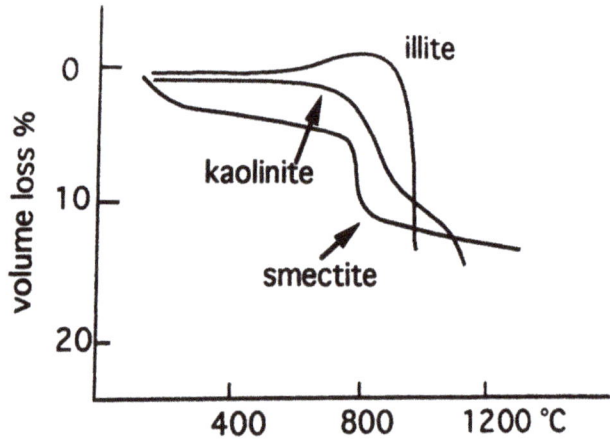

Fig. 3.11. Volume loss upon heating for 10 min per 10 °C for illite, smectite and kaolinite.

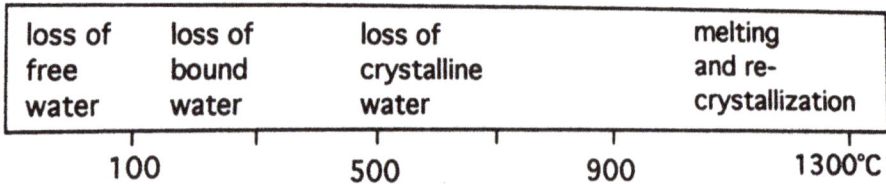

Fig. 3.12. General range of temperatures at which different clay transformations occur. As temperature is higher, more time is necessary to accomplish the reaction indicated. Total loss will be a function of time and temperature.

3.5
Kinetics

The rate at which water is lost from clays and the transformation from crystalline to amorphous material occurs is due to several factors. These are factors governing the kinetics, or speed of reaction. The first variable is the thickness or size of the clay grain. The thicker or larger the grain, the more slowly the entire mass will react. The length of time needed to reach a certain temperature is important also. The more slowly a temperature is reached, the more time is available for the clay minerals to react. The final product will be more "cooked". Finally, the length of time that a maximum temperature is maintained is of still greater importance. The longer the time a material is held at high temperature, the more it will react and the more completely it will react. Thus, it is evident that the heating rate (rise in temperature in the object) and the duration of the heating to which a clay mineral is subjected will determine its reaction. The reasons for this can be explained as follows.

3.5.1
Grain Size

The thicker an object, clay particle here, the more slowly it will release its water. Water release occurs by diffusion of the H_2O molecules from the inside of the particle to the outside. The thicker the particle, the further the water has to travel to reach the surface of the grain and its final liberty as a gas. Hence, under the same conditions, thick grains will be destabilized and lose their water more slowly than small grains. In a large crystal, the distance that a water molecule must travel to escape the structure is greater than in a smaller grain. Other things kept constant, the time necessary for a crystal to be destroyed will be roughly related to the square of the radius of the grain. Figure 3.13 indicates this relation for muscovite or illite. In this clay mineral, only crystalline water is present in great amounts, about 6 % by weight. Thus, the size of the clay particle can be a factor in the rate or speed with which it reacts to temperature.

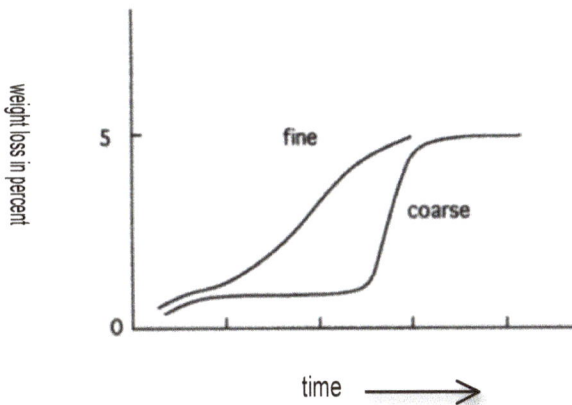

Fig. 3.13. Effect of grain size on the rapidity of water loss. Smaller grains lose water faster because the water has a shorter distance to follow in order to be liberated.

3.5.2
Heating Rate

The next variable is the rate at which heating is effected. Does temperature go up rapidly or slowly? Is it maintained for a long period or a short one? The effect of the duration of a heating event on a clay mineral and its water retention is illustrated in Figure 3.14. It should be noted that laboratory determinations of mineral stability are based upon heating rates of about 10 °C min. This is a greater rate and faster rise in temperature than in many archaeological contexts. In the examples given here, the temperature is maintained for periods up to 100 h. The reaction of the clays is not the same concerning their weight loss (and therefore volume change). Heating at high temperatures changes the weight of the clay rapidly while heating at lower temperatures changes the weight at different rates. Temperatures of 100 °C or so will not change the weight of the clay even after long periods. High temperatures, above 1050 °C, change the weight of the clays very rapidly and thereafter not much change is seen. These are the extremes of heating rates. At intermediate temperatures, the effect of time is very great in the weight loss. For example, at 650 °C heating for 3 h changes the weight by 2%, but after 10 h the weight loss is doubled. In these regions of time- temperature variables, the heating rate of the length of time that a furnace temperature is maintained is very important to the evolution of a clay miner- al in a ceramic paste. One can see that the length of time that a firing operation is maintained can effect much the same clay mineral transformation as that of temperature. These are the two variables of clay mineral change.

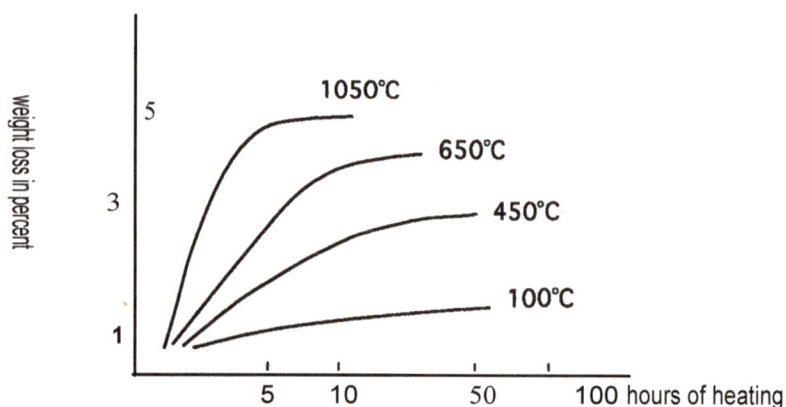

Fig. 3.14. Illustration of the effect of time and temperature on the loss of water (weight loss) from a clay sample. Note that the time scale is on a log scale, the distance between 50 and 100 hours is much smaller than between 10 and 50 hours. This is necessary in order to show the relative losses. On a normal scale the changes would be difficult to see graphically.

3.6
Summary

Clays can be seen as small, sheet-like particles which attract water to their surfaces. This attraction gives a plastic property to the clays. They can slide over one another as they are deformed, pressed or pushed to make an object of a desired shape. Hence, clay paste is closely identified with water content. The water thus added to clays to make a pliable, plastic paste must in the end be evacuated in order to produce a solid ceramic. The drying process which liberates the added water is slow but can be done at normal atmospheric temperatures.

Different clays attract different amounts of water to their surfaces; the smaller the grain, the more water is attracted. Smectite clays have water molecules inserted within their structures, associated with cations such as calcium (Ca). This water is evacuated at higher temperatures, above 110 °C. Much water is present and this changes the volume of the smectite clays greatly. Smectites are not very good clays for ceramics in that respect.

Crystalline water is present in all clays. It is firmly held and much thermal energy is necessary to loosen it. Depending upon the clay mineral species, this water is liberated between 400 and 800 °C. The loss of crystalline water destroys the clay mineral structure and renders the clay into an irreversibly rigid structure with rigid shapes determined by neighbouring particles. Loss of crystalline water is the initiation of the formation of a rigid ceramic.

At still higher temperatures, above 1100 °C, the clay material which has lost its normal crystalline structure but which is still in the rigid solid form melts to become a silicate liquid. In this liquid, new crystals begin to form and the process of recrystallization has begun.

One can hasten or prolong this process by changing the parameters of time as well as temperature. The succession of dehydration steps can be controlled by heating rapidly and initiating each step in rapid succession. Otherwise, it is possible to heat at a lower temperature but for a longer time. These are the parameters of kinetics, the function of time and temperature.

There is an equivalence of time and temperature. One can substitute one for the other to a certain extent. A long time of heating can often be the equivalent of a high temperature. A high temperature can do the work, transforming clays, of a long firing period. This is the key to kinetics. Time, temperature and grain size can play the same roles or give the same effects. Thus, they provide several variables in the potter's bag of tricks. It is possible to arrive at the same result with different methods of firing. This is a key to archaeological interpretation.

Bibliography: Clay Minerals

Bailey, SW (ed) (1988) *Hydrous phyllosilicates*. Mineralogical Society of America, Washington DC, pp 725

Bohn, HL, McNeal, B, L, and O'Connor, GA (1985) *Soil chemistry*. Wiley, New York, pp 341

Brindley, GW, Brown G (1984) *Crystal structures of clay minerals*. Monograph 5, Mineralogical Society, London, pp 496

Dixon, JB, and Weed, SB (eds) (1989) *Minerals in soil environments*. Soil Science Society America, Madison Wise, pp 1244

Fripiat, JJ (ed) (1982) *Advanced techniques for clay mineral analysis*. Elsevier, Amsterdam, pp 235

Gard, JA (ed) (1971) *Electron optical investigation of clays*, Monograph 3, Mineralogical Society, London, pp 383

Greenland, DJ, and Hayes, MHB (1981) *The chemistry of soil processes.* Wiley, New York, NY, pp 714

Grim, RE (1962) *Applied clay mineralogy.* McGraw Hill, New York, pp 422

Mackenzie, RC (1957) *The differential thermal analysis of clays.* Mineralogical Society (GB) Monograph, London, pp 456

Newman, ACD (ed) (1987) *Chemistry of clays and clay minerals.* Mineralogical Society (GB) Monograph 6, Longman Scientific and Technical, London, pp 480

Velde, B (1985) *Clay Minerals: A physicochemical explanation of their occurrence.* Elsevier, Amsterdam, pp 427

Velde, B (1992) *Introduction to clay minerals.* Chapman and Hall, London, pp 198

Velde, B (ed) (1995) *Origin and mineralogy of clays.* Springer, Berlin, pp 334

Weaver, CE (1989*) Clays, muds and shales.* Elsevier, Amsterdam, pp 819

Wilson, MJ (ed) (1987) *A handbook of determinative methods in clay mineralogy.* Blackie, Glasgow, pp 308

4 Origin of Clay Resources

In this chapter we wish to indicate the different origins of clays minerals, clays and clay resources used by potters. For the most part, this information forms a portion of geological science and geomorphology. First the clay minerals are created; they are concentrated and deposited in a "final resting place" where potters can find and extract them for use. Understanding these processes will help to understand the variation found in ceramic materials.

4.1
Segregation of the Elements by Weathering

The origin of clays starts for the most part with rocks. Rocks have their origin generally somewhere below the surface of the earth. They are hard and compact because they have been compressed by the weight of sediments or other rocks at some depth at significant temperatures. When these dense materials are brought to the surface at the air-water-rock interface through mountain-building forces, or volcanic eruption, they are unstable, having been brought out of their natural "habitat".

Rocks are unstable in the rain. Rainwater, combined with atmospheric carbon dioxide (CO_2) becomes slightly acidic, containing an excess of hydrogen (H+) ions, and this attacks the minerals in the rocks. Under these atmospheric conditions, rocks become "hydrated", overall they exchange hydrogen ions for other similarly charged elements (cations) in the crystals which compose the rock. This phenomenon is called chemical weathering when it occurs at the surface of the earth. This weathering is essentially an exchange of hydrogen for dif- ferent ions such as sodium (Na), potassium (K) or magnesium (Mg) in minerals. In effecting this exchange *new minerals are formed*. They are generally *clay minerals* when they are silicates. Some elements, especially iron (Fe) tend to form *oxides,* i.e. they combine directly with the oxygen in the air to form a phase or mineral of the element and oxygen. The elements easily expelled from minerals are absorbed into the slightly acidic water as ions, such as Na+, for example. These are *soluble elements.* The end result of these chemical alteration processes is to produce clay and oxides, which is the basis of mud, and to produce ion-charged water. This is shown schematically in Figure 4.1. The interaction of acidic water and rocks, weathering, is one of *segregation* of the major elements into new minerals. The cation elements which are found in new clays of weathering origin are silicon (Si), aluminium (Al), hydrogen (H) and some iron (Fe) and magnesium (Mg).

One also finds some potassium (K) permanently fixed in the mineral. The oxides are mostly iron (Fe) forms.

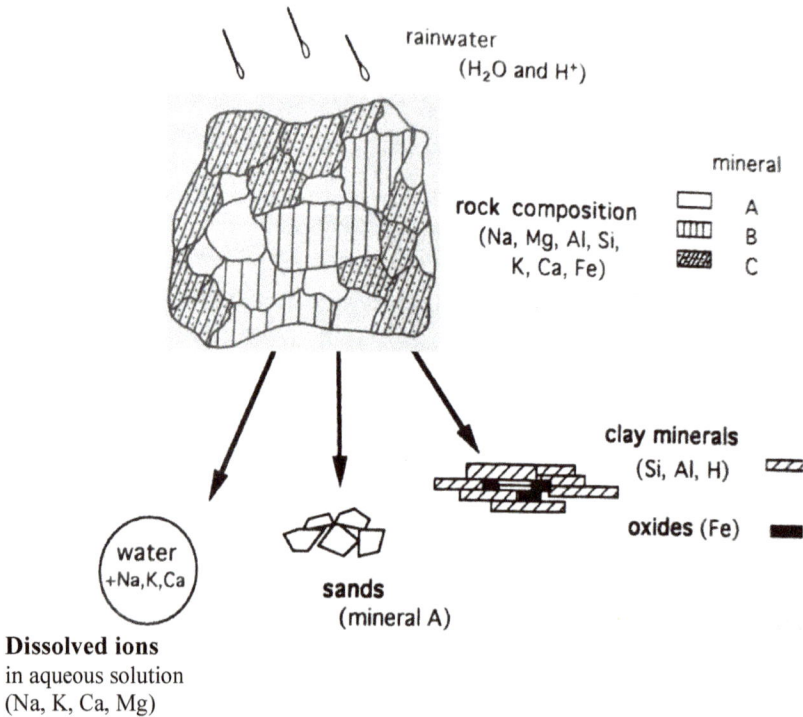

Fig. 4.1. Schematic illustration of rock alteration. Mineral components of the rock are illustrated as minerals *A* to *D*. Hydrogen-rich rainwater falling on the rock alters the minerals by introducing hydrogen ions in the solids in place of ions such as sodium *(Na)*, potassium *(K)*, calcium *(Ca)* and magnesium *(Mg)* which become dissolved ions in solution. Some minerals do not alter readily, such as mineral *D*, and become sand grains or of smaller size. Some elements form oxides (notably iron, *Fe)* and others clay minerals *(Al* and *Si)*. Three materials are produced by the alteration process, solutions containing soluble elements, those exchanged for hydrogen ions, unaltered solids and new solids which are clays and oxides. The clays and oxides are predominant in soils.

Calcium (Ca), sodium (Na) and, to a slightly lesser extent, magnesium (Mg) and potassium (K) are taken into aqueous solution. The fate of most of the solution is to eventually find its way into the ocean where a large portion of the cations are used by animals to make their carbonate-rich shells (Ca and some Mg). Sodium remains in the sea, giving it its salty character. Soluble elements leave the site of weathering.

During the interaction of rainwater and rocks to form clays and oxides, some of the mineral grains in the rocks do not react to the rain. These are part of the alteration product, they are leftovers. They are found in granular form as sand or grits. The segregation operated by chemical weathering then produces two of the major components of ceramics, clays (new minerals) and sand or grits (old minerals). Soluble elements leave the site of weathering.

4.2
Weathering of Minerals

It has long been noted that the different minerals interact with acidic rain during weathering at different rates. In the same soil developed from a rock with different minerals in it, some minerals will be more transformed by the weathering interaction than others. The stability of minerals during weathering is guided by two principles, temperature of formation and chemical composition. In a general manner, minerals which have a high-temperature origin will be less stable than those formed at a temperature nearer that of the earth's surface. Thus, minerals in volcanic rocks formed at temperatures above 1200 °C will be less stable than minerals formed in volcanic rocks whose origins were near 800 °C. Basalt minerals will be less stable than those of granites. Olivine and pyroxene will be less stable than potassium feldspar and quartz.

Minerals rich in magnesium and iron (divalent Fe^{2+}) will alter more rapidly than minerals rich in aluminium. Thus, chemistry has an influence also. However, both chemistry and temperature often go hand in hand, making high-temperature rocks rich in iron-magnesium minerals, while those formed at low temperatures are richer in aluminium (Al), which increases the stability of the mineral.

In the same rock, an assemblage of several mineral species, each miner- al will respond at a different rate to weathering. In granites, where potassium feldspar, plagioclase (Na-Ca feldspar) and quartz are present, the plagioclase is the first mineral to alter. Next one sees potassium feldspar which is affected by weathering. The last mineral is quartz, a very stable one. It is possible to rank the stability of minerals relative to each other. This is done below in Table 4.1. One can match the stability of individual minerals with the mineral assemblages in different rock types in order to obtain an idea of the effects of weathering on rocks and their constituent minerals.

Table 4.1 Mineral stabilities during weathering

Increasing Stability ---------- >

Least   Most

Olivine
   Pyroxene
      Amphibole
         Plagioclase (Na-Ca)
         Biotite mica (Mg, Fe-rich)
            Potassium feldspar (K)
            Muscovite (Al)
                  Quartz

It is important to realize that the impact of weathering is variable, depending upon the rock and the minerals in it. Thus, in a soil developed upon granite, for example, the biotites and plagioclase will normally be highly transformed into new clay minerals where the potassium feldspar will be little affected as well as the muscovite. The quartz will be unaffected, at least as far as one can see with an optical microscope.

4.3
From Rocks to Soils to Sediments

Chemical weathering produces clays and sands or grits, as indicated above, which are the major materials in ceramic production. These are found in soils which are the products of weathering. Usually, soils contain some organic matter also which colors them and affects their chemical response to firing. However, soils are not static. They tend to move in the geological landscape.

The forces of geological action, mountain building and erosion, give rise to the common features we see at the earth's surface, our habitat. Rivers and lakes, flood plains and beaches, river deltas and coral reefs all are important sources of geological variation and they act on the major materials found in pottery, clays and sands. The geologic cycle of forming rocks and eroding them is important because there is a great segregation of elements and materials, and often a higher concentration of elements in different assemblages produced than that found in the rocks which were eroded. The following section is designed to give an idea of the interrelation of geological forces and the formation and accumulation of pottery-making materials, clays and sands or grits. In the diversity of formation one can see that different sources will give different materials and thus a correct identification of each will aid in identifying the different sources of materials in a ceramic.

4.3.1
Weathering Profiles

The interaction of rainwater and rocks produces clay minerals and oxides, as mentioned above. One characteristic of the water-rock interaction during weathering is that it is rarely complete, i.e. some of the rock-forming minerals are not transformed. The fragments left behind tend to be monomineralic. Sand grains usually contain only one mineral crystal. The larger the grain size, sand to gravel, the more multimineral clasts, or rock fragments, are present. The most common mineral which is not chemically reacted during chemical weathering is quartz, SiO_2. This mineral is common not only in many types of rocks, sedimentary, metamorphic and igneous, but also in soils.

Weathering profiles show a concentration of new and fine-grained minerals near the surface where rain-rock interaction is the greatest. The proportion of unreacted material increases with depth. The amounts of these different size materials is then a function of depth in a weathering or soil profile as indicated in Figure 4.2 In the upper soil horizons clay is predominant and the proportion of larger grains, silt then sand then gravel, increases with depth until one reaches the bedrock or the unaltered rock source. The rainwater coming from above has its greatest impact near the surface where its acid content (hydrogen ions) is highest. As the water moves downward, the hydrogen ions are exchanged for other ions which enter the solution. The rainwater is gradually impoverished in its hydrogen ion stock and then it interacts less with the rock and altered silicate material. As the water moves down- ward in the profile, it loses its power to alter the minerals in the rock. However, the solutions moving to the groundwater table in the bottom of the profile are charged with dissolved rock in the form of soluble ions. Overall, the weathering process has the effect of reducing the amount of solids, as much rock enters directly into solution. Because the clay minerals formed by water-rock interaction are very small (< 2 μm) they tend to remain in physical suspension in water, "floating" due to thermal agitation or Brownian motion, even though they are denser than water (density = 2.5). Clay particles are what give a clouded, muddy color to rivers and some lakes. Thus, rainwater in weathering removes not only dissolved ions but also some solids in the form of clays in suspension.

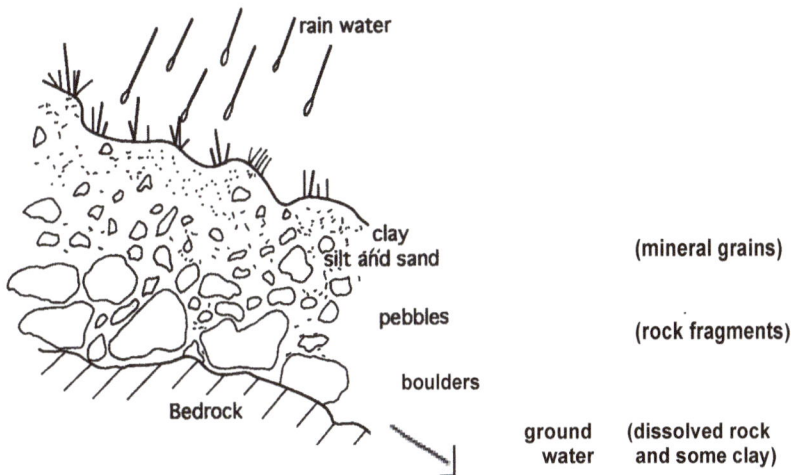

Fig. 4.2. Weathering of rocks showing the distribution of different elements such as rock boulders, pebbles and other fragments along with the sand and clay materials as a function of depth in the weathering zone.

A very important part of the soil profile is its very top. Here, one finds the organic remains of plants in various stages of decomposition. At several centimeters depth this organic matter is intimately mixed with the fine clay and sand particles. This material gives a black or brown aspect to the clay-rich top part of the soil profile. In soils with much organic matter, one will find black or grey clay fractions. In many ceramics one sees, especially in the center of the body, a black to grey color. This is the sign of an organic-rich soil source of the clays used to produce the ceramic. Upon firing, depending upon the variables of the firing program (see Chap. 5.4), the organic matter is destroyed and the clay-sand mixture of the core becomes grey and eventually brown or red. Such changes indicate the firing technique used.

Some of the variables which affect soil thickness are climate, age and topography. Weathering profiles are affected by the amount of water available (rainfall) and the length of time over which there is water-rock interaction. For example, rocks in the Arctic or on mountain tops are little weathered because water is most often ice. Ice cannot interact with the rock to ex- change hydrogen for rock cations. In the same way, desert soils are not very thick. Little rain falls, and when it does it is so sudden and in such great quantities that there is not much time for it to interact with the rocks. These torrents erode most of the soils that could have developed.

Weathering profiles are affected by temperature, as one would expect, chemical reactions are speeded up by temperature increase. Therefore, weathering reactions are faster in the tropics than in the Arctic. Hence, cold climates, dry climates and rocks on steep slopes have shallow soil profiles. By contrast, in equatorial jungles, it rains much of the time and the water can penetrate well into the soil-rock substratum. Here, alteration is extensive and soil profiles can be tens of meters thick. On stable, flat continents where soil profiles have formed over long periods of time, the weathered material is very deep. Low slope as well as time favour soil formation. These relations of soil thickness are shown in Figure 4.3.

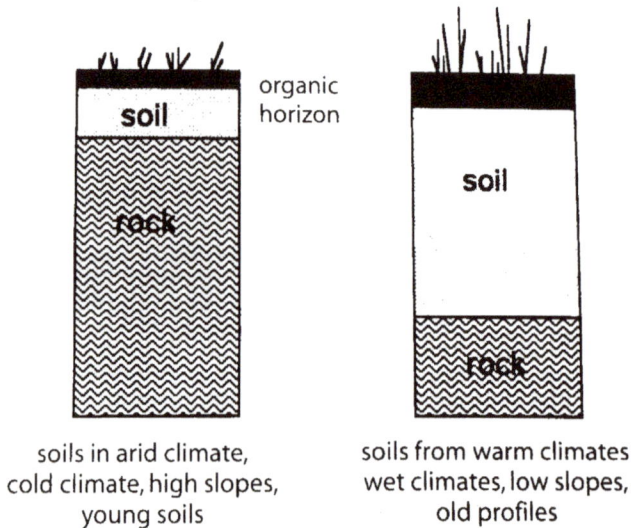

soils in arid climate, cold climate, high slopes, young soils

soils from warm climates wet climates, low slopes, old profiles

Fig. 4.3. Weathering profiles as a function of climate, age and slope where the amount of the different components of weathering (clays, sand and rock fragments) are shown to be dependent on the weathering intensity which is, in turn, a function of climate, slope and age of the weathering process, which is dictated by the type (chemistry) of the rocks involved.

Weathering profiles are affected by slope. On mountain sides, the water, though it may be abundant, will tend to run off rapidly and hence the soil is not very thick there. Lower, more gentle slopes allow more water-rock interaction and soil profiles become thicker and more clay-rich. Hence, topography is a factor in the production of clay minerals by weathering.

Weathering profiles which are very old will be thick. The action of alteration has much time to accomplish the production of clay minerals in old weathering profiles. Soils developed on old, flat and stable continents, such as western Africa, tend to be very deep and clay-rich, reaching tens of meters of depth.

Weathering profiles are also affected by or dependent upon the type of rock upon which they form. Some rocks are difficult for water to interact with. For example, basalts which are hard, dense and have many cracks tend to give shallow soil profiles. Limestone, in which the constituent carbonate dissolves in acidic water, leaves shallow soil profiles because there is little in them which is not dissolved. Most of the rock dissolves, leaving little clay behind. Soils developed on rocks which already contain much clay will obviously contain much clay in their turn. One can see then that weathering "intensity", the thickness of a soil and weathering profile will depend on several factors: *climate, rock type, topography* and *age.*

The products of weathering are clays and oxides, sand and silt grains and gravel. The proportion of these products will depend on the three factors listed above, the more intense the weathering the greater the pro- portion of clays compared to sand and gravel-sized rock fragments and old mineral grains. In weathering, a soil will normally contain clay-size particles, sand and silt grains.

4.3.2
Transportation by Water Flow, Grain-Size Sorting

If one takes the product of weathering, soil, and puts it into a beaker or glass, then stirs it up, a mechanical sorting is effected. The lightest and, more importantly, the smallest grains settle the most slowly. As most silicates have about the same density, around 2.5 times that of water, grain size is a very important factor in settling. The smaller the grain the more friction is effected on its surface as it falls through the water. This is basically controlled by the ratio of surface of the grain, compared to its volume. As clays are the smallest of the grain-size material, they tend to stay "afloat" longer and can be separated from bigger grains. Figure 4.4 illustrates this point. If one pours the liquid from the beaker, the clays are separated from the sand and gravel. When the remaining material is stirred again, and allowed to settle, one can extract the sand fraction from the gravel, and so forth. In nature it is a fundamental process.

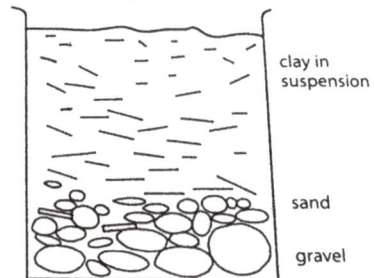

Fig. 4.4. Size fractionation in water suspension. Example: a beaker and sediments in water which, after being stirred, are allowed to settle. The larger grains settle first and the smaller, flat, clay grains settle last.

In mountains the sediments tend to be boulders and gravel, on plains and low hills the sediments tend to be sandy with some clay, along beaches sediments tend to be sand while offshore in more quiet waters sediments tend to be muddy, i.e. clay. The winnowing effect of water transportation separates the components of weathering by grain size. This is a geographic distribution; the flatter the terrain, the smaller the grain size. Along rivers one can find mixtures of sand and clay, especially on flood plains along flat-lying rivers. Those rivers which wind their way through the countryside will give mixed deposits of clay and sand. Often in one spot there will be a concentration of sand while in another there will be a concentration of clay. The effect of transportation is to homogenize the grain-size distribution of a sediment. The higher, mountain rivers contain grains of varying size, the sands of a beach are much more homogeneous in grain-size distribution.

As the sand grains and other sized particles are transported, they tend to knock against one another. The net effect is to round the edges of the grains and to round them. Hence, transportation (and wind) rounds grains and sorts them into more homogeneous classes. This is shown in Figure 4.5.

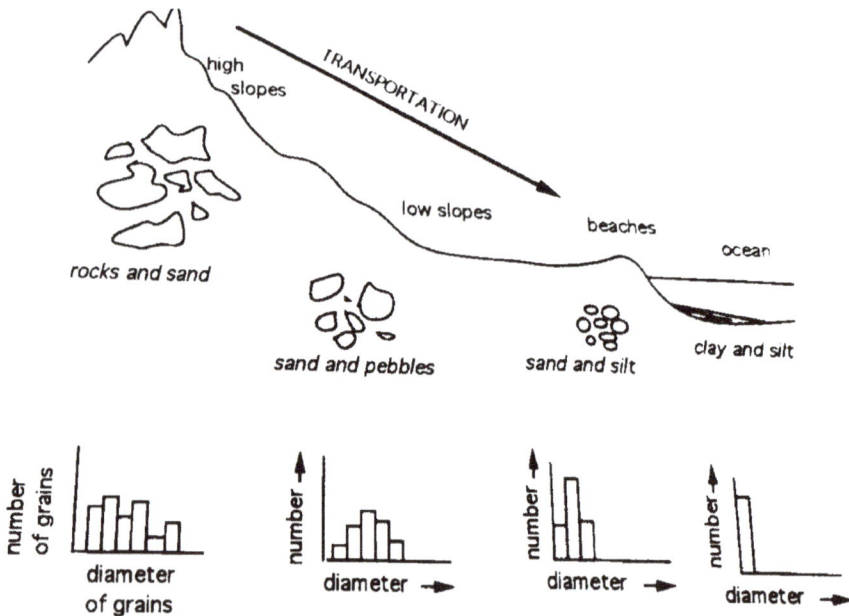

Fig. 4.5. Grain size and water transport of sedimentary materials in a landscape. The smaller, flatter materials (clay grains) are transported further, to the ocean, and the larger grains are carried less far and deposited in streams or on the beach, depending upon the energy of transport. The small histograms indicate the grain-size distribution to be expected in each of the transport zones. Not only are the grains rounded, but the grain size changes with transport and the distribution of the sizes changes also.

4.3.3
Transport and Deposition of Clays

Rivers are the major agents for transportation of clay materials. If the river flows with great speed, on high slopes, most of the clay-sized materials will remain in suspension in the water. As slopes change and the river flow is slower, grains begin to sediment out. The slower the flow, the more material drops out of the river. As a river flows into a body of water, lake or ocean, the flow is decreased and water speed decreases. Sediments tend to be deposited. One finds clays deposited from waters moving at slowest speeds. Hence, lake and ocean bottoms tend to accumulate clays as sediments.

A special, and important case of this system occurs during certain periods of the year along the edges of rivers. During the year, river flow changes speed; the rainy season tends to increase river flow more rapidly since they carry more water. They often have so much water to move that they overflow the normal riverbed. In doing so, they make temporary lakes along their edges. These are the flood plains of rivers. Here, water which was flowing rapidly and charged with material in suspension is suddenly slowed because it has nowhere to go. Clays are then deposited on river flood plains. Therefore, one finds clays transported by rivers to lakes or the ocean during normal months and one finds them depositing clays along their banks during rainy periods.

When clays move, they contain different componants of the materials from which they formed, clays (fine particles forming plastic materials) and grits (non-plastic materials). The relative proportions of clays and grits (sand grains) determine the working properties of the clays. Natural clay-grit mixtures show unimodal distributions of non-clay materials (Fig. 4.6a). A mixture of two clay sources (often) shows a bimodal distribution of non-clay particles (Fig. 4.6b). Use of grits to modify the clay properties shows a strong irregular distribution of more coarse particles (Fig. 4.6c). These three methods of modification of the physical properties of clay sources can be seen in using computer technology. Of course, there are counter examples to this rule. One example is from a silty clay material from China, from a plaine south of the Yellow River naturally loaded with fine to very coarse granitic fragments that could be mistaken for added temper. So Figure 4.6c could be an example of a natural clay. Furthermore, the potters often refine the raw materials used, homogeneizing grit sizes (or eliminated the coarser ones).

Recognizing the presence of more than one sediment (raw material) requires many threads of evidence. Conducting geologic survey to obtain comparative samples is a very important step to have an idea of the types of sediments present in the area of study.

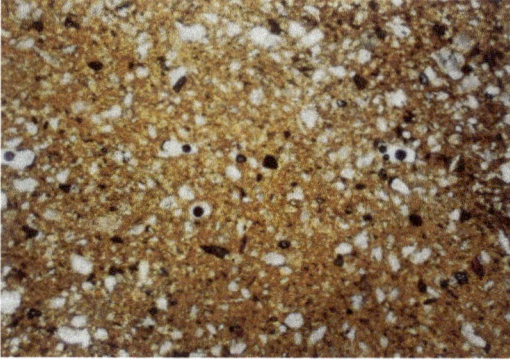

Fig. 4.6a Clay rich deposit with almost homogenous grit size. Transported material.

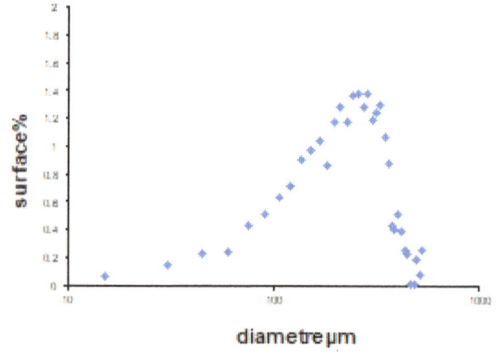

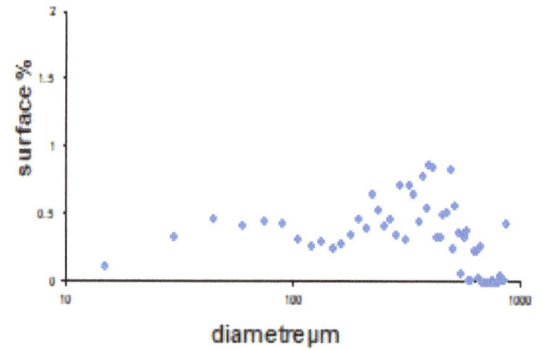

Fig. 4.6b.Mixture of sediment deposits, two populations.

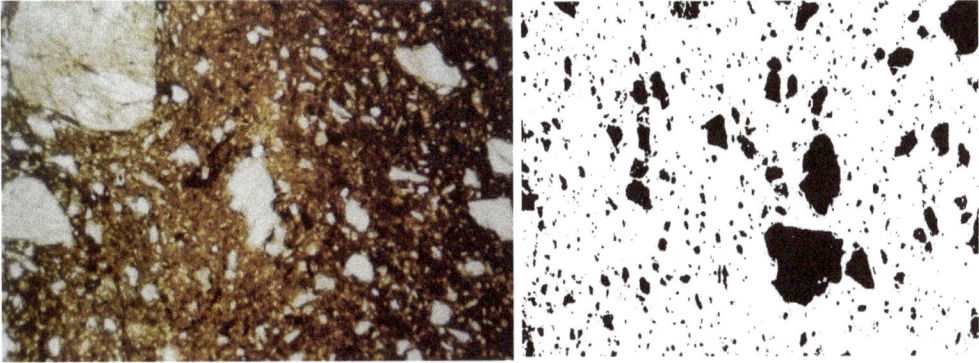

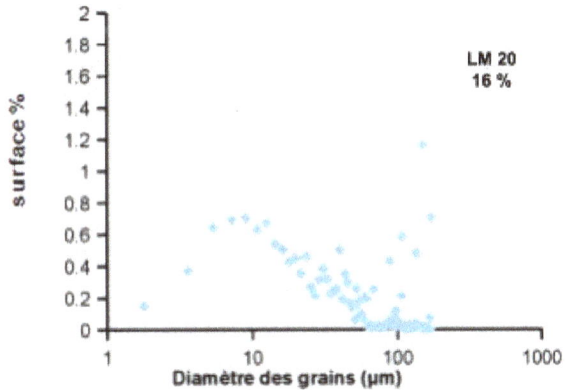

Fig. 4.6c Addition of coarse grit to sediment population.

4.3.4
Wind Transport

A special type of transport and deposition is that of the wind. This is mostly concerned with fine material above clay size (greater than 0.002 mm). Most people know about sand dunes, large accumulations of sand along sea coasts or in desert basins. These materials are largely moved and deposited by the effect of wind. A second type of deposition is through the action of wind acting on glacial outwash plains. During the time of continental glaciation, large amounts of fine materials were deposited on the edges of the glaciers by streams. This material was, in turn, swept around in great wind storms and it accumulated further from the glacier's edge. These outwash plains and silt deposits (silt is a fine-grained sand size and is carried further than sand, due to its smaller size, of course) cover typically hundreds of miles from the glacial edge. They could subsequently be concentrated on the flood plains of rivers. The most striking examples of this type of deposit (called loess) are found in China. However, in large portions of North America (especially in the Midwest of the United

States) and in Northern Europe, these loess deposits are very frequently present. The loess layers vary from tens of centimeters to meters in thickness. This fine-grained material, in fact a mixture of silt and clay, can be used readily in the production of simple pottery. It typically turns up in Neolithic pottery in Northern Europe, identifying the river banks or flood plains where it accumulated.

4.3.5
Burial of Sediments

As sediments are deposited in basins or on the edges of oceans, they tend to be buried by other sediments. This implies that the floor or basement on which the sediments are deposited descends as new sediments are added on them. This is roughly true, though sometimes, sedimentary basins become filled and sedimentation stops, and sometimes the basement subsides faster than the sediments can fill the basin. The filling basin is the most commonly evoked in geology, where there are just enough sediments to keep up with the subsidence of the bottom of the basin. In a sedimenting basin, each layer goes deeper in the earth, and as is normally the case, the ambient temperature increases with depth. Also the ambient pressure increases, which tends to favour the more dense phases, silicates have a density 2.5 times that of water, and hence the water of sedimentation is expelled. The sediments become drier. Upon sedimentation, i.e. deposition on the floor of the ocean, clay-rich materials have a free water content of 80 %. That is to say that there are about 80 % holes in the sediment. As this sediment is buried, its free water content decreases, the holes or pores decrease and they become about 15% of the sediment at depths of 3 km. This changes the physical properties of the sediment.

Also as burial is greater, temperature increases and this effects change in the minerals present producing metamorphic rocks. Before metamorphism, much of the clays is not drastically transformed and can still be used as a plastic material by potters.

4.4
Hydrothermal Alteration

If a molten silicate mass, magma, becomes "stuck" on its upward ascension, it will in time cool to form an intrusive magmatic rock, as we have said earlier. In many cases the magma in its liquid state can contain up to several weight percent water. When the water content is great enough, it leaves the crystallizing mass and enters the surrounding rocks or acts upon previously crystallized magmatic material. The minerals of the rocks which have formed from the magma react with the water, as do rocks at the earth's surface during chemical weathering, and they produce clay minerals. A main clay mineral produced by these reactions is kaolinite.

The temperatures of these reactions are on the order of 300-500 °C well above those of surface weathering, to be sure. Although the chemical changes effected by these hydrothermal alteration reactions are quite similar to those observed in weathering, the hydrothermally altered rocks which contain clay minerals most often do not contain iron oxide. The iron is kept in a reduced state (Fe+) and it is carried away in the altering solutions unlike the case of weathering alteration. Therefore, hydrothermally altered rocks are often whitish in color and the clays do not have color and they do not give color when they are made into ceramics. From this characteristic, it is evident that potters would search for these clays of hydrothermal origin when they wished to make a clean, white ceramic object. This is the case for porcelain, invented by the Chinese and sought after for more than several centuries by European craftsmen and entrepreneurs.

The clay minerals formed by hydrothermal alteration are most often composed almost exclusively of aluminium (Al) and silicon (Si) cations. Along with hydrothermally produced clay deposits, one often finds ore deposits. They contain such metals as lead (Pb), silver (Ag), tin (Sn) and copper (Cu), for example. Although these metals do not interest a potter in the first instance (to make a pot), the discovery of useful, white clays associated with the precious ores was undoubtedly often made as a result of mining activity. This is the case for white slip ware from Cyprus produced during the second Bronze Age, for example.

The ore metals are often used in decoration effects on pottery.

4.5
Sources of Materials Suitable for Ceramics

4.5.1
Clays

Clays and clay minerals are formed for the most part in soils and concentrated in the top of soil profiles near the surface. Through erosion they are moved and eventually concentrated by river transportation in lakes or off-shore environments along oceans or along river banks on flood plains. These deposits are frequently buried as a function of time, and eventually compressed and rendered more rigid, they become rocks. Clay-rich rocks are not likely to be the first choice for a potter in his search for raw materials. However, rock clay resources can be softened when put into contact with water after they have been crushed in mills (or by hand) to produce a useful material for potting.

For most potters in ancient times, sources of clays tended to be those easily available, those at the surface. Soils are a likely candidate, to be sure, but if one needs a large quantity of clay for a large number of pots, the resources need to be important. Soils tend to be about a meter or so thick at most, not a great resource. If a clay-rich rock is subjected to weathering on a slope, a much larger amount of soft clay can be found. If a sediment is young, and not deeply buried, it will not be as hard as a rock and can be used for ceramics. If a rock is highly altered by volcanic gases, it will be clay-rich and soft. These are some of the most common clay resources used in the pre-industrial era.

Hydrothermally formed clay deposits are more rare but of better quality. They are found in irregular masses, frequently they are mined at some depth. Some rocks are altered to a clay-rich composition and they occur over large surfaces. These rocks are found in volcanic or igneous rock areas.

Figure 4.7 shows diagrammatically the types of environments where clay resources are likely to be found.

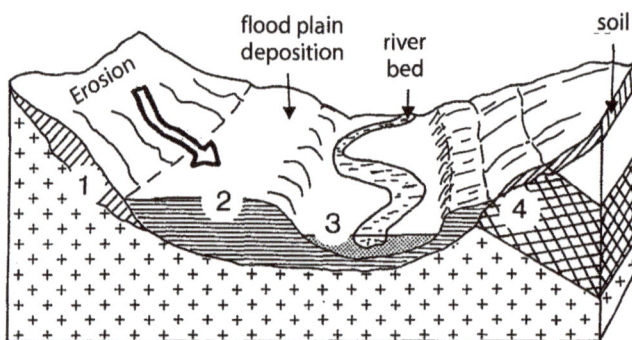

Fig. 4.7 Clay sources. Soils formed at the surface of a rock by weathering can be extracted as a clay source (indicated by 1). Clay-rich sediments, due to transport and deposition of the soil materials by a river on a flood plain can be extracted as a clay source (2). Sediments in a river bed will tend to have more sand and silt components than clays (3) an hence they will be used as temper materials if needed. In certain cases, clay-rich sedimentary rocks can be used as clay sources for ceramics. These materials occur as "outcrops", places where the sedimentary layers reach the surface and can be exploited (4). The youngest clay resource material is the soil, the next oldest is the sediment along a stream, then one can find older sedimentary rocks (clay-rich).

4.5.2
Non-Clay Grains

Up until now we have considered two types of material, clays and non-clays. Non-clays are of grain sizes greater than clays. The non-clay materials are usually divided into grain-size categories of silt (diameter of 2-50 μm), and sand (50 μm to 2 mm in diameter). These two size categories are usually of non -clay material and hence they have a relatively small attraction for water to their relatively small surface area compared to their volume. These materials are for the most part the non-plastics of a clay paste. In nature, one finds various mixtures of the different fine grain fractions sand, silt and clay. Figure 4.8 shows how different proportions of these grains give different names to such materials in either geological or soil science vocabularies.

Given this rich variety of mixtures of the different fine-grained components, it is clear that in many places natural deposits of these elements can be used directly by a potter for different purposes. It is evident that some soils and sediments will be well adapted to form a ceramic without further treatment. By their natural mixture of plastic and non-plastic components, they can be used to form pottery. If the proportions of clay, silt and sand are not adequate, a potter can mix the soil with large amounts of water and wash out the clays, thus leaving sand behind, as shown in Figure 4.4 in the beaker example, used to illustrate the sorting of weathered material in stream transport. The separation of the different elements in soils or clay-rich sediments has often been practised in the past in order to correct inherent imperfections in clay-rich materials as the needs of a potter are perceived. Thus, one resource can be treated by a potter to give another mixture of materials from the same source.

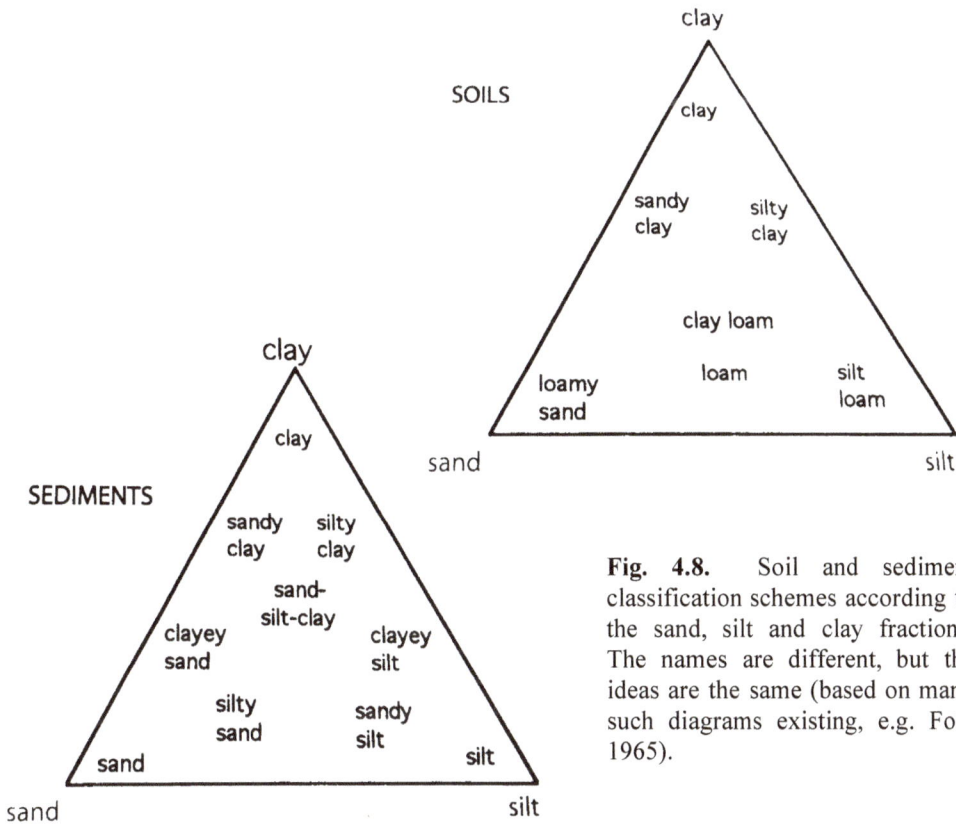

Fig. 4.8. Soil and sediment classification schemes according to the sand, silt and clay fractions. The names are different, but the ideas are the same (based on many such diagrams existing, e.g. Folk 1965).

As a reference, Stoops et al. (2010) offer an invaluable source of information for soil materials and micromorphology, with 27 chapters and many color plates covering colluvial deposits, pedofeatures, siliceous and anthropogenic features, archaeological materials (soils used for construction, floors, walls, etc.), paleosols, to cite a few.

4.6
Factors Favorable for Clay Formation and Concentration

Geology
Rock type (factors favorable for the formation of ceramic clays)

1 Formation of clays from volcanic glass materials in volcanic rocks
   These rocks have a large proportion of silicate glass which becomes basically
   clay minerals when in contact with surface water.

2 Transformation of low grade metamorphic and diagenetic clays from shales
   This material is basically clay-rich and becomes more so under surface
   conditions.

3 Alteration of igneous minerals to clays
   These rocks contain minerals highly unstable under surface conditions but most
   of them remain unchanged under surface conditions.

These three geological factors produce clay-rich deposits but of different types with
different proportions of clay minerals and non-clay materials which can become sand
and eventually grit material. Clays tend to resemble each other in composition but grits
often tell of their geologic origin.

Physical factors (climate)

1 Water – rock ratio (rainfall, residence time in rock, and altered rock system)

2 Temperature

These variables determine the chemical activity of the interaction of rocks with surface
conditions.

Areas of production and accumulation

1 At depth: hydrothermal action by water and water vapor, tens to hundreds of
   meters depth. Frequently contractions of very clay-rich material are produced.

2 Surface interaction: soils, interaction with plants producing clay-rich zones at the
   surface. Rock alteration interaction with rainwater below soils.

These variables determine the type of clay mineral produced and its relative abundance
compared to non-clay materials.

Factors of accumulation

1 No movement (hydrothermal subsurface deposit), relative clay abundance
   depending upon the intensity of the interaction between the rock and aqueous
   materials.

2 Soils and surface erosion (displacement and accumulation of clays). Controlling factors: topography, rainfall, wind (loess)

These variables determine the relative concentration of the clays and sand (grit) materials due to the initial alteration process. Clays and sand grits are less heavy and more likely to be moved massively by wind or water movement in concentrated deposits. However, one deposit can be covered or cover another of a different nature and hence the grit–clay ratio and grit size will vary from one deposit to another. For example, along a river in the sediments at its edge one can find different relations between clay and grit content in each layer. Thus ceramists needed to find clay-rich deposits with fine grits but had to be aware that these contents could vary from one layer to another.

Movement of Clays

Clay particles at the surface are joined to one another through chemical bonding with water molecules. This fixes the edges of the crystals into a larger more complex mineral structure which is still largely two dimensional. The two dimensionality gives the clay particle resistance to gravitational action in water. Hence clays tend to be transported by surface water due to their structure. They remain in suspension for long periods of time. All one needs to see is the water in a river or stream to see that clays are moving.

Grits (largely quartz) are not fixed together by water-induced chemical interaction and tend to remain unidimensional and not transported in suspension by water unless it is moving with great energy. Then clays and grits tend to be separated in sedimentation processes except when the grits are small enough to be incorporated int the clay assemblages.

4.7
Sediment Types of Use for Ceramic Making and Their Mineral Characteristics

When looking at the mineral composition of a ceramic, at the quartz and other minerals, and non-plastic grains, one important question to answer is: what type(s) of material(s) was/were used to produce this object? We have presented the weathering process, how clays and non-clay materials are accumulated and transported, but we want to examine here the characteristics of different sediments, the type of mineral assemblage, size, angularity, that are expected in, say, alluvial sediments versus glacial-till, what can be expected from clastic and detrital materials, coastal versus up-valley materials, etc.

Alluvial soils are not consolidated, not yet transformed into a rock, young in age, eroded and redeposited elsewhere (if it is consolidated we speak more of a deposit). They are multicomposite (different types of minerals and rock fragments), with different angularities and sizes, with clay, silt, sand and even gravel. They often fill flat areas, and basins.

Fluvial sediments would be, of course, deposited by rivers, and found along their course. According to the strength, flow, and slope, the material will be graded, with finer sand material where it is calmer. Pockets of very fine sand can be found in meanders, the banks can present horizontal and vertical stratification, with layers of different sand grades. River banks have often been visited by traditional potters to gather tempering material. Not much clay is usually found there at is has been washed away. The composition usually reflects surrounding geology and can change at each branching of an affluent. Often, the higher up stream, the less composite, coarser and more angular the material.

These sediments types can be affected by flooding, heavy rains, monsoons, that not only may mix materials, but reshape the landscape. In Peru, after the big Niño disaster of 2017, where rains rushed downhill cubic meters of sediments, mud and clay, the types of deposits in valleys and fields changed. Gravel can be seen where fine sand used to be, the fine material was washed away, pans of hard clay caps some areas. Even in river areas up-valley, the fine material was washed away, leaving behind only coarse material, pebbles, and river boulders. The consequence for potters is that the places where they used to collect material are not available anymore. They have to find other sources, go further up, and experiment with new material.

Glacial-till as the name implies comes from glacier environment. The materials would be angular, of sand-size, often derived from granitic sources, thus constituted of much clear angular quartz, plagioclase, biotite, hornblende.

Pyroclastic material results from volcanic activity. As for glacial till, the quartz, plagioclase, biotite, hornblende, are 'clear' (un-altered), angular to subangular, the quartz and plagioclase show a volcanic origin (embayment, zoning of plagioclase, as seen in thin section with a petrographic microscope. The shape of the biotites and hornblendes are often euhedral, coming from porphyritic rocks. Along with this, some of the rock fragments may be rounded due to alteration (rhyolite, andesite, dacite, basalt), one can also see pumice and glass shards.

At the bottom of mountains or hillslopes is found what is called colluvium, a type of unconsolidated sediment that was not transported much (beside downward). As expected, the composition reflects the different geological and soil layers up above, sometime with a heavier presence of the constituent of main outcrop (say granite). It is also less rounded than alluvium or fluvial sand, and may contain clays and grit of up to gravel size. This was a source of material often targeted by ancient potters.

Coastal sand displays a multicomposition, this is a key characteristic. It often presents mix materials of volcano-sedimentary, metamorphic, and intrusive origins, mineraloclasts and rock fragments, usually fairly rounded and weathered, but the angularity of the grains could be mix, up to angular for certain fragments or crystals. Organic material, microfossils, bioclasts (minute fragments of biological origin such as shell, corals), ooids (rounded coated grains), material of marine origin, are often present. Granulometry can be varied or sorted. Wind can round grains as much or even more than water.

A clastic sediment is simply one that is composed of fragments (of rocks and/or minerals), usually not referring to clays. A clastic rock would be made up of fragments of different rocks or minerals cemented together (a sandstone is a good example).

Soil clays (produced in the organic-altered rock zone) and alterite clays (produced in the rock-draining rainwater alteration zone) show no distribution of grain sizes in that all of those possible due to surface alteration are present to a large extent. In upper zones, there are more clays and the grits show no selection of sizes. These materials may be used as is by ceramists and if it becomes not useful, another deposit is sought.

Detrital sediments result from the movement of clay rich altered surface materials by water or wind movement.

We illustrate below what sands and different sediments look like up close. These types of materials have often been used in ceramic making, as is, or, more often, refined, sieved, ground, or even heated (e.g. rocks, mussels) to break them up more easily. Sampling sediments, clays, soils, sands in the region of study helps assess local geology, natural compositions and granulometry to be expected in potting materials. Proximity of a source is not necessarily a pre-condition of use, as traditions and other constrains may influence potters' choices and behaviours. Also, the proportion of clay/grit in the source, exact location and level where materials are mined, introduces variability in paste composition that the analyst must consider. How much within-group variability should be 'accepted' in a compositional group of archaeological ceramics partly depends upon the overall variability of the entire corpus of study, and knowledge of variability in local materials.

Some of the clays and sands illustrated have been collected in the same area, a few kilometers apart. This shows the diversity that could exist within a small region -in this case a high valley in the north-central Peruvian Andes- and how important it is to prospect and sample comparative materials to have an idea of the available potting materials in the region of study (see Figs. 4.9-4.11). Figures 4.12-4.14 show fairly 'clean' clays with low presence of natural silicate grains of silt to medium sand size, used as is to produce ceramics.

a

Fig. 4.9. Coarse (**a**) and fine (**b**, next page) clays from the Peruvian Andes, collected few kilometers apart. AT-C1 and AT-C5.

b

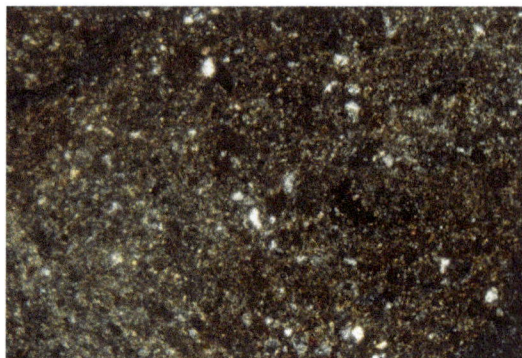

Fig. 4.9. Fine (**b**) clay from the Peruvian Andes, collected few kilometers from the coarse clay 4.9a.. AT-C1 and AT-C5.

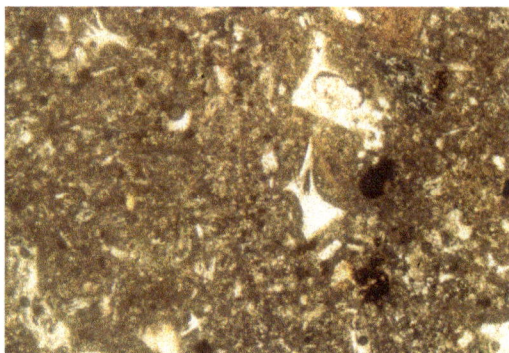

Fig. 4.10. Volcanic clay, rich in glass shards, fragments of wall of gas bubbles (white hemispherical clasts or cuspicles, ppl view. In cross-polarized view, most of the thin section appears opaque as glass is isotrope (the light only vibrates in one direction and when you introduce the polarizer it blocks it. This type of clay is a good choice for producing cooking pots, as it withstands well thermal chocks.

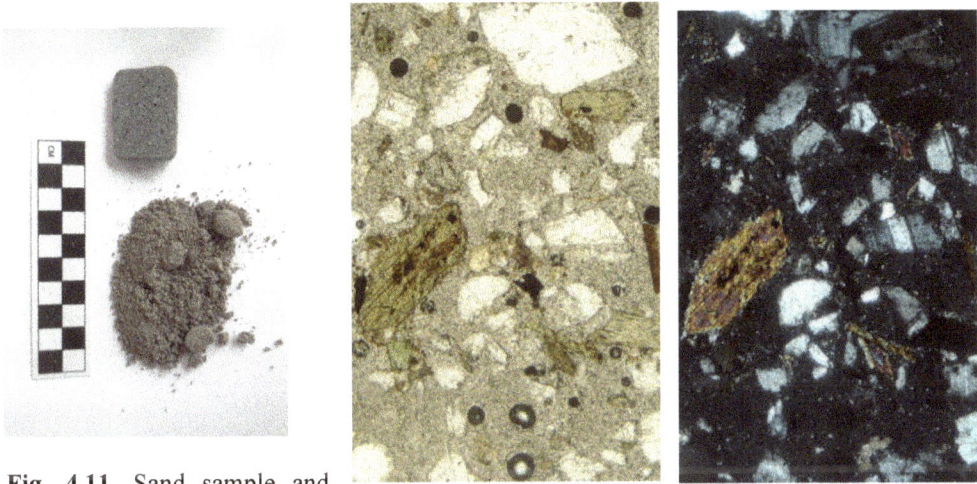

Fig. 4.11. Sand sample and slab (epoxy consolidated) from which the thin section was made, showing the composition of this sand (ppl and xpl).

Fig. 4.12. Unprepared yellow clay: low amount of silt and very fine sand-size quartz grains and micas, iron nodules, and larger shale clasts in the matrix. 25x, xpl, Yaquia, PR67.

Fig. 4.13. Clay with natural non-plastic inclusions (mostly quartz), TT16, Liangchengzhen area, China. Reflexive light, 90x DinoLite portable microscope. (See Druc et al. 2018 for more details).

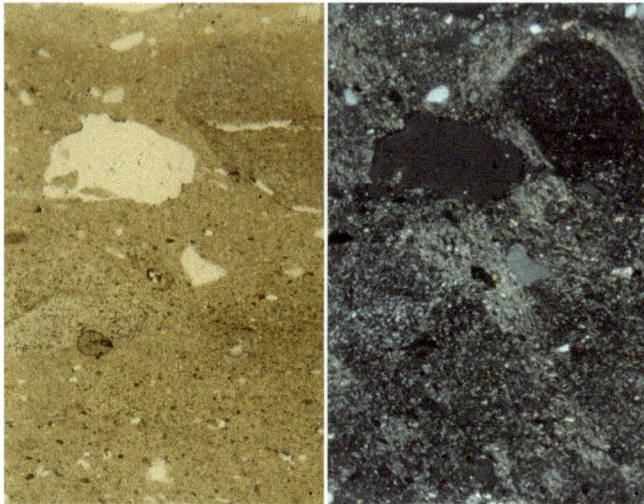

Fig. 4.14 Paste of a white ware vessel (handle). China. Kaolin-rich, very fine clay. The light color in ppl (a) and grey in xpl (b) are typical of this type of clay. LCZw43-1, claystone is 0.85 mm long, 4x.
(See Druc et al. 2021)

Bibliography: Clays and Soils

Bohn, HL, McNeal, BL, and O'Connor, GA (1985) *Soil chemistry*. Wiley, New York, pp 341

Daniels, RB (1992) *Soil geomorphology*. John Wiley, NY, pp 236

Dixon, JB, and Weed, SB (eds) (1989) *Minerals in soil environments*. Soil Science Society America, Madison Wise, pp 1244

Druc, I., Underhill, A., Wang, F., Luan, F., and Lu, Q. (2018) A preliminary assessment of the organization of ceramic production at Liangchengzhen, Rizhao, Shandong: perspectives from petrography. *J of Arch Sci Reports* 18: 222-238.

Druc, I., Underhill, A., Wang, F., Luan, F., Lu, Q., Hu, Q., Guo, M., and Liu, Y. (2021) Late Neolithic white wares from southeastern Shandong, China: the tricks to produce a white looking pot with not much kaolin. Results from petrography, XRD and SEM-EDS analyses. *J of Arch Sci Reports* 35 https://doi.org/10.1016/j.jasrep.2020.102673

Ergenzinger, PJ (1994) *Dynamics and geomorphology of mountain rivers*. Springer, Heidelberg, pp 326

Folk, RL (1965) *Petrology of sedimentary rocks*. The University of Texas. Hemphill's, Austin, Texas

Foster, RJ (1988) *General geology*. Merrill, Columbus OH, pp 507

Gerard, J (1992) *Soil geomorphology: an integration of pedology and geomorphology*. Chapman Hall, London, pp 269

Gill, R (1989) *Chemical fundamentals of geology*. Unwin Hyman, London, pp 291

Greenland, DJ, and Hayes, MHB (1981) *The chemistry of soil processes*. Wiley, New York, pp 714

Julien, PY (1995) *Erosion and sedimentation*. Cambridge University Press, Cambridge, pp 280

Lutgens, FK, and Tarbuck, EJ (1989) *Essentials of geology*. Merrill, Columbus OH pp 378

Martini, IP, and Chesworth, W (1992) *Weathering, soils and paleosols*. Elsevier, Amsterdam, pp 618

McManus, J, and Duch, RW (1993) *Geomorphology and sedimentology of lakes and reservoirs*. John Wiley, NY pp 278

Nahon, D (1991) *Introduction to the petrology of soils and chemical weathering.* John Wiley, NY pp 313

Ollier, C (1996) *Regoligh, soils and landforms.* John Wiley, NY, pp 316

Panizza, M (1996) *Environmental geomorphology.* Elsevier, Amsterdam, pp 268

Pye, K (1994) *Sediment transport and depositional processes.* Blackwell, Oxford, pp 397

Rapp, G, Jr, and Hill, CL (1998) *Geoarchaeology. The earth-science approach to archaeological interpretation.* Yales University Press, New Haven, CT, pp 274

Reading, HG (1996) *Sedimentary environments: processes, facies and stratigraphy.* Blackwell, Oxford, pp 688

Robinson, DA, and Williams, RBG (1994) *Rock weathering and landform evolution.* John Wiley, NY, pp 519

Selby, MJ (1985) *Earth's changing surface: an introduction to geomorphology.* Oxford University Press, pp 607

Skinner, BJ (1992) *The dynamic earth: an introduction to physical geology.* Wiley, NY, pp 570

Sparks, BW (1986) *Geomorphology.* Longmans, London, pp 561

Stoops, G, Marcelino, V, and Mees, F. (2010*) Interpretation of micromorphological features of soils and regoliths.* Elsevier Science, Amsterdam, pp 752

Summerfield, MA (1991) *Global geomorphology: a study of landforms.* Longmans, London, pp 577

5 Physical and Chemical Processes of Making Ceramics

In the preceding chapters we have dealt with the materials which are found in ceramics, clays, silts and sands. Now we wish to look at the processes which come into play as one uses the raw materials to form and fire them into a ceramic body.

The process of making a ceramic from natural raw materials, clays, sands and so forth is the reverse of the weathering process or hydrothermal alteration. In weathering and alteration, a rock is divided into different chemical components by the affinity of the different elements present under the conditions of interaction of silicates and acidic water. The rock is a solid and the alteration products are soft, plastic or finely granular. The task of a potter is to take these different components, to shape them into a desired form and fire the object until it becomes hard and unalterable. The potter metamorphoses sediments into an artificial rock, in a way. In order to do this, there are several steps and choices made to obtain the desired object. These steps and choices are dictated by the materials, availability, tradition to which the potter pertains to, and the type of objects to produce. We will deal here only with the material aspect of the equation.

5.1
Plasticity

The object of pottery-making is to produce a desired shape out of a plastic material. This shape is "frozen" by heating to form a stable rigid shape. The plastic phase of the paste is destroyed by the firing process. However, it is not all that easy to obtain a useful plastic state in the first place. This is the initial and one very important part of ceramic production. The shaping and forming of the plastic paste is the step most observed and commented upon. It gives the aesthetic qualities, functional properties and formal attributes which can be used in a rapid examination to identify a pottery production. These are often used as a criterion to establish a hierarchy of progress in pottery making, as well as a commentary upon the society in whose context the pot was made. The study of the shape and details of construction, surface treatment, decoration can lead to typological classification of pottery productions. We focus here only on the physical changes observed in the material used.

Attaining a plastic state using clays is the first step in producing pottery after fetching and preparing the material(s). It is necessary to obtain a fine-grained clay powder, usually from dry clay lumps, which are pounded or ground in some way, or settled in water to separate the coarser from the finer fractions and eliminate impurities (rootlets, plants, etc.) and dried. The resulting powder may be sized, sieved to assure homogeneity, and is mixed with water. Other material(s) may be added to regulate the plasticity of this clay base, give it more strength, limit retraction upon drying, etc. As seen earlier, this added material is often called temper in the archeological jargon. According to need, tradition, or level of production organization, these initial treatment(s) can vary, from rudimentary to elaborate.

Often at this stage, the clay-water mixture is left to stand and "mature" for a few hours to several days (or months). The clays frequently contain organic matter, which evolves in a water-soaked state, and putrefies. As a pond of stinking mud is not the best neighbour and due to the smoke, toxic fumes at times when firing, and danger of fires, potters were often banned from carrying out their trade within the city walls in Medieval Europe. Workshops may also be grouped in neighborhoords at the outskirt of a city.

The matured clay-water material is next kneaded to obtain the desired plasticity and homogeneity. Schematically, there are three physical states, which can be obtained in such a procedure. These states are shown schematically in Figure 5.1.

1. The *brittle state* occurs when not enough water is added or too much non-plastic temper is present. The material cracks upon bending or folding. The structure will have planes of weakness in it. The brittle state hinders the smooth formation of a pottery object, it is likely to break, and it limits the thinness and curves made possible with a well-wrought plastic state.

2. The *plastic state* occurs where the mixture can be moulded or formed without breaking the paste material. Once the forces of shaping are relaxed, the material maintains its structure and shape. This is the plastic state.

3. The *fluid state* occurs when the mixture of clay resource, tempers and water flows under the force of gravity. This is not plastic but viscous. It is only desirable in pottery production for casting into a mould. For other uses, it needs to be tempered by adding non-plastic materials.

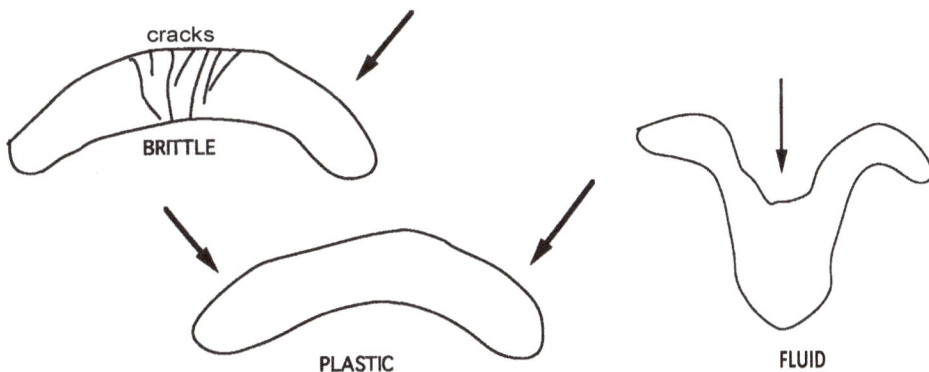

Fig. 5.1. Representation of the brittle, plastic and fluid states of a prepared paste. The illustration is a rolled lump of paste which, when bent (arrows represent pressure), cracks in the brittle state, is homogeneously deformed in the plastic state or flows in the fluid state.

5.2
Mineral Temper Grains

The plastic state depends upon the proportions of clay, water and non-plastics, broadly called here 'temper grains'. This term for the purpose of our discussion simply refers to a non-plastic (non-clay mineral) material which does not take much water onto its surface and which gives rigidity to the paste assemblage. Mineral temper grains can be natural, i.e. present in the clay assemblage in the state of extraction of the clay resource from its natural context (stream bank, soil, soft rock deposit), or temper grains can be present in the added material when preparing the paste. These can be of mineral, vegetal, and/or organic nature. The important concept is that temper grains modulate the tendency of clay and water to form a mixture that is too "runny" or liquid. In some cases, temper has a role during the firing process also (see Chap. 6.3.3).

In the analysis of the internal part of a ceramic, we are interested in observing, identifying and classifying the non-plastic grains.

5.2.1
Natural Mineral Grains

First, it is very important to remember the sources or origins of clays or clay deposits.

a) Some result from rock weathering, where clays form from the decomposition of certain minerals in a rock but other minerals are left intact as sand- or silt-size grains. These non-plastic mineral grains are not clays, and because they have less water adhering to them, they modify the sediment plasticity of the clay source and hence could be called natural temper, or natural mineral temper grains.

b) In sedimentary clay deposits, clays are also usually accompanied by other minerals, non-plastic in nature by their shape and size, which also can be considered to be mineral temper grains, as in the case of soils, and are natural as they are present in the geological deposits.

c) In hydrothermal alterations of rocks, clays are produced by high-temperature chemical reactions, similar to those in weathering. As would be expected, certain minerals, such as quartz, resist the chemical attack and remain intact. They are non-plastic and natural mineral temper grains.

In fact, most clay sources naturally include plastic (clay minerals) materials and non-plastic mineral grains, which when the material is mixed with water, form a paste. If the natural mix of plastic and non-plastics is appropriate for the use of the material to form a ceramic (and reacts well to drying and firing), the paste is made of natural mineral temper grains and clays. If it is necessary to add other non-plastic material to the clay resource when mixed with water, the non-plastic material is an added temper. Clays can also be 'refined' and in that case, some mineral material can be substracted. When non-clay mineral grains are present, they have a tempering function, whether they are natural to the clay resource or added by the potter. For an archaeologist or ceramic analyst, what is important to identify are the original materials used and the act of tempering, if done by the potter, to modify the physical properties of the paste, as discussed in further in this chapter and the next.

5.2.2
Decantation and Separation of Natural Mineral Temper Grains

Preparation can also involve taking a clay source and dividing or separating it into its natural components, clay minerals and non-plastics, either silt-sized grains or sands. This can be accomplished by passing a suspension of the raw material diluted in water into a series of pans. The heaviest or biggest grains fall out first, the lightest or smallest last. The process is called levigation in ceramic production. This is a classical process of concentrating clays from a less clay-rich source (Fig. 5.2). If the material is left to settle in water, in a tin for example, with no transfer from one pan to the other, it is called decantation (Fig. 4.4).

Let us assume that the clays contain too much non-plastic natural temper. Then the potter will decant out the coarse material; but in the end the clay concentrate might be too clay-rich or too plastic when mixed with water. The potter will then need to add some temper to the clay (or some of the fractions retrieved).

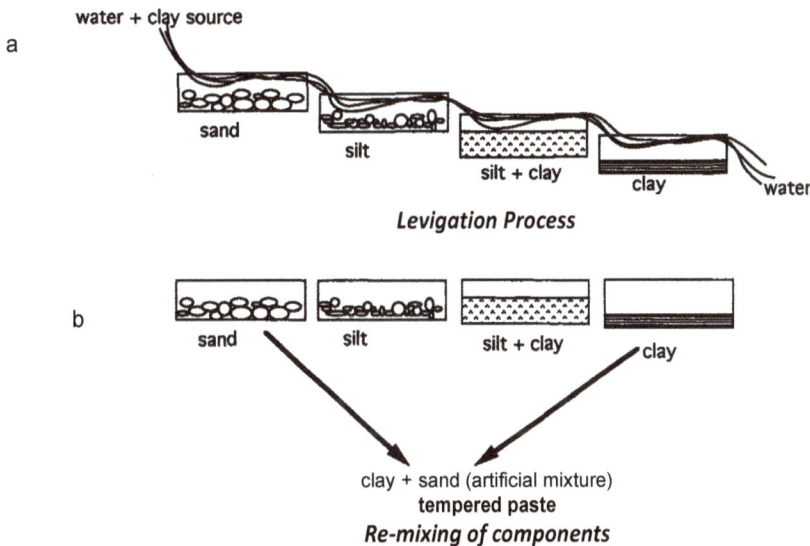

Fig. 5.2. Effects of levigation on a clay source material. (**a**) Separation of different size materials by water flow into recipients at different levels. The heaviest materials and coarse fraction fall out first and the clays last. (**b**) Remixing of the levigated materials can be performed by using the sorted fractions during the preparation of a paste. In this way, clays can be tempered with their initial temper grains sorted according to size by an action of the potter. The sand grains and clay are from the same natural source, be it a soil, sedimentary rock or whatever. They represent a natural accumulation of materials but their proportions have been modified by the potter to meet his/her needs.

Since the potter has sand or silt grains which can be used as temper at hand from the decantation of the clay source, this natural temper extracted from the clay can be mixed with the decanted clay in controlled amount. Thus is produced an artificial mixture from the same natural clay-temper source. The potter just modifies the proportions of the end members. This is illustrated in Figure 5.2b. The advantage of this process is that the potter can vary the proportions of clay and temper (and size of it) according to the type of vessel to produce. This is useful when a range of products is made in a work site from the same resources when the different objects need different proportions of clay and temper, either for structural reasons (thickness of the walls of the object, needs of thermal conductivity, etc.) or aesthetic ones (smoothness of surface or, in the case of certain Japanese ceramics, need of small grains or other irregularities indicating "rustic" or unsophisticated qualities, a paradox for a sophisticated potter).

An example of varying proportions of natural non-plastic silicate temper grains and clays coming from the same source which when used in varying proportions, correspond to the structural needs of a ceramic can be found in some common ware from the St Marcel site near Argenton sur Creuse (Creuse) France. In some of the common or domestic ceramics from a medium-sized Gallo-Roman agglomeration, certain characteristics are present which are indicative of sources adapted to the potter's use. Common kitchenware samples studied are of different types, jars, bowls and cooking ware. Observation of the clay and non-plastic temper grains of the objects indicates very similar materials. The clays are mica-rich, and the temper grains are almost exclusively quartz and feldspar grains of different sizes. Thin-walled vessels with smoothed surfaces or coarse, thick-walled jars are made of the same basic geologic components, mica-rich clay and quartz-feldspar temper grains in the silt-and sand-size ranges. However, if one makes a numerical analysis of the temper grains and makes histograms of these grain populations, it is apparent that the thick-walled ceramics contain grains of greater diameter than the thin-walled samples (Fig. 5.3a). Further, the total amount of non-plastics or temper grains (counted as the surface visible in a thin section (a two-dimensional cut through a three-dimensional object) is related to the thickness of the wall of the objects (Fig. 5.3b). This suggests that the potters used not only more coarse grains for coarse, thick-walled pots but that they used more of the tempering material (non-plastic) in total. Thus, there must be a relation between the need for temper and wall thickness of these common ware objects. It would also seem that for these potters it was easier to gain in bulk tempering by using larger-grained material. Perhaps bigger sand grains behave better under fabrication (mixing the paste and forming the object) and firing of coarse, thick-walled jars. These relations will depend upon the type of material available (clay source plasticity, sand-grain size, etc.) but there is a very interesting relation between the amount of tempering material used by Gallo-Roman potters and the object that they wished to produce.

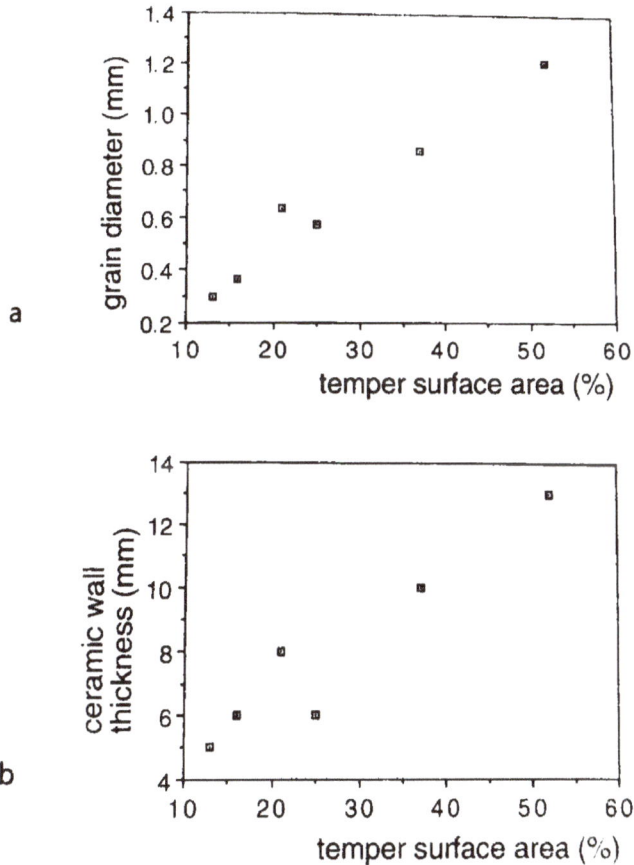

Fig. 5.3. This is an illustration of the relation of grain size and the volume they occupy when used for temper in some common ware from a Gallo-Roman site in central France. Observations are made from thin sections of sherds, which are a two-dimensional representation of a three-dimensional object. However, the extrapolation is not too hardy. **(a)** If one considers the relation between average diameter of temper grains and the total surface occupied by the temper grains in the thin section, it is clear that there is a relation. The larger the average grains the bigger the proportion of the paste they occupy. **(b)** If one considers the wall thickness of the ceramic object and the surface area of the temper grains present in the thin section (volume occupied), there is also a clear relation. The thicker the wall of the ceramic object, the more temper material is present. In summary of these two diagrams, the more temper, the greater the grain size on average and the thicker the wall of the object. In these samples there is a clear use of temper of different sizes to produce ceramics designed for different use. Wall thickness is often related to use.

In the case cited above, it is important to note that the temper grains are natural in the sense that they were accumulated with the clays in the clay source in a natural manner, most likely on flood plains along the banks of the local river. Both clays and sand temper grains come from the alteration of granites. The precise clay source used for the objects is not known, be it one or several. Whatever the method of obtaining the paste necessary, the potters obviously manipulated their resources to their needs. However, as far as the geologist is concerned, the materials are from the same source, only their proportions have been changed.

A similar process can be observed in ethnographic context, one example being that of Mexican potters of a small village in Puebla State, decanting their clay to extract the coarser fraction. The latter is 'reinjected' into the clay base in different amounts for the production of cooking griddles or other coarse vessels (see Druc 2000).

The act of subtracting coarse grains from the clays and adding them to the clay material in different proportions in order to obtain the desired paste texture is also called tempering, even if all of the materials come from the same natural source. As expected, it can be difficult in some cases to determine the act of tempering with a search for naturally occurring materials of different proportions, clay minerals and non-plastic grains, to produce the right paste for the right pot.

5.2.3
Tempering by Mixtures of Materials

Tempering, the act of adding material to modulate the clay plasticity or change the paste properties, changes the clay and non-clay proportions to make a paste of a desired consistency. This can be accomplished by different means (see also Chapter 7).

5.2.3.1
Mixtures of Clays and Non-Clay Grains from Different Sources

Temper material or temper grains can be found as coming from entirely different sources from the clay materials used by a potter. For example, clays from one source can be mixed with beach or river sands. In this case, the sands can be of a very different origin from the clay source materials. The clay resource itself will also most likely contain some non-plastic, associated grains. An easy example to identify the mixing of different materials would be shell-filled beach sand and a clay bank material. The sand would contain shells but the clay from the bank source would not. Another possibility is crushing rock or sands to make a temper material suitable for tempering a given clay source. Tempers which are added to a clay source will then often show different characteristics, size distributions, shapes or mineralogy, from those of the clay source. Mineralogy (and more specifically petrography) will be the easiest method to identify multiple source materials (see Fig 5.4 for quartz sand added to clean clay).

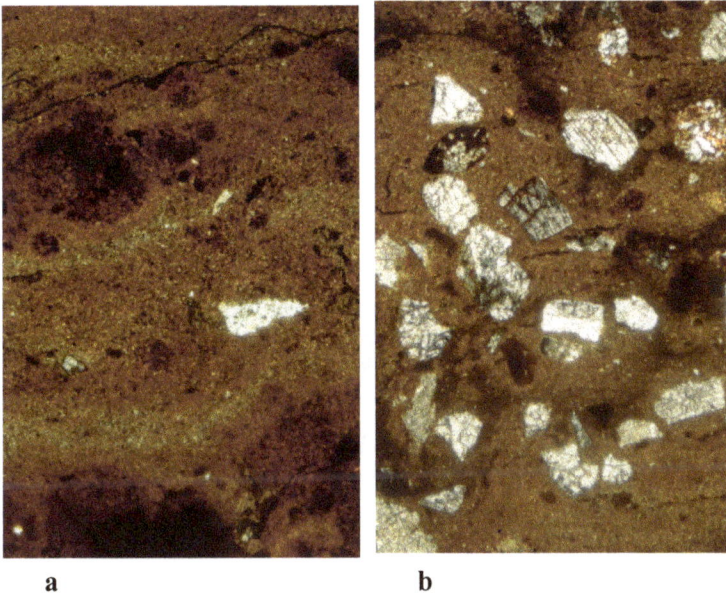

Fig. 5.4. (a) HV136 beach clay and **(b)** HV136/138 mixed with beach sand (1 sand /4 clay), central coast Peru. Test tiles, fired in experimental open kiln, oxidizing atmosphere.

a   b

The use of non-clay materials of different sizes is not restricted to silt or sand grains. For example, some potters use small claystone fragments, as temper. This use is quite logical. The claystone behaves as a large grain. The clays do not react with water because they are in the rock fragment. The claystone fragments have a non-plastic behaviour, but they have thermal properties of clays which most sand grains do not. Upon heating to high temperatures, they will mimic clay behaviour. Shale, phylite, slate, schist, volcanic rock, tuff, ash, pumice, etc. all can be used as temper, crushed by the potter or collected as weathered material in primary sediments, eroded from the nearby parent rock or mixed in sands. Thus rock fragments can be used as temper to moderate the swelling, and plastic qualities of clay materials. The importance for provenance analysis is then to identify the type of material and what types of sediments were used (primary, secondary, alluvial, fluvial, glacial-till, etc.). These types of sediments and their characteristics are discussed at the end of Chapter 4.

Another source of temper, one that is anthropogenic, is grog. Grog refers to fragments of previously fired ceramic materials, usually crushed to be added to the clay base (Fig. 5.5). These materials, though of clay composition, do not have the grain sizes characteristic of clays because the mineral properties have been destroyed by firing. They behave as tempers. Grog has been often used as a tempering agent, depending the potting tradition. In some cases in Thailand, it appears that grog was and is now produced by firing balls of clay at low temperatures. These dried and slightly transformed materials were/are then used as a tempering agent in ceramic production.

Vegetal material may be used as temper, as adding straw, rice hull, dung, or other organic material will change the plasticity of the clay-water paste. It can also be accidentally incorporated into the clay mix and in that case, it would be in very low percentage (Fig. 5.6b-c). In some instances, the clay itself is rich in organic material, or the temper (sand or other) may have rootlets, pellets, broken snail shells or other occasional inclusions. Upon firing, the organic material is partly or entirely destroyed, according at which temperature and in which conditions the pot is fired. This leaves holes (organic-derived voids, Fig. 5.6a-b) in the ceramic, with sharp boundaries and sometimes remains of charred cellular tissue, which will allow the analyst to identify (or guess) what the original plant or organism was (the void could have derived from an animal shell and calcium can leach out, for example, Fig. 5.6c-e). These accidental or natural plant fragments and bioclasts seen in the raw material(s) give valuable information as to the provenance or geological environment where the materials could have been formed or collected. Oxides and clay can fill in some of the voids left by burned off organic material or some of the carbonaceous material may still be visible after firing as in Figure 5.7.

The temperature at which organic carbonaceous material starts degrading is around 200 °C, to burn off totally below 800 °C. But experimental studies have shown that certain materials under reducing conditions or incomplete oxidation can partly withstand temperatures up to 900 °C (see the excellent article by Van Doosselaere et al. 2014 on this regard, as well as Dumpe and Stivrins 2015). Many voids in a ceramic may present some structural problems, or they may be desired; it increases the porosity and decreases its capacity to hold liquids if the vessel is low fired and the surface untreated. Some studies have also shown that the desired effect could have been to diminish the weight of a vessel, a jar for example, for shipment purposes. In that respect, adding pumice would also provide structure, good heating expansion rate, lower the plasticity, with less weight than the same amount of quartz sand for example.

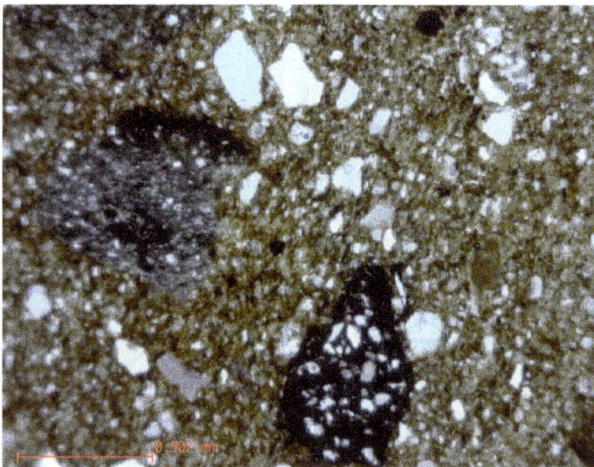

Fig. 5.5. Examples of grog in a ceramic thin section from a Louisiana sherd from Mallard Bay, 225x ppl.

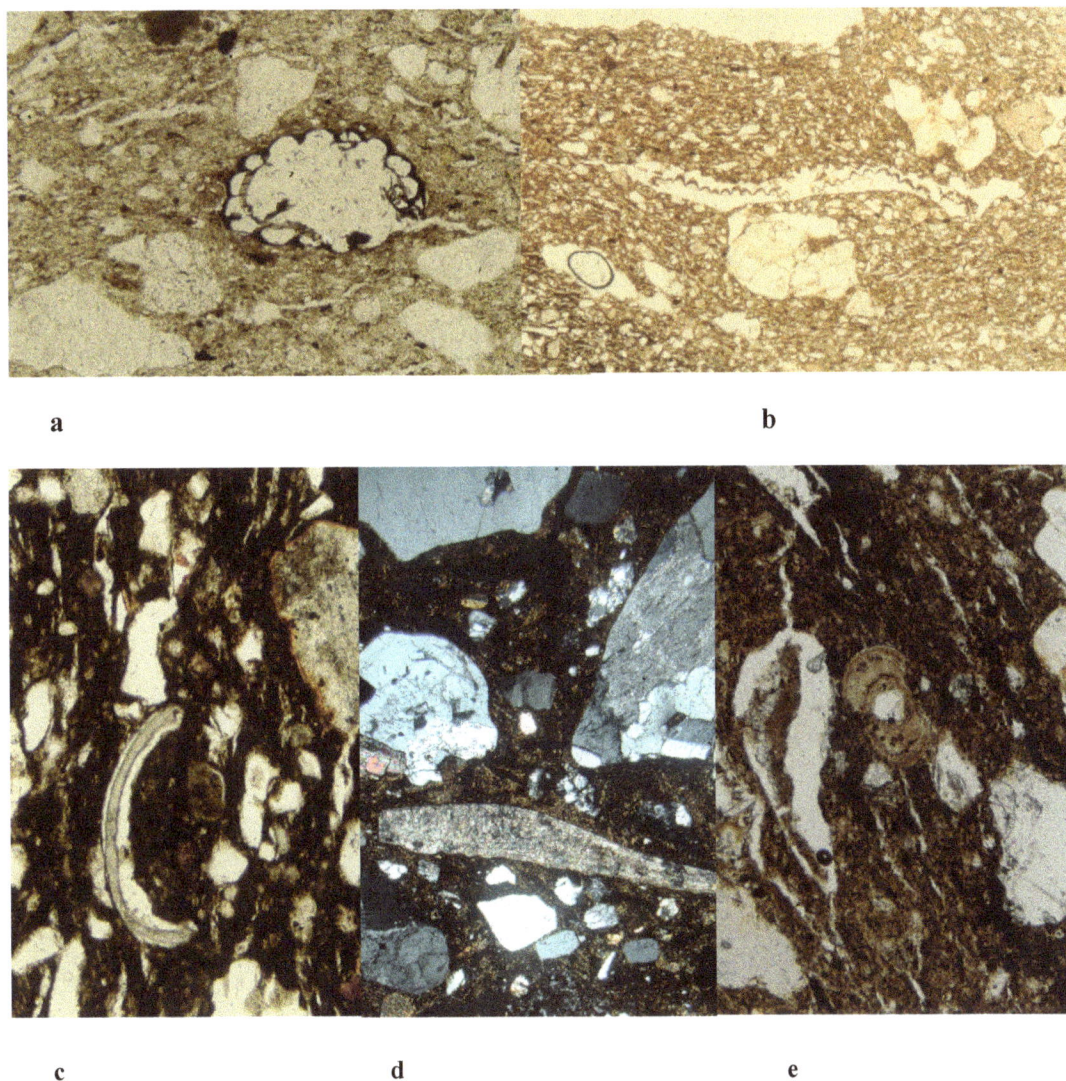

a   b

c   d   e

Fig. 5.6. Examples of plant remains and bioclasts in ceramic thin section. **(a)** Charophyte gyrogonite pseudomorph void (0.34 mm long) with remains of cellular tissue, from old lagoon-clay deposits, seen in a ceramic paste, PU128, Puemape, north coast, Peru 10x, ppl, reproduced with permission from Druc 2015, fig. 9b; **(b)** void with plant tissue (possibly rice hull), from a Chinese Longshan ceramic, accidentally incorporated into the paste; **(c)** ostracode (bioclast) fragment, KW41, Kuntur Wasi, Peruvian Andes; **(d)** shell fragment (1.62mm) in granitic-rich paste, HM25, lower coastal Moche valley, Peru; **(e)** ooid (circular concretion around a grain, here a quartz) and left of it, possibly a shell void resulting from the leaching of Ca from post-depositional processes, CA16, Puemape, Peru.

Fig. 5.7. Organic material (rootlets) in natural clay, TT21, Liangchengzen area, China. reflective light, digital Dino-Lite microscope.

Clays may also contain natural soil aggregates (pedofeatures), ferruginous nodules and pellets left by worms, etc. They are typically rounded or oval and may vary greatly in size. In figure 5.8a we see a large iron-clay-silt concretion juxting the wall of a ceramic vessel, while in 5.8b iron nodules and concretions are within the paste as natural inclusions. Concretions or pedofeatures are not added by the potter. These natural inclusions are not to be confused with grog (crushed ceramic fragments used as temper). Ian Whitbread (1986) and Patrick Quinn (2013), among others, building upon sedimentology and soil nomenclatures have discussed this issue and propose different names for these iron-rich nodules according to how they look and if they can be considered natural or added inclusions. One clue as to their presence being natural or not is their percentage in the paste. It is possible that the potter may add dry, powdered clay, to his/her slurry to reach proper consistency and these clay pellets or nodules will be numerous and show a darker color, mostly because they will not have been rehumidified (will not have the same water content) when used as temper. Iron may be present also as nodules and stains. The presence of iron and organic material has an impact on the firing of the ware, as will be seen later on.

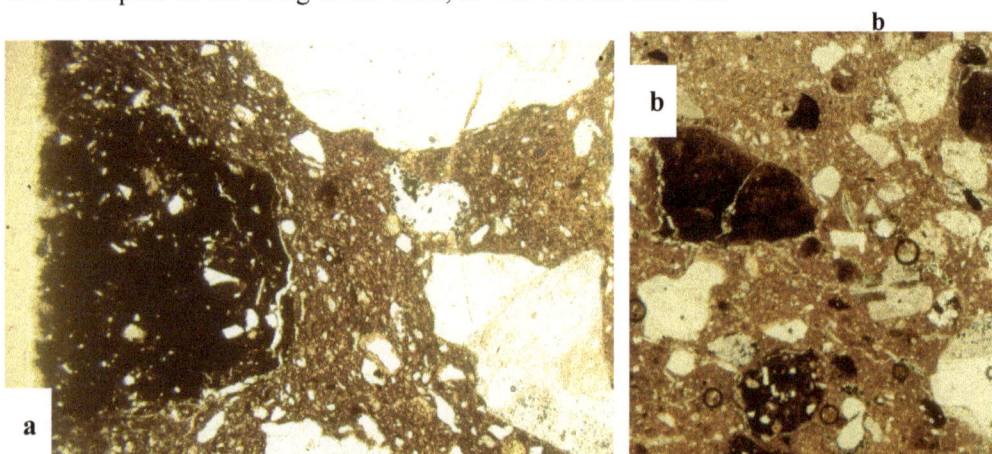

Fig. 5.8 (a) Coarse pedofeature (Fe-clay-silt concretion 0.95mm large) in ceramic paste, HM5 4x, ppl. It may have been incorporated into the paste by accident, but other similar ones can be found as natural inclusions in the clay along with iron nodules as in **5.8 (b)** (HSC21 small nodule = 0.375 mm diam., 4x, ppl).

In summary, the plastic properties of natural pure clays are often too great to allow their heating. Pure clays often are not physically stable, they are too plastic and flow with time. They have too much water associated with them in most instances and, upon firing, the ceramic paste swells as the water becomes vapour. The clay paste cracks and becomes useless for producing ceramic. In order to modulate the plastic physical properties of a clay-water mixture during the process of forming and drying a ceramic paste, a non-plastic material such as sand or silt should be present. This tempering material reduces plasticity and improves paste workability.

The importance of silicate mineralogy of non-plastics for archaeology, and in particular provenance studies, can be understood when one realizes that silicate sand grains are most often little affected by the heating necessary to produce a ceramic. They tend to retain their mineral identity. Thus, they can be used, when identified, to trace the rocks from which they came. Microfossils are another material naturally present in certain clays, which may not fully disappear upon heating and can yield a great deal of information about the environment from which the clay was collected. Clay minerals, on the other hand, tend to be completely transformed during the firing process (See Chap. 6). Clays lose their identity in space, forming a sort of glass or unstructured material. Therefore, sand-size grains can be used to identify the geological source of that portion of a ceramic while clays cannot. For this reason, sand and silt non-clay mineral grains have a great use in archaeological research.

In archaeological descriptions, the visible or identifiable portion of a fired paste will be mineral grains of silt and sand size which are non-clays. They provide a tempering effect of the natural plasticity of the clay minerals. The identity of the non-plastic material can often be ascertained by mineralogical observation. Hence, the non-plastic silicate (temper) portion of a ceramic paste can be used to indicate provenance of the materials as the mineral species indicate rock type and an assemblage of grains will indicate whether or not several rocks produced the sands found as temper.

5.2.3.2
Mixtures of Clay Sources

A more complicated, but perhaps common, practice is to use two or more clay sources as a base to form a paste. The addition can be considered as the tempering of a more plastic clay with a less plastic one. In some cases, the more non-plastic material can, in fact, be a clay-rich rock. Further, tempers of this sort can be of different plasticity. They can contain different proportions of clay compared to non-plastic grains or different proportions of clay minerals (kaolin, illite, smectite, etc). Such a usage is still practised in traditional potting in Japan, China and South America, among other areas.

It must be stated that many clay sources used in traditional and ancient ceramic making present a mix of clay minerals (interlayers of smectite, illite, kaolin, etc.) and are very rarely pure (one clay type only). In fact, clay-mineral layers can stack different types. Thus, an XRD analysis of the raw clay material will yield a multi-component composition. The archaeologist will label the clay according to the highest amount present, say a montmorillonite clay, even if small amounts of smectite and kaolin are also there. These different clays are formed from different parent rocks and environments, and identifying the clays may help identify the resource area -if the pottery has not been fired too high and only incipient vitrification is seen. Potters, on the other hand, usually choose clays according to which color they turn when fired, which also depends upon the amount of iron present in the paste and firing conditions (temperature-time-amount or oxygen present). Pure kaolin clay will turn white, but a slight presence of iron will bring hues of pink, for example. The whole idea is that the more non-clay material, the less plastic the total paste mixture.

The above information gives some idea of the natural mineral associations which one should expect in a rock. However, these simple relations are rarely encountered in ceramics. The problem is that the sands of a river reflect the rocks in a wide geographic and hence geologic area. It is rare that a river drains a region made up of one type of rock only. Hence, one can expect that rivers will produce deposits of different mineral grain associations (e.g. Fig. 5.9).

Fig. 5.9 Sandy paste with mix composition: volcano-sedimentary, metamorphic and intrusive rock fragments and derived minerals, subround to subangular fine grains, sorted material, in a clay with many carbonates (oval, elongated, brown grains), PU178 4x, ppl and xpl, Peruvian coast. Many voids developed parallel to the ceramic walls due the retraction in relation to the clay type used, but also compaction of the paste during building, drying and firing of the ceramic

The simplification of mineralogy into a pure mono-mineral assemblage (e.g. quartz grains) can occur on beaches, but only rarely. When it does, provenancing can yield great results as the study by Dickinson et al (1996) of Lapita ware shows. Usually, some other minerals are present, though they may be there in only small quantities. The rule in ceramics is one of multirock sources when the materials have been transported. Hence, when one finds, for example, only minerals common to a granite in a ceramic sherd: quartz, potassium feldspar, plagioclase, and mica, for example, one should suspect that the material has not travelled far before it was collected by a potter. An opposite example, where there is a mix of rock fragments and minerals from different origin, plus the presence of carbonate is a good indication of a beach setting (see Fig. 5.9).

Simple mineral associations will reflect either systems of little transport or those of important transport and weathering. The contrast of temper grains, representing different rock types, can be of great help in archaeology. If one finds some special minerals, indicative of a special rock type, this might be an index of provenance or site of production (e.g. Pailes et al. 2015). Anna Shepard (1965) already pointed this out. Examples of this sort of detective work are given in Chapter 9.

5.3
Decorations and Surfaces

The mixture of clay and temper grains forms the core, aka the ceramic body, which one sees normally only when the object is broken. This body can be masked by a surface treatment, for aesthetic or functional reasons. In fact, a high proportion of archaeological ceramic objects have a surface treatment which modifies or hides the aspect of the basic materials inside. One objective is to make a ceramic object attractive, one which makes a user prefer one producer over another. In other cases, ceramists wished to imitate objects made of other materials, metal objects for instance, and so tried to modify the earthen aspect of the initial materials. Reducing the porosity of a ware to better hold liquids, limiting abrasion and wear, are other reasons to apply a surface treatment. Surface treatment is often a criterion used by archaeologists to distinguish different cultures, communities, villages, or even individual producers from one another.

These surface treatments are of several types, which can, of course be combined in different ways to give the desired effects or results. The basic methods are grouped into the four techniques below. Our emphasis, again, will be on what can be observed at the level of the materials.

5.3.1
Surface Smoothing and Polishing

Smoothing is the simplest method to obtain a surface effect. This technique is effected by pressing a smooth, hard object on the surface of a clay-rich ceramic. This causes the particles of the clay matrix to orient themselves parallel to the surface of the ceramic object. In pushing on the plastic material of the body, the temper grains are pushed into the matrix, enhancing the proportion of clay at the surface of the paste. The clay-particles' orientation gives a lustrous shiny effect similar to that produced by a pure clay slip. Polishing is going a step further, rubbing over and over with a smooth hard object (a pebble for example) the leather-hard surface of a ceramic (before firing). The process requires more pressure than for smoothing, and tends to give a more dense state of the paste at its surface; it not only orients the clay particles but compact them more than just smoothing. This is time-consuming but it brings a metallic shine that is not achieved by smoothing.

This is shown in Figure 5.10a. It illustrates a thin section (a) and cross section view (b) of an experimental tile made with sigillata-clay. The tile front face has been lightly polished (with no addition of a slip) and let to dry, then fired with barrel-reduction at the end to obtain a black surface (Valley and Druc 2017). Polishing the surface while still slightly wet brings up a thin layer of very fine, wet clay; the clay particles can then easily align themselves parallel to the surface. Both the silt and very fine-size of the material and the orientation of the particles are responsible for the nearly opaque layer that we see in thin section. This layer operates as a water barrier, a functional gain in addition to the visual effect of having a polished surface. The other side of the tile, unpolished, does not show such a layer of aligned or fine particles and carbon penetration is greater on that side, with a diffuse line. In comparison, Figure 5.10b illustrates the application of hematite powder mixed to a sigillata-clay slip. The surface has then been burnished with a soft pebble, and the pot fired. However, we can still see silt- and sand-size grains close to the surface, and the composition is similar to that of the core. The same clay was used for the core and the slip, somewhat more watery for the later and acting as the carrier of the mineral pigment. The aim of adding hematite in the slip here was to bring out a deeper surface color be it red or black according to the firing atmosphere. The difference is mostly visible in terms of surface color (see Fig. 6.6).

Next page, Fig. 5.10 (a) Thin section photomicrograph (left), 4x plain polarized light of an experimental tile of sigillata clay, with uncoated polished surface. (Right) The same tile, seen in cross-section with reflective light, showing both sides (back on the left and front on the right). The polished side (right of the image) is smoother, with sharper boundaries and less grains closer to the surface than the back side. The dark layers on both sides result from more intense reduction at the surface, while the core is not as reduced.

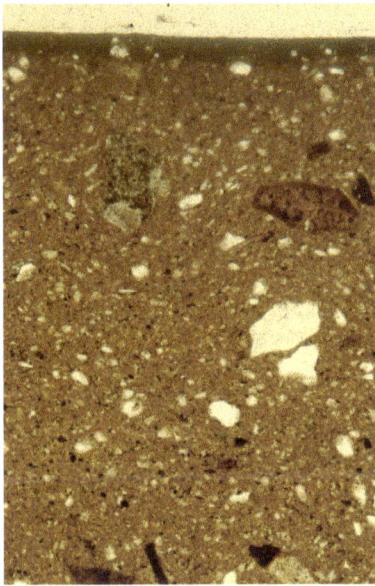
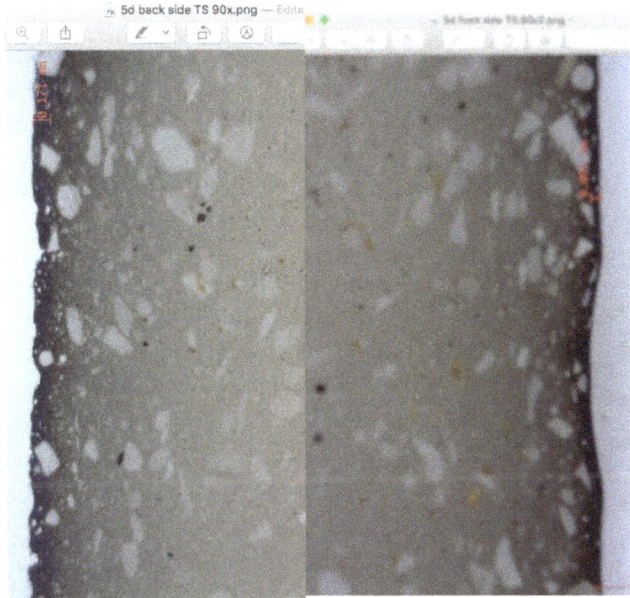

a

Back of tile   Front of tile

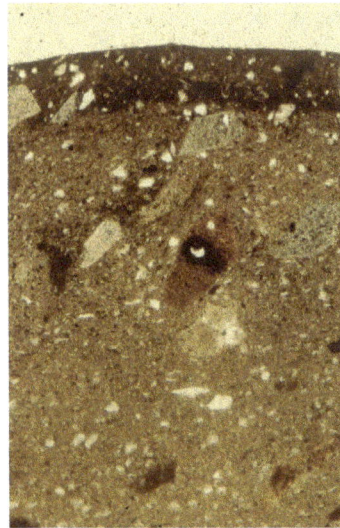

b

Figure 5.10 (b) ppl, 4x. Left: Hematite + sigillata slip over sigillata body. Right: hematite + mineral oil. Neutral firing, 4x ppl, experimental tiles produced and fired by Andrée Valley, ceramist, see Druc et al. 2019).

5.3.2
Slip

A slip is a layer, normally of clay-rich material, applied in a liquid state to the air-dried ceramic body. The ceramic can be dipped into the clay-water mixture slip, or the slip be applied onto part of it with a brush (or other). The dryness of the ceramic extracts water from the slip which concentrates the clay particles on the surface of the unfired ceramic body, producing a clay coating. This clay coating does not run but adheres to the ceramic when fired. One objective of a slip is to cover, visually, the surface of a ceramic, hiding defects or unevenness of the body (Fig. 5.11). The covering effect occurs because the slip has a strong tendency to melt or lose its structure before (at a lower temperature-time stage) the core of the ceramic. If temperature is high enough, the material becomes glassy or hard while the ceramic core maintains its competence. The adherence of slip to core is an important technological step in ceramic production (see Fig 5.12 for a case of bad slip adherence). The clay-rich slip usually shows a sharp border with the surface below; its composition should also be different from the body, with finer and less non-plastic inclusions. Slip layers are usually on the order of several tens of micrometers thickness to a millimeter or perhaps more.

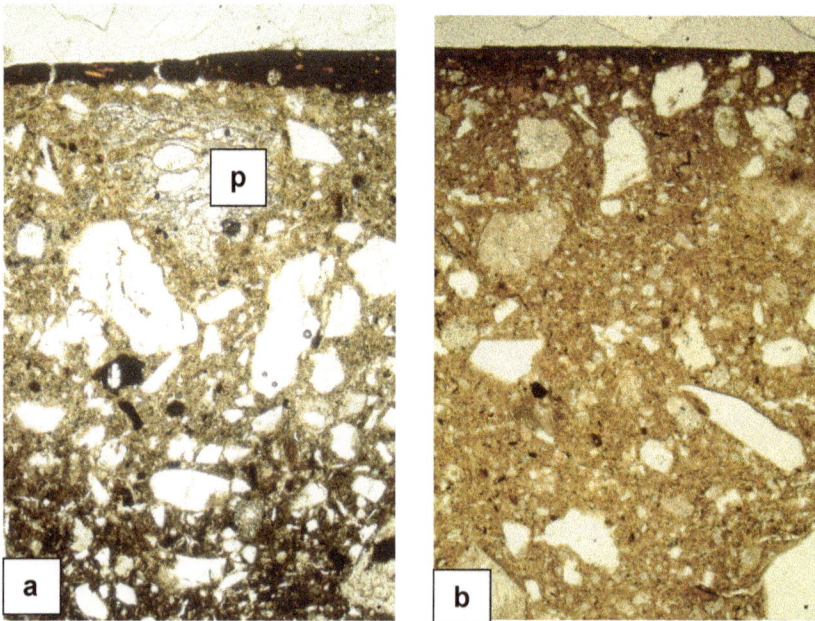

Figure 5.11 (a) shows a slip covering the uneven surface of a bowl (ID5). The paste is volcanic and a large pumice (P) is seen just below the surface. The slip layer is also uneven. Contrast this to (**b**) where the ceramic (KW24) has been smoothed before firing, may be with a wet cloth, bringing the clay (and iron oxides) particles to the surface. This is however not a slip. There is no sharp limit between the slip layer and the body of the vessel, heat can penetrate more into the ceramic, oxidizing fully the body. Both photomicrographs are at 4x, ppl view.

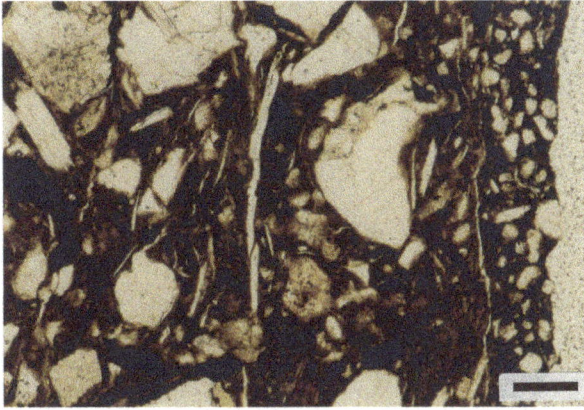

Fig. 5.12. Here, one can see the effects of bad adherence of slip to the clay body. The slip layer shows finer temper grain-size material (right side of image) than in the body. It contains many fine quartz, feldspar and mica grains, while the main body has medium to large quartz-muscovite schist clasts and derived minerals. Slipped bottle, pre-Hispanic Andes, 4x, ppl.
Bar= 0.18 mm

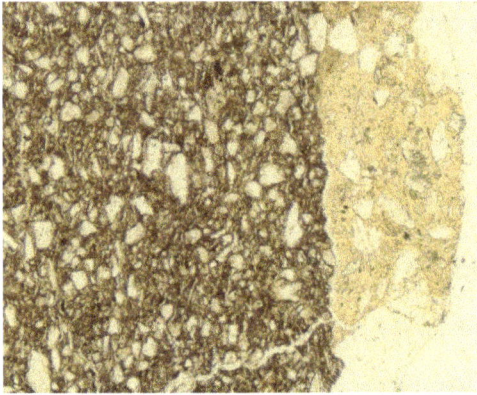

Fig. 5.13. Slip (light color in ppl) on fine *bei* cup, Liangchengzhen, China, slip is 0.22 mm thick, 10x, ppl (see Druc et al. 2018).

By contrast, see Figure 5.14. The surface is oxidized, with no slip. The cross polarized view shows no composition or granulometry difference.

Fig. 5.14. Thin section photomicrograph of oxidized surface of white ware, LCZ-wh108 4x ppl and xpl. The mineral composition is the same as in the body (see Druc et al. 2021).

According to the firing cycle, the color of the slip may change. An oxidizing atmosphere will produce a red color in a clay- and iron-rich slip, a reducing one, a dark to black slip with the same base material. In other cases, the slip can be multiphase (i.e. clay and silt grains), showing irregularities in its texture. When the slip is a mixture of clay and other minerals, fine quartz and so forth, its gives a thicker texture and the slip layer is thicker also (see Fig. 5.12).

Generally, slips are clay-rich materials -but they could contain many inclusions, as seen above. The clays are easily kept in suspension in water because of their small grain size and shape, sheet form, which keeps them from falling rapidly to the bottom of the slurry. The clay slurry can be applied with ease. The clays will usually melt at a lower temperature than the ceramic core paste due to the addition of a fusing agent or perhaps by their own chemical composition should they be alkali (potassium or sodium)-rich. In any event, the clays must melt or be more strongly transformed than the core paste in order to give the cover sought for and to adhere to the object. Some clay slips become glassy when well fired. Roman sigillata ware (red) and pre-Roman, Campanian ware (black) from Italy are examples. Their clay-based surface materials are opaque when fired. The slip is a very clay-rich, potassic clay which melt more easily than the minerals in the clay body. These same clay materials are used as paints as well.

Difference in granulometry and mineralogy can help identify the true presence of a slip. So, if the grain sizes of non-clay materials is different in the slip and core material, the slip is not the same material as the core but could be derived from the core clay resource by levigation.

5.3.3
Paint

Paints are usually metal oxide-rich assemblages which are high in color, or lack of it (black) contrasted to the slip or body of the ceramic. Usually, paints are applied locally to a surface with some sort of a brush. Painting indicates a decorative act, designs of some sort. Paints are of two kinds, (1) those which are heat-resistant and (2) those which are not resistant to heat (post-firing paints).

Heat-resistant paints should hold their color when fired. It is not necessary that paints keep the same color when in the "raw" state and in the fired state. What counts is the final result. Paints are usually applied in much thinner layers than the slips, on the order of less than 10 μm. Paints can be applied to the surface of a ceramic body and not covered by the protective coating of a glaze. These paints are generally formed of iron oxides giving red, brown or black colors, or of manganese oxides giving black to brown colors, or of iron-titanite (ilmenite) giving black colors or perhaps other metal oxides. These metal oxides are not greatly affected by the normal oxidation occurring during the firing process.

Other transition metal elements can be used, but they need to have been employed below or within a protective glaze coating. These are cobalt (giving a blue color), copper (green or, at times, red), lead-tin (white), among many others. In fact, these paints are usually in a glass matrix which preserves their color during (and after) firing.

Or they can be applied post-fire, thus are not fired. This is reserved for wares not used for cooking or subjected to a lot of wear.

Some paints have a different chemical history from the rest of the ceramic material during the firing cycle. These paints change oxidation state, being reduced during firing while the core of the ceramic is oxidized.

A very classic example is that of Greek ceramics produced in the early Hellenistic period. These ceramics are for the most part large drinking vessels and jars with black-painted figures in a red, orange or pink background. The brush strokes of the painter are frequently thinned near their edges and show the superposition of black designs on red or pink clays. The paints are produced from iron-rich clays which produce reduced iron oxide phases that are very stable upon further firing under oxidation conditions, i.e. normal firing conditions at high temperature (Noble 1988). The process of producing these black, iron oxides under oxidizing conditions, which gives pink oxidized clays in the core of the ceramic, is one which can only be deduced. However, it seems probable that the initial paints were charged with an organic material which gave off reducing gases during the initial stages of firing. This reduced the iron oxide in the clays to produce the black iron oxides early in the firing cycle.

A similar process was used to reduce precious metal ions dissolved in solution in lustre ware of the 19th century Staffordshire Jersey ware and Hispano-Mauresque painted wares of the 15th-17th centuries. In these wares, gold, platinum or silver metals were dissolved in acid solutions and combined with organic-rich liquids such as turpentine. The concentrated paint solution, when fired at 700-800 °C, reduces the metallic ions in solution as they are condensed and deposited on the surface of the ceramic. The resultant coating of metal (not an oxide but in a metallic state) as a thin film gives a shiny surface which makes a reasonable imitation of a metal object. The reduction is local while the entire ceramic is subjected to a normal oxidizing atmosphere during the firing cycle. The surface coating is reduced while the core clay is oxidized.

Sometimes paints do not resist heating. They are applied to ceramics of ceremonial or decorative use only. For example, many funerary ceramic statues in Aztec Mexico and Classical Greek times used paints which were applied to the ceramic when the firing cycles were finished. Aztec blue colors were obtained by the absorption of organic dyes onto clay substrates which were mixed with metallic oxides in aqueous solution (Jose-Yacaman et al. 1996). The paint was stable with time but not with temperature. Greek funerary statues frequently were colored with inorganic minerals. For example, jarosite was used, which is a brown mineral composed of potassium, iron, sulphur and water. It is unstable at temperatures above 100 °C. However, its place in a tomb did not require conditions above the stability of the minerals involved. At times, paints of low thermal stability were applied to ceramics which had some colors applied during the firing cycle. This is true of certain large drinking or liquid recipients (craters) produced in Classical Greece. Iron oxide or kaolinite-lime were applied on the surface of otherwise normally fired color patterns of vitrified clays in the black and red range.

If paints are applied with a brush, locally to give a design, one could conceive using a slip material, the clays for the Roman sigillata ware for example, to paint a ceramic. If one looks at Greek Hellenistic decorative ceramics, it is evident that the material is the same as the Campanian or sigillata ware slips. A liquid, low-firing clay material can be used to cover the whole object or only a part of it. Hence, one must be careful to define what one means by a paint. It could be an oxide, never used as a slip, or it could be a material used under other circumstances for other purposes.

5.3.4
Glazes

Glaze is usually a term used to describe a glassy, silica-rich layer which has been applied onto the surface of a ceramic. When fired at a certain temperature, depending upon the materials, the glaze vitrifies and the surface becomes impermeable. Normally, glazes contain only small amounts of alumina. According to recipes, they may melt before the core materials of the paste of the ceramic, and are often applied on bisque fired bodies in a second, lower temperature firing. The fusion of silica in glazes at a low temperature is produced by the addition of a fusing agent. Most glazes are prepared as a glassy material before being applied to a ceramic object. The glassy state allows them to react rapidly to become liquid under heating.

The most common and earliest glazes contained lead or another heavy element. Such glazes melt at low temperatures, slightly above 700 °C, depending upon their composition. Some glazes are made of alkalis (sodium, potassium) and silica. These glazes usually melt at higher temperatures than lead glazes. Figure 5.15 shows a glaze that is alkali but may contain some lead as the firing temperature was much lower than alkali glazes need to mature (c. 800-850 °C instead of 950-1000 °C, Ownby et al. 2017, 620). Glazes can be opaque or transparent.

Fig. 5.15 Alkali clear glaze over marl (calcareous) clay body with quartz and feldspar grains, fired between 800-850 °C, Medieval ware from Elephantine, Egypt, blue on white splash decoration. Glaze (clear layer on top of ceramic body) shows some sand and rare bubbles from gases not being able to escape due to the viscosity of the material. Glaze is 0.6mm thick. Photo: Mary F. Ownby, Ownby et al. 2017, Fig. 4a. Reproduced with permission from M. Ownby.

Glazes can be charged with coloring agents so that even when they remain transparent they give a general color tint to the object. The early glazes used in the Middle Ages in Europe were a lead-based silica composition charged with several percent copper, giving a green color, or with iron, giving a brown or yellow color. Cobalt (tinting blue) was also a common, but more expensive, coloring agent, and as a result it was more rare. Depending upon the amount of metal used, these glazes are transparent or opaque. When opaque, they give a homogeneous color to the surface of the ceramic and mask the color of the core. A common opaque under-glaze in faience ware from Renaissance ceramics in Europe is a tin-lead-silica base (Sn-Pb-Si). It provides a white opaque layer, with a general. The thickness of glazes is often on the order of 30 to 300 µm (0.03 to 0.3 mm) but could be more.

The lead-tin opaque glazes give a white color that masks the color of the ceramic body. This is the basis for what is often called faience ware, used to imitate the porcelain of the Chinese in Europe in the Renaissance period and later. One can paint on this glazed surface and fire the object again at lower temperatures in order to bake the paint into the white glaze so that it will be durable under use. A second coat of glaze is often used to protect the paint from oxidation in the furnace. This gives a double-glazed structure with one opaque layer next to the ceramic core and another above it at the surface of the object. Thus, glazes can be used to mask entirely the clay body of the ceramic, they can be used to color the surface and mask the clay body or they can be used to protect colors below their surface when the glazes are transparent. Glazes also have a functional aim: that of protecting or impermeabilizing the pot's surface.

Some glazes show interaction with or penetration into the ceramic body beneath them. The best glazes will enter into the clay-temper grain matrix by diffusion to form a strong bond which keeps them intact. They will not flake off. At times, there is an interaction between the glaze and the ceramic body to produce small crystallites.

Fig. 5.16. Jar neck fragment (**a, c**) and cross section (**b**). Bar at corner left of cross section view is 0.5mm, Dino Lite portable microscope, reflective light. Reproduced with permission, Druc 2015, Fig. 4.33.

Figure 5.16 shows the cross section of the neck of a big stoneware jar produced in PhuLang, Vietnam. The plant-base glaze (bamboo leaf ash + river mud) applied on the outside surface is clearly visible. Due to the glaze, the surface below did not receive as much oxygen during firing and remained reduced (Fig 5.16c), while the unglazed face (interior of the wide neck) was oxidized. These big jars are fired once, in dragon kilns at about 1200° C. Figure 7.21 in Chapter 7 illustrates a similar ware as seen in thin section.

The surface treatments described above can either be applied to the dried object before the first firing or they can be applied after a bisque firing and be subjected to subsequent heat treatments. This second instance is common when the surface material needs a lower time-temperature of firing to obtain the desired effects than necessary to fire and transform the core materials. For example, porcelain is fired at temperatures above 1030 ºC for more than 8 days, while melting of glazes are commonly obtained at temperatures below 950 °C. Further, many glazes tend to melt rapidly, given their thinness, and hence will need shorter firing times than does the ceramic body. Some complicated decorations can require a glaze undercoat, a paint cycle on this undercoat and a subsequent transparent glaze over the paint and undercoating of glaze. Other glazes are applied on top of an initial opaque or transparent glaze coating. These are called over-glazes. The trick in all of the operations is that the last coat fires and melts at a lower temperature-time span than the under layer. Thus, successive firings are of lower and lower temperatures-time cycles. Different combinations of slips on air-dried ceramic (green ware) with or without paints applied later with different firing cycles are, of course, possible.

There are then three large groups of surface treatments with materials added to or deposited on the surface of a ceramic: slips, paints and glazes. There can be different combinations of these materials to produce decorative or artistic effects on the ceramic. Other surface treatment and decorative techniques (incision, excision, etc.) are not "materials", hence they are not presented here.

Bibliography

Dayton, J (1978) *Minerals, metals, glazing and man.* Harap, London, pp 496

Druc, I., Bertolino, S, Valley, A, Inokuchi, K, Rumiche, F, and Fournelle, J (2019) The Rojo Grafitado case: Production of an early fine-ware style in the Andes. *Boletin de Arqueologia, PUCP*, Advances in ceramic and pigment analysis in archaeology, Part 1, 26, 49-64.

Druc, I., Underhill, A., Wang, F., Luan, F., Lu, Q., Hu, Q., Guo, M., and Liu, Y. (2020) Late Neolithic white wares from southeastern Shandong, China: the tricks to produce a white looking pot with not much kaolin. Results from petrography, XRD and SEM-EDS analyses. *J of Arch Sci Reports* 35 https://doi.org/10.1016/j.jasrep.2020.102673

Dumpe, B, and Stivrins, N (2015) Organic inclusions in Middle and Late Iron Age (5th-12th century) hand-built pottery in present-day Latvia, *J of Arch Sci* 57: 239-247.

Gebhard, R, El-Hage, Y, Wagner, FE, and Wagner, U (1988/1989) Early Ceramics from Canapote, Colombia, studied by physical methods. *Paleoethnologica Buenos Aires* 5:17-34

Jose-Yacaman, M, Rendon, L, Arenas, J, and Serra Puche, MC (1996) Maya blue paint: An ancient nanostructured material. *Science,* 273: 223-225

Kingery, WD, and Vandiver, PB (1986) *Ceramic masterpieces; art, structure and technology.* Free Press, NY, pp 339

Levin, EM, Robbins, CR, and McMurdie, HF, (1964) *Phase diagrams for ceramists.* American Ceramics Society, Columbus, OH, pp 601

Mason, RB, and Tite, MS (1997). The beginnings of tin-opacification of pottery glazes. *Archaeometry* 39: 41-58.

Harman, CG (1973) [Parmelee, CW (1951)] *Ceramic glazes*, 3rd ed. Cahners Books, Boston, pp 612

Noble, JV (1988) *The techniques of painted attic pottery.* Thames and Hudson, London, pp 216

Ownby, MF, Giomi, E, and Williams, G (2017) Glazed ware from here and there: Petrographic analysis of the technological transfer of glazing knowledge. *J of Arch Sci Reports* 16: 616-626.

Parmelee, CW (1973) [1951] *Ceramic glazes.* Completely revised and enlarged by Harman CG. Cahners Books, Boston, 3rd Ed. pp 612

Pailes, M, Killick, D, Ferguson, TJ, and Mills, B (2015). Diabase temper as a marker for Laguna ceramics. *Kiva* 80 (3/4): 281-303.

Quinn, P (2013) *Ceramic petrography.* Archaeopress, Oxford, pp 254

Rice, PM (2015) *Pottery analysis, second edition, A source book.* The University of Chicago Press, Chicago, pp 592

Rye, OS (1981) *Pottery technology.* Manuals on archaeology 4, Taraxacum, Washington DC, pp 150

Sentance B (2004) *Ceramics a World Guide to Traditional Techniques.* Thames and Hudson, London, pp 216

Shepard, A (1965) *Ceramics for the archaeologist.* Carnegie Institution of Washington, Washington, DC, pp 414

Tite, MS, Freestone, I, Mason, RB (1998) Review article: Lead glazes in antiquity. Methods of production and reasons for use. *Archaeometry* 40: 241-260.

Tite, MS, Wolf, S, and Mason, RB (2011) The technological development of stonepaste ceramics from the Islamic Middle East. *J. Archaeol. Sc.* 38: 570-580.

Van Doosselaere, B, Delhon, C, and Hayes, E (2014) Looking through voids: A microanalysis of organic-derived porosity and bioclasts in archaeological ceramics from Koumbi Saleh (Mauritania, fifth/sixth-seventeenth century AD). *Archaeological and Anthropological Sci* 6(4): 373-396. DOI: 10.1007/s12520-014-0176-5.

Valley, A, and Druc, I (2017) Experimental replication of black pigment on Andean ceramics. What's left after firing? Colloque d'archéométrie du GMPCA, Rennes, France.

Wagner, U, Gebhard, R, Murad, E, Riederer, J, Shimada, I, Ulbert, C, Wagner, FE, and Wippern, AM (1991) Firing conditions and compositional characteristics of formative ceramics: archaeometric perspective. 56th Annual Meeting of the Society of American Archaeology, New Orleans

Whitbread, IK (1986) The characterization of argillaceous inclusions in ceramic thin sections. *Archaeometry* 28: 79-88.

6 Physical and Chemical Processes in Firing Ceramics

6.1
Firing and Material Transformations

The process of making mud into a pot involves the transformation of the minerals in the plastic clay-rich paste into a solid object. The means of transformation is one of mineral phase change, the destruction of certain minerals and at times the creation of others. These changes are accomplished by firing, i.e. bringing the clay-rich materials to a high enough temperature for a sufficient period of time to effect the needed changes in physical state. The firing process is one of hardening of the material or object. The degree of hardness depends upon the length of time and the temperature used to transform the material (and type of clay, fusing agents, particle size, among other factors). This is the firing process. The hardness or the degree of transformation is due to the methods of heating. This heating is determined by the needs of the potter to make a desired product and the stage of technical development of the society in which the potter works.

6.1.1
Variables of Transformation to Make a Ceramic

The same degree of transformation (often considered to be the quality of the ceramic product) can be due either to a choice among a number of options or due to a limit given the period and technology known to the culture the ceramic was produced in. For example, the Roman potters were capable of making very fine ceramic materials but they did not always do so because many of the objects they produced did not need such a high refinement of technical prowess. Roof tiles, amphora or cooking pots were not in the same category as the sigillata ware found on the tables of local notables. The society as a whole, considering its technical capacity, was capable of making items of very high quality, but these methods were not always employed if they were not needed. However, Neolithic potters used more or less the same methods to produce the ceramics used as offerings as well as cooking pots. Details of decoration and form differentiated the more prestigious materials from those of everyday use. Hence, the type of material produced is often a conjuncture of means or capacity and need. Some pots are better made than others, but it is up to the archaeologist to find out why. One should always remember the proverb, "he who can do the most can do the least".

6.1.1.1
The Firing Process: Time and Temperature

The transformation of a slightly humid clay-rich shape into a ceramic then passes by the firing process which is one of heating the object until the clays are destroyed or heated enough to retain the shape they have been given for significant periods of time and under some mechanical stress. If one cuts a steak on a plate and the plate breaks, the heating process was poorly done.

Heating and mineral phase transformations are done in the framework of two variables, *time and temperature.* These are the key variables to the firing process. Each can be varied independently of the other. Other factors influence too the firing process which we will see further below. As well, the type and dimension of the kiln influences the time taken to achieve the desired temperature. In a large kiln chamber, it may take more time than in a small chamber to have the kiln temperature homogenized, allowing all the pots, be they placed in the center or in the periphery, to be correctly fired. A third, and equally important variable in the mineral transformation process of ceramic making is the *chemical composition* of the ceramic material. This variable affects the time-temperature relations of firing.

The interrelation of time and temperature is extremely important in the production of ceramics. Taking the example of baking a cake, the free water from the batter needs to be expelled, as also the water from the flour, eggs and other organic matter to allow the raw materials to transform into something edible. In ceramic production, the heating process allows to expel free water and water from the clay mineral structures, and this at specific temperatures.

Given the firing temperature and cycle, a ceramic will be more or less fired (transformed), depending upon the time it resides in the furnace. If one raises the temperature, the time necessary for the chemical and physical transformations necessary to form a solid, stable ceramic will be lessened.

Time and temperature relations can be seen in Figure 6.1 (a), where the heating time (log scale) is plotted against weight loss (a measure of mineral transformation) for muscovite mica. At high temperatures (1050 °C) the mica is transformed rapidly, whereas from 110 to 450 °C the transformation does not occur in a significant manner. Att 650 °C one can see hope of transformation, but it will take a considerable time.

The same is true for the clay minerals in a ceramic body. The temperature of firing is not the only variable in producing ceramics. A practical illustration of the time-temperature approach to producing ceramics is illustrated by the use of ceramic indicators such as the Wedgewood cones to tell the state of firing that a ceramic charge has attained in a hot furnace. Since the measure of attainment of transformation or ceramic formation is difficult to define in a strict sense, the method used in the 19th c., and still used today, is the Wedgwood scale 'measuring' the degree of transformation or melting of a cone of ceramic material. Cones of clay, feldspar and quartz are made, with different mineral proportions, which melt at different temperatures given the same time of heating or different times of heating given the same temperature of firing (Fig. 6.b).

Fig. 6.1 (a) Change in weight of muscovite as a function of the temperature and time of firing. The curves show the importance of both temperature and time. In all but the highest temperature curve one sees a continuing weight loss as a function of firing time up to 100 h, which is more than 4 days time. . (**b**) Example of a cone that has been fired with ceramics within a kiln.

Several cones are placed among the ceramics to be fired. When one cone is deformed, the degree of transformation has been attained for the given mixture of clay, feldspar and quartz. When cone number three, of a given Wedgewood composition, has been deformed, for example, a potter knows that his specific ceramic materials are sufficiently fired (Fig. 6b). The potter then compares the state of transformation of the Wedgewood cones to his specific ceramic materials to be transformed. The deformation of a Wedgwood cone is determined by the temperature and the length of time that is heated. These are the two variables of the firing program.

In forming porcelain, the firing cycle takes a number of days (8 to 10 or so) at the same firing temperature (near 1300 °C). Since different furnaces or portions of a furnace do not maintain exactly the same temperature over such a long period, the use of Wedgewood cones became widespread in order to assess the change effected in different parts of the furnace. It is quite clear that if temperature were the only variable in firing, no reasonable potter would fire his material for 8 days at 1300 °C to make porcelain instead of firing it for several hours. These relations of time and temperature as factors in the transformation of ceramic raw materials are, of course, well known to ceramists, and they have been studied for archaeological ceramics since a long time.

6.1.1.2
Stages of Transformation in Time-Temperature Coordinates

In order to understand the process of clay to ceramic metamorphosis, one needs to define the stages of this transformation and their bearing upon the quality of the finished product. In transforming a humid clay paste into a ceramic, several stages can be recognized. They follow the transformation of clays given in Chapter 3.3. They are:

1. Drying and loss of free water. This is the stage normally accomplished under atmospheric or not heated conditions. A potter puts his shaped ceramic piece on a rack in a well-ventilated place to dry. This stage is done slowly in order to avoid cracks or deformation due to loss of the considerable amount of free water. Shrinkage of the initial object can be up to 15 %. Initial drying occurs over periods of days or a week.

2. First stages of firing are those of loss of bound water on the surfaces of the clay and temper materials. This is accomplished at temperatures above 100 °C usually and for a reasonably long period of time, hours. Not all pottery-making processes use this step.

3. Temperature is raised, and heating occurs so that the clay minerals lose their crystalline water. This is an irreversible process. Once the clays have lost their water, the ceramic becomes rigid. Often, if the ceramic is withdrawn at this stage, the material has a low durability and is soft, i.e. it can be easily scratched and eventually corroded. Much earthenware and faience are fired at high temperature, but before reaching stage 4, of the partial melting of non-plastics.

4. The next stage of transformation is the total transformation of the clays and partial melting of the non-clay particles and the clays. Often, the smaller non-clay particles are melted to the clay mass. Larger temper grains are still identifiable.

5. The most advanced stage of ceramic formation is that of a porcelain. Here, in true porcelain, the clays and the temper are melted and a new mineral is crystallized. It is mullite, an alumino-silicate. The clays and much of the temper form a glass and the mullite forms a network of fine, needle-shaped crystals which holds the structure together.

In common pottery terminology, the *porcelain stage* (5) is the highest fired material, stage 4 is that of *stoneware*. Stage 3 is the most common one -in archaeology-, producing *earthenware*, and one that includes well-fired examples and poorly fired samples, depending upon the amount of clay transformed and the bonding of transformed clay with smaller temper grains. Stage 2 is essentially an unfired state which is more stable than stage 1, which is often called the green state. The porcelain stage gives a translucid material which is very easy to identify. It allows one to make an object which is very thin-walled (but this is not only the case for porcelain) and extremely hard. Stoneware ceramics are hard but not translucid.

It is important to remember that the stage of firing is affected by the time-temperature relations of the firing program and the composition of the ceramic material. One can describe the firing program of a potter in the framework of time and temperature: the initial heating period, which can be the longest part of the process, the maximum temperature and the cooling stage. Each stage can vary from one firing process to another. Two examples are given in Figure 6.2.

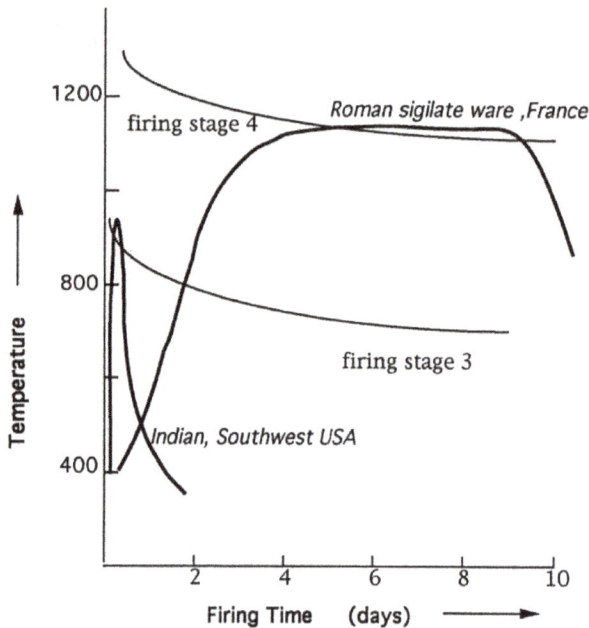

Fig. 6.2. Illustration of two different types of firing programs, Roman sigillata ware (La Graufsenque, France) and Zia Southwestern American (USA) Indian ware. The Roman ceramics are produced in well-constructed kilns where temperatures can be maintained for days whereas the Indian ceramics are produced by the bonfire method where a pile of fuel is ignited and left to burn around the ceramics on the surface of the ground. The Roman sigillata ceramics reach a stoneware transformation <*firing stage 4,* see text) and the Zia ware is well fired with moderate transformation *(firing stage* 3). The firing stages are represented by *lines.* The position of these lines is not an absolute value in time-temperature coordinates in that it is affected by composition, grain size and other factors. It is clear in this Figure that the two different types of firing environments give very different results due principally to the firing time and temperatures reached. (Combined data compiled by Velde from Shepard 1963 and unpublished material of Roman ware study).

The typical firing program of southwestern Indian potters (Shepard 1963) is on the order of several hours at most and the maximal temperatures are in the range of 850-950 °C for surface firing or in pit kilns. It can take long to reach the desired temperature. Then temperature is high for a short period of time. The pottery attains a stage of destruction of the clay mineral matrix but not the stage of glass formation of stoneware. By contrast, the Roman potters making a highly durable, very dense and nearly porcelain-type pottery fired the material near 1150 °C for periods of up to 8 days. The sigillate ware produced is near stoneware in its texture and stage of mineral transformation.

The firing programs are highly different, as are the products. Japanese traditional pottery producing wares similar in degree of transformation of the clay and tempers use programs similar to those of the Roman producers. Nineteenth century porcelain firing needed temperatures above 1300 °C for a period of nearly a week. In general, one finds that the lower temperature firing programs are of short duration. However, it is sometimes necessary to fire a ware for a long period at a low or medium temperature to allow a better adherence of the slip or melting of the glaze. A low firing is also important if one wants to conserve the shiny effect of polishing.

If we take the five basic stages of ceramic transformation outlined above in Section 6.1.1.2 as being 1 through 5, we can represent how to attain them in coordinates of time and temperature as in Figure 6.3. The effect of the two variables in furnace use can be foreseen by the rates of weight loss of clays. Higher temperatures effect more rapid weight loss of clays and hence a greater structural change. Lower temperatures take much more time to change the structure of clays. The stage of firing is influenced by the length of time and temperature of the firing program. However, composition is important also. For this reason, we indicate the stages of firing in time-temperature coordinates in a general way with lines which are only indicative of the physical conditions necessary to transform the paste of the ceramic. The actual stage of trans-formation of a given pot can vary to either side of these indicative lines in the figure. For very short heating periods, enormous temperatures are necessary to transform a paste into a ceramic. Clearly, more than seconds or several minutes are necessary at any furnace temperature. Of course, there are limits to the process.

The lowest temperatures of transformation to stage 2, for example, will be at 110°C, where water becomes vapour overcoming the attraction of water to the clay particle surfaces if the temperature is maintained for long periods (days). The stage 3 will occur probably at a minimum temperature of 500 °C in most instances, depending upon the clay minerals present. Ceramics of porcelain grade need temperatures above 1300 °C. Most ceramics of reasonable hardness and durability are made at temperatures above 900 °C. The longer the firing time, the harder the material and the better the grade of firing. But fusing agents and even firing atmosphere can change these relations.

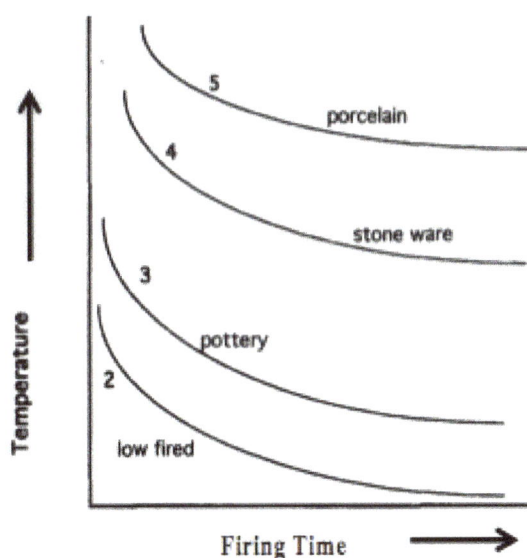

Fig. 6.3. The four stages of ceramic transformation induced by firing are schematically indicated as a function of general time and temperature conditions of transformation. No absolute values can be given because these will depend upon the materials used to make the ceramic. The stages are indicated by *lines* where the stage of transformation is reached in time-temperature space. Notice that the different stages are not at equal distances from one another in the diagram.

6.1.1.3
Paste Composition and Fusing Agents

If one uses only refractory clays and a refractory temper such as quartz, firing temperatures will need to be very high in order to destroy and melt the clay to form a sturdy, hard and impermeable ceramic, i.e. temperatures well above 1050 °C. However, if one adds calcium oxide, potassium or sodium in other forms, the melting temperature is greatly reduced. These are fusing agents. Figure 6.4 illustrates this principle.

The most common fusing agent is calcium oxide (lime or CaO), which can be present in the clay itself or added as such, or in the form of calcium carbonate ($CaCO_3$), the mineral calcite, which degasses, releasing CO_2 upon firing in the range of 700-900°C. This range of temperatures depends upon the firing program (time-temperature) and grain size of the crystals. For example, one frequently finds different calcium contents in Roman sigillata ware which reflect the use of a fusing agent and hence can show the production area as well as the period when the ceramics were produced. A very good chapter about calcium-rich clays, their behaviour, and their use in the Meditarrenean world, for example, is given by Daniel Alvaro Santacreu (2015).

A preheated source of calcium is lime (CaO) obtained by roasting calcite until the CO_2 gas has been liberated. The advantage of such a material is that when in the paste, it combines directly with the clay material without the bubble-forming gas CO_2. The action of the fusing agent occurs at lower temperatures than those when calcite is destabilized. Calcite-rich pastes are actually quite common, often using microcrystalline, sparite calcite, which presents lesser risks than using coarser calcitic material. Maritan et al. (in prep) report the use of this type of temper in Italy (4th-14th c) for producing cooking pots, derived from the grinding of marble. This reminds us that co-occurring industries can provide tempering materials to potters as well.

In Figure 6.4 the left-hand part of the diagram shows that quartz (SiO_2) melts at 1800 °C. The line separating liquid from solids intersects the temperature axis at this temperature for pure SiO_2. As small amounts of CaO (lime) are added to the quartz, one moves to the right on the diagram, and the line separating pure liquid and quartz moves down, to lower temperatures. Thus, as more calcium is added to the mixture, the temperature at which one can produce a liquid with no quartz becomes lower and lower. Thus, the temperature of destruction of quartz is decreased as the composition becomes more calcic. This is the principle used in ceramics to decrease the fusion temperature of clays.

Of course, potters try not to melt quartz but to melt or at least partially melt clays. This occurs at lower temperatures than the melting of quartz. The range of melting is effected between about 900 and 1500 °C. The effect of composition on changing the state of clays and temper grains in a paste changes the firing temperature. Composition change in a paste can change the position of the lines in the temperature-time diagram. This chemical effect on firing characteristics is indicated in Figure 6.5. Hence, one cannot give a specific series of time-temperature conditions to define a given stage of ceramic transformation. Composition will shift the physical conditions necessary to change the clay-temper assemblage into a solid ceramic of a desired quality.

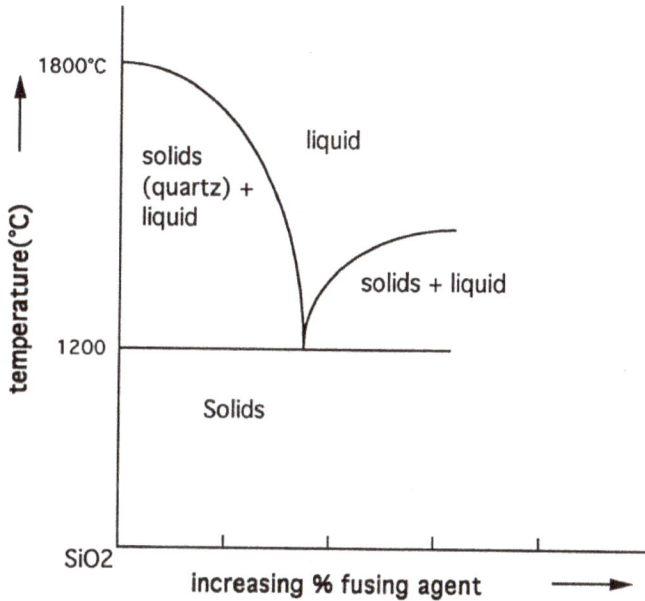

Fig. 6.4. This diagram illustrates a hypothetical fusing agent diagram. The effect of a fusing agent on the melting temperature of silica (quartz) and associated phases is shown. The area where liquid alone occurs reaches a minimum temperature at near 2 % of the fusing agent. In the composition near the SiO_2 pole, the high-temperature assemblage is quartz plus liquid. The proportion of liquid varies with the amount of fusing agent present. If one wishes to achieve a ceramic with fusion of the solids (quartz), the best composition is with 2% fusing agent where the temperature is 1200 °C compared to 1800 °C in a sample without a fusing agent.

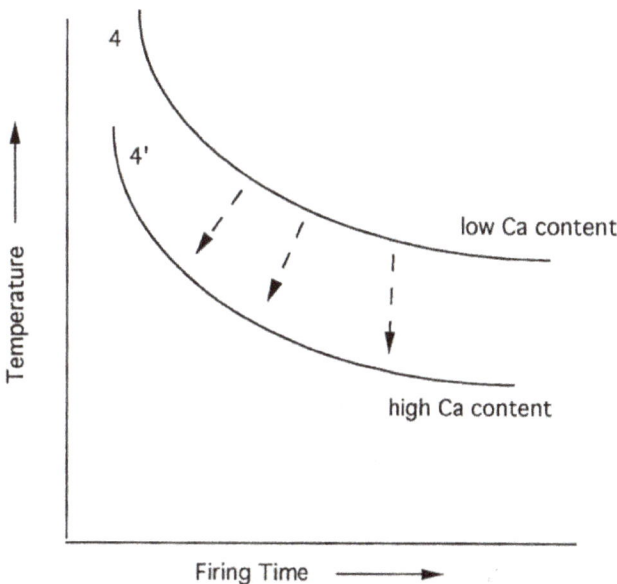

Fig. 6.5. The effect of chemistry change is illustrated here as it affects the firing stage of a sample. When CaO (lime or calcium oxide) is added to a ceramic paste, it tends to change the values where a stage of firing is achieved in time-temperature variables. The example is for the firing stage 4 of Fig. 6.3, which becomes modified to 4'.

6.1.1.4
Reduction of Iron

Most common clays contain a reasonable amount of iron oxide, about 1-3 %. The iron in these clays is usually oxidized, and very often in its oxide form of Fe_2O_3, regularly accompanied by some water bound to the oxide in crystalline form. The iron cations are Fe^{3+}. This oxide can be transformed into a less oxidized form, FeO for example, by heating in a reducing environment. A reducing environment is one which changes Fe^{3+} to Fe^{2+}, reducing the positive charge on the ion. Reducing environments can be created in a furnace by improperly burning the fuel, for example by putting wet, organic material, such as tree bark or wet straw in the furnace where the fuel is burning. This often produces smudging on a pot. This process may be sought to achieve impermeability (see Chap. 7). In this case, the carbon deposit on the surface blocks the pores, hence lessening permeability. Reducing fires tend to be less hot than oxidizing ones because the fuel does not give its entire heating capacity. However, in a reducing environment the iron will be reduced from an oxide to a more active form.

This reduced iron tends to combine with the minerals present, especially the clays, and it forms a glass destroying a portion of the clay structures. The reduction of iron changes the melting temperature of a clay body much as does the addition of lime (CaO). This lowers the time-temperature curve of ceramic transformation as shown in Figure 6.5. There is a tradeoff in this process between lower firing temperature, as the fuel burns less efficiently, with one of lowering the melting temperature of the ceramic. One must make the melting temperature of the clay-oxide mixture lower than the reduction in temperature of the fire. However, reducing the iron does produce a nice highly glassy material. More details about oxidation-reduction effects and firing cycles are given further below. The firing atmosphere, firing cycle and fluctuation(s) bear also consequences for the surface color of a vessel, as illustrated in the example discussed below.

6.1.1.5
Firing Programs and Surface Color

Surface color is affected in great part by firing atmosphere and oxygen access in particular at the end of the firing cycle or when opening the kiln, at which time re-oxidation can occur. An example will illustrate the behavior of mineral pigments in slips according to surface treatment in relation to different firing programs, when 'playing' with oxidizing and reduction effects. It is taken from the experiment with sigillata tiles by Valley and Druc (2017, Druc et al. 2019). The tiles were coated on their front surface with a mixture of terra sigillata (the medium) and mineral pigments such as graphite, hematite, and manganese (MnO_2), or with the mineral pigments mixed to mineral oil, or simply left uncoated. Made from commercial earthenware clay with 17% grog, the coated tiles were bisque fired to cone Δ010 (887°C) to prevent them from blowing up in subsequent firings due to ambient humidity.

Fig. 6.6. Experimental firing cycles with different mineral pigments and sigillata slip (Tsig). (**a**) Reduced and reoxidized, (**b**) Reduced and cooled in H20. (**c**) Reduced in sawdust + H20 (Valley and Druc, 2017).

In the firing program (**a**) (Fig. 6.6, first column), the tiles were fired in a Raku kiln (a short firing), with a reduction flame then re-oxidized (opening up the air source) at the end of the firing. In the (**b**) program (Fig. 6.6 second column, the tiles were fired in a Raku kiln, in a neutral atmosphere, but heavily reduced at the end of the firing, and cooled off quickly in water. In the (**c**) program (Fig. 6.6, third column), the tiles were again fired in a Raku kiln, then quickly transferred to a barrel and carbonized using sawdust. Being red hot, the tiles ignite the sawdust. A lid is then put on the barrel, which promotes carbonization on the surface of the tiles. Oxygen in the clay is drawn out to promote combustion and the drawn-out O is replaced by CO_2 in the clay body. This produces a nice jet-black carbon surface.

After the tiles are in the closed barrel for about 10 minutes, they are transferred to a pail of water to 'freeze' the carbonization process. If the tiles were left to cool naturally in the air, they could re-oxidize. They could also be left in the barrel with the lid closed until cool with the same effect. In Figure 6.6, note the difference in shine according to the mineral pigment applied. Also, even if a tile is fired in a reducing atmosphere, a re-oxidation at the end of firing will yield a light color surface. A dark core will show if the firing has been long enough. When exposed to reduction, the extent of reduction on the non-coated side was greater and more intense than the coated side, which did not allow for as much carbon penetration. In fact, the coating seals the surface: the finer grain size of the particles brought to the surface by burnishing and their alignment parallel to the surface oppose perpendicular circulation of air and carbon. As for surface color, unless the piece is carbonized so heavily the surface is fully reduced to black, iron reds are expected in a reduction atmosphere, as well as in atmospheres that have been reduced then re-oxidized. Oxidation, light reduction or even heavy reduction with a time of re-oxidation will allow for iron reds and darker manganese blacks. (Andrée Valley is a ceramist who teaches, ceramic firing processes).

6.1.1.6
Particle Size

Most clays are of less than 2 μm in diameter, as their initial definition states. However, some are smaller than others. Some are on the average half the size of others. Natural clay-rich deposits often contain grains of other sizes, which are not of the same mineral type as clays. These differences are important in the firing characteristics of the clay-rich paste used to make a ceramic. The finer the grain size on average, the lower the temperature at which the clay will be transformed given the same firing time or the more the material will be transformed during short firing cycles. This is the case of the mica shown above, and also the case for other minerals such as feldspar or calcite. Grain size is also a factor in the mineral transformations necessary to transform a clay-temper material into a ceramic.

Summary

All of these variables, grain size, mineral species, oxidation-reduction relations, fusing agents etc. contribute to the transformation of a clay matrix into a durable ceramic material. The physical variables of the firing cycle, time and temperature are used to reach the desired firing state according to the materials which make up the ceramic paste. For these reasons, the simple determination of a firing temperature will not be adequate to describe the state of ceramic transformation or the type of firing cycle used to obtain the ceramic transformation state. Measurement of temperature is only one of many variables in the production of a ceramic object.

6.2
Firing Practices

There are almost as many furnaces and types of firing as there are archaeological investigations. Details are unnumbered when it comes to the type, shape, time-temperature programs, atmospheres, etc. of ceramic firing. Here, we will give only the basic variables which give very different results in the production of ceramics. These are the different types of firing conceptions.

6.2.1
Firing on the Ground (Bonfire)

The easiest type of furnace construction is one where there is no furnace. If one piles the green ceramics to be fired on the ground on fuel and covers them with fuel or other (wood, dung, grass etc.), one can fire the materials to a very good degree of transformation. Temperatures of above 940 °C can be reached easily, but they will not last very long unless new fuel is added on to the pile, after about half an hour at most. This is the technique which costs the least in labour and material as far as furnace construction is concerned. However, the cost in fuel is the highest. Figure 6.7 illustrates this principle. The heat which is produced in the surface firing method is very largely lost to the surrounding atmosphere. One needs fuel which burns rapidly and gives a high heat. Cotton wood is used instead of oak, for example, straw instead of coal or charcoal. The mass of material per item fired is not favourable economically.

Second, the firing can be of good quality, but it never reaches stage 5, that of porcelain and rarely that of stoneware (stage 4). The structure of a fire in the open air does not lead to the constant high heat which needs to be maintained to melt feldspars and kaolinite. The rate of heating is very great also, which leads to irregularity in the firing process. A misplaced branch of wood or a layer of straw which is too thick can momentarily retard the heating and will not give the same results on the whole pot. These two factors lead to uneven firing and a significant loss of ceramics due to poor conditions producing underfired objects, over fired ones and irregularly fired objects.

One often has parts that are overfired (partially melted and deformed) or those that are underfired, giving irregularities in their solidity or transformation state. It is possible, nevertheless, to vary the firing period by adding more fuel on to the pile during the firing cycle. This gives some latitude to the potter in his use of the firing program to transform his material to a ceramic.

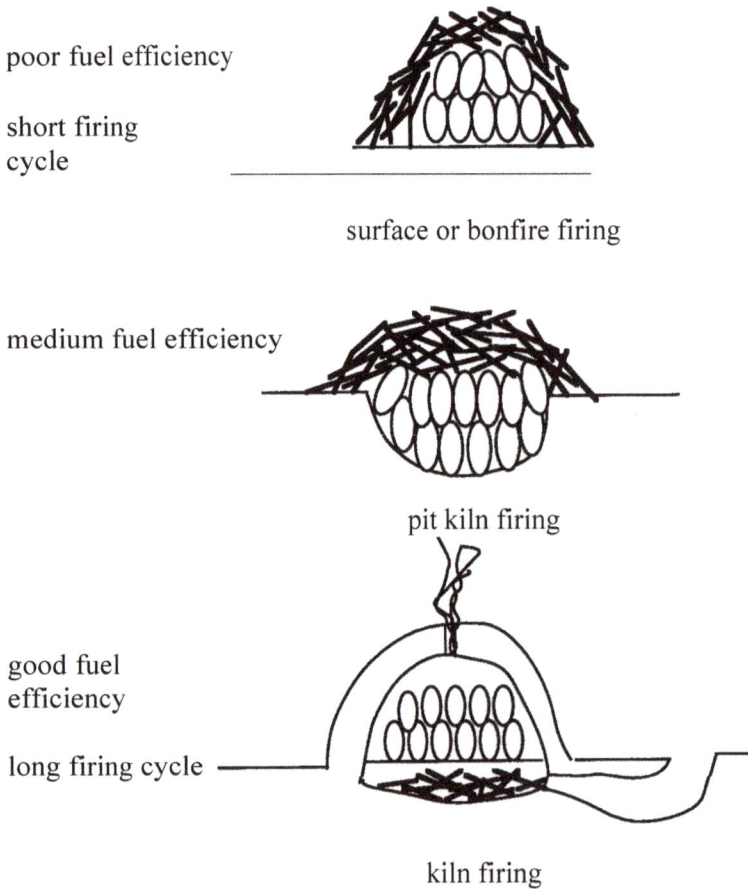

poor fuel efficiency

short firing
cycle

surface or bonfire firing

medium fuel efficiency

pit kiln firing

good fuel
efficiency

long firing cycle

kiln firing

Fig. 6.7. Schematic description of the different general categories of firing practices. The more the ceramics are protected from heat loss, by the earth in a pit or in a furnace or kiln, the longer the firing cycle can last and the higher the efficiency the system will be as far as fuel (represented by *heavy black lines)* is concerned. Pit or surface firing lasts hours and kiln firing generally lasts days.

6.2.2
Pit Firing

Digging a small hole in the ground can increase the efficiency of a firing greatly. The earth of the pit walls will keep the heat in and release it over a longer period than, of course, the air will in a ground surface firing. However, the pit kiln, by its structure, limits the amount of fuel used and it is difficult to add more. Once the fire has started and burned, it is not possible to modify the firing program. Pit firing is, then, more economical, it gives a more consistent temperature over longer periods with a given amount of fuel, but one cannot lengthen the firing time. The structure of the pit walls regulates the firing temperature and gives more consistent results. Firing times at maximum temperature are on the order of a half hour or more.

An intermediate, ameliorated pit-type kiln can have walls but not a covered roof. Walls are built of stones, laid around the pit or even on a flat surface. The effect is similar to that of a pit kiln, one where a rudimentary thermal barrier is built to hold in the heat of the firing process. This permits a longer firing cycle, up to 24 hours in some cases.

6.2.3
Kilns

A good system is a kiln where a fire is made under the material to be fired and the heat is kept under thermal control by the walls and roof of the kiln. It is the most economic and flexible method of producing ceramics, using almost any type of fuel, and the firing cycle can be constantly modified by adding different types of fuel. This is the ideal type of firing system. It has been and still is used by many complex societies. Firing times in kilns can be from hours to a week or more. The atmosphere, oxidizing or reducing, can be modulated during the course of the firing cycle by controlling the temperature and changing the material which is to be burned. The principle of a kiln is to have a fire apart from the ceramics. Usually, it is below or to the side of the charge to be fired.

The necessity for a draft to regulate the combustion of the fuel has led to different types and shapes of kilns. A second problem is putting the ceramics in the kiln and then closing it up so that it maintains a homogeneous temperature. One should close all the doors. Sometimes, the ceramic charge is sealed in by making a temporary wall in the entry, at other times, different methods of introducing the charge are used. These constraints also lead to different sorts of kiln construction.

In general, the longer the firing cycle, the more control one has over the quality of the end product. Thus, it is generally considered that advanced civilizations use closed kilns to produce ceramics where the rate of transformation of the clay-rich paste is slower and more consistent. In a closed furnace, the temperature tends to be more homogeneous, and all the objects will be transformed to the same degree. This is, of course, true in a general sense, but one should not be fooled. Some very fine, well-fired and homogeneous pots can be produced with rudimentary bonfire, pit or open kilns (for example the Moche and Sican ceramics, and those of Indian cultures in the Southwestern United States, Shepard 1963).

It is evident that firing systems on the ground or in a shallow pit cannot maintain a temperature for any important length of time. The maximum temperature will not be imposed for more than a half hour or so. The more insulation used (pit walls, kiln wall and roof), the longer the maximum temperature will be maintained. Pictures and more details on kilns are presented in Chapter 7.

6.3
Summary of Factors in the Formation of a Ceramic Body During Firing

From the above sections, it is evident that the formation of a ceramic, i.e. transformation of a plastic paste into a rigid, stable form, is controlled by numerous factors. Not only the variables of the furnace-firing program, but also the composition of the paste and the grain-size distribution of the minerals in it can affect the process of ceramic formation. The variables of the transformation process are then *temperature* of firing, length of *time* of the firing, *oxidation-reduction* atmosphere of the furnace, *composition* of the paste mixture, and at times the *grain size* of the paste minerals. Given these factors, a search for the highest temperature that was attained by firing a ceramic will not give very much pertinent information as to the process of ceramic production. Firing temperature will vary, as do the minerals in the paste, the length of time that the material was heated and the oxidation-reduction conditions of its firing. The conjugation of the above variables will determine the quality of the fired ceramic obtained. A mastery of these different variables leads to a well-made product, whereas a poor use of the variables leads to objects of lesser quality.

6.4
Structure, Porosity and Density of Ceramics: Non-Plastics, Clays, and Pores

The different steps of ceramic production, the formation of the plastic state by addition of clay, tempering agents and water, shaping and finishing the ceramic object and then the firing program, all contribute to the final internal structure of the ceramic. Why is this important? In stylistic studies, one looks at the shape of the ceramic object, the surface finish, such as lustre, a high or low polish, the decoration, paint colors, use of glazes and so forth. This is important for classification purposes, typologies, in iconographic studies or for identification of the producer. But there are other features that characterize the internal aspect of a vessel and may influence its function. A storage pot was usually designed to hold something, keeping rodents or insects from it or keeping it from changing its properties such as liquidity or humidity. A cooking pot was used to hold liquids in a high-temperature environment. Pots may also have non-functional uses (funerary, offerings, status symbols etc.). As well, decoration and shape of a ceramic object had to fulfill certain aesthetic or quality requested by the user. The secret of the use-parameter is often found by investigating the inside of the ceramic material. This can be found in studying the structure, density, porosity part of ceramics.

In order to investigate the inside of a ceramic material, the easiest method is to look at it. Since most of the elements in a ceramic body are in the microscopic or submicroscopic scale, it is necessary to use a hand lens or a microscope. A binocular microscope, or now a portable digital one allows examination in the range of 3 to 50 in magnification. One can see the types of grains (i.e. their color, relative size, and to some extent their shape). One can determine if there is much non-plastic material on the scale of observation used, its distribution, and in many cases, identify some of the minerals in the coarser temper. This first step can be used in the field for initial classification or description. It is useful in a general sense.

However, a more careful investigation can be made by using a true optical polarizing microscope which gives a resolution of up to 2 μm, the definition of clay size particles. In order to do this, one needs to have the ceramic object cut into a thin section, as it is known in geological studies, one where the ceramic is near 30 μm thick. In this state, light passes through the majority of the grains present and using transmitted light as petrographic microscopes do, one can identify the minerals, rock fragments and other inclusions (with a trained eye). Mineral identification is most often for an expert, a geologist or petrographer. However, much can be gained by a careful examination of the relations between the clay matrix (unidentifiable for the most part due to the resolution of that type of microscopes) and the temper grains present. One can obtain a great amount of information about technology and possible provenance of the raw materials in that way, and classify the materials into clay matrix, texture, and types of non-plastics. It should be remembered that the temper grains can be in natural association with the clays or added or modified (subtracted) by the potter.

6.4.1
Pores

6.4.1.1
Primary Pores

A ceramic piece is made up of solids and pore space, filled by air. These pores can be due to several causes, one being the natural interstitial space of the clay material worked by the potter, called here *primary pores*. This is most apparent in ceramics made from clay sources which have been little treated or hastily assembled. It is normally a sign either of lack of technique or of rapid execution. The example shown here, (Fig. 6.8) the lip area of a Roman amphora, shows pores which are due to incomplete packing or working of the clay materials, they show in black, as does the gap where the lip was folded.

Fig. 6.8. Direct scanned image of a Roman amphora lip cut and polished. Primary pores *(black areas* in the paste) are due to incomplete working of the clay. White temper grains are evident. The juncture of the material folded to form the lip shows a large linear pore where the folded material did not get welded to the central body by the pressure of a potter's hand. *Lighter areas* near the edge of the sherd are due to oxidation of its surface during firing. Sherd length is 7 cm.

6.4.1.2
Secondary Pores

Secondary pores can have two major causes, gas release and shrinkage of the clay substrate. Normally, decantation, kneading and mixture of the paste are designed to eliminate pore space. Hence in well-elaborated ceramic materials the pore space will be at a minimum. However, even though a potter has elaborated a clay-temper-water mixture largely exempt of pores or holes, the firing process can create new ones.

As we know, when clays are fired they release significant amounts of water. Also, fusing agents, such as calcium carbonate ($CaCO_3$) will release gas when destroyed by heating, and then they become fusing agents. If this water vapour or gas (CO_2) escapes when the clay body is plastic and if the materials do not let the water escape as a gas, pockets of gas can be created, producing *secondary pores.* These pores will be irregular and oriented to some extent. These types of pores are frequently visible on the surface of a ceramic body. Secondary pores can also be the result of the drying and dehydration of the clays. This is especially evident when the clays are highly oriented. as seen in Figure 6.9 where the temper grains are large.

When clays dry and are fired, they shrink as the water leaves the body. These shrinkage characteristics create different sorts of shrinkage cracks which are usually visible in thin section as long, linear or nearly linear pores very often parallel to the surface of the ceramic body. Such shrinkage cracks can be a good indicator of the working process. They are *work-induced pores.* For example, one finds them in handles or lips of vessels where the clay matrix has been compressed more than in other areas. These cracks result from the compression of clays, which tend to align parallel to their sheet structures. Shrinkage accentuates the clay texture, giving a line of shrinkage with respect to less oriented parts of the object. The shrinkage pores are not usually connected, at least as seen in thin section (two dimensions of a three-dimensional object), and they are slightly wavy, finishing at their ends in thin slits.

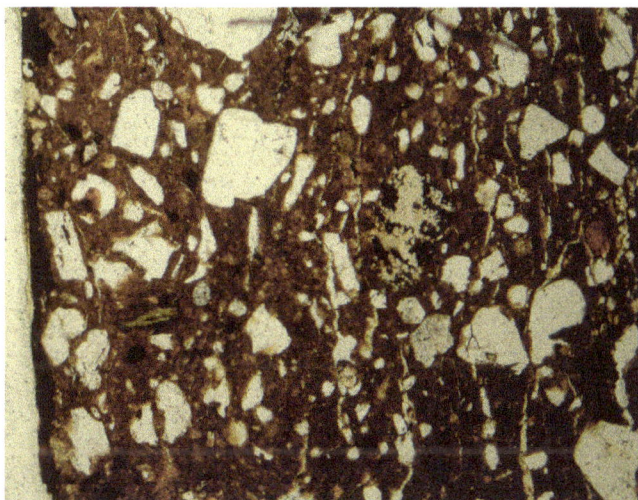

Fig. 6.9 Irregular secondary shrinkage cracks in earthenware from Peru, 1st mill B.C. Shrinkage of the clay mass upon firing makes wavy pores, which indicate the orientation of the clays due to working of the paste. KW82, 4x, ppl.

Pores are classified according to their form and size, as spherical (or vesicles), planar, channel, or vughs following terminologies found in micromorphology (See Rice 2015 or Quinn 2013: 97-98 for examples) and can be used to classify pottery. Pores, here, are not voids caused by Ca leaching, plant or bio-organisms destroyed during firing.

Pores can be accidents in production or they can be purposefully introduced. Common ware ceramics intended to contain water and keep it cool are designed to have pores which allow some water to leak out of the container to the air. The water in contact with the air evaporates, taking heat from the pot and cooling the water in it. The movement of water through the pores is called *permeability. Porosity* is different and relates to the percentage of pores within a ceramic. There is a great difference between pores that are connected to provide pathways through a ceramic and pores that are isolated in the core of a ceramic. The isolated pores are insulators. They keep the wall of a pot from transmitting heat. The connected pores, creating permeability, let fluids flow and reduce the thermal resistance making heat transfer greater. These relations are shown in Figure 6.10. The result of permeability (linked pores) in a vessel is to delay heating of the liquid inside. The leaking water moves out of the pot and cools it as it heats. This is, however, what is expected of a good liquid storage container, to keep the water cool. Potters may keep the surface of a pot from letting the water out by sealing it in one manner or another. They can apply an impermeable layer of material (glaze) on the inside or outside of the pot. They can polish the surface, which orients the clay particles perpendicular to the walls, which produces a barrier and close the pores that could have been present. In any event, porosity and pore structure are important factors in thermal properties.

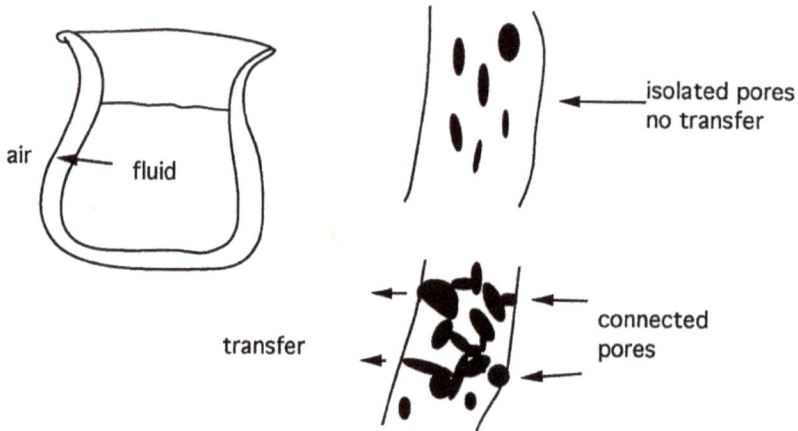

Fig. 6.10. Pores and permeability are diagrammatically represented for ceramic sherds as they affect the passage of water through a ceramic. A high porosity of unconnected pores is, in fact, an insulator because air is a very poor conductor of heat. When pores are connected, not only can water pass but heat can be transferred by the exchange via the water or liquids flowing out of the vessel. Thus, connected pores are to be avoided in cooking ware. However, unconnected pores producing high porosity are not necessarily a bad thing in conserving heat. Air, unconnected to the outside (air or water in the pot) is a good insulator, and will conserve heat. However, good insulation will prevent a pot from heating its contents rapidly. This is why modern pans are made of metal where heat in transferred inward, and water does not leak outward. Modern ceramics also fulfill this function.

6.4.1.3
Microporosity

This porosity is not seen by examination under the optical microscope but can play a great part in determining the physical characteristics of a ceramic body. The microporosity is a function. of the total force exerted on the clay paste, the relation between temper grains and clay minerals, the types of clays used and certainly the existence of fusing agents. Microporosity is the very fine pore space which is usually not connected to larger pores. This pore space forms an isolated body of very small air chambers. They act as insulators in the clay body.

6.4.2
Temper Material and Firing

Temper material, as defined in Chapter 5, is used in the formation of a paste which can be modeled into a convenient or desired ceramic shape. A second, and not minor, function of temper is to keep the object in its desired shape and dimensions during the firing process. If a clay-rich object fires too rapidly, producing a glass, it will lose its shape. Temper is often used as a sort of building block that will keep the plastic phase from running away. Sand grains, non-plastic, are less likely to melt under high temperatures. Clays are the first phases to be affected by heat and the silicate, sand-grain tempers the last. Thus, when one looks at the body of ceramic containing temper grains, one sees sand grains, which have not been modified by the firing cycle. Thus, one role of temper is to serve as a backbone for the ceramic during firing.

Other silicate tempers can be composed of grog, which is already fired ceramic material added in sand-grain size or greater. Grog is an ideal temper because it has the same physical properties of thermal expansion and chemical properties as the clays and temper of the new ceramic. And it changes the properties of a paste. It has been used extensively in various cultures in the Americas for example, and several samples of 19th century hand-made roof tiles from western France show a high proportion (more than half) of grog in their core.

Some special tempers can be seashell fragments. Further, they have the characteristic of having shapes which can lead to their identification under the microscope by a specialist. Seashell (carbonate) tempers have the unhandy property of disintegrating near 900 °C. This leaves gas (carbon dioxide) and the highly active calcium oxide or lime (CaO). Lime combines with the clays to create a glass (Sect. 5.4.1.3 on fusion agents). However, the loss of gas is very disrupting to the ceramic, and thus the firing temperatures should not destabilize the shell material to a great degree. Gas bubbles in a ceramic body lead to leaks and uneven surfaces. One remedy to the inconvenience of shell temper is to add salt (NaCl) which surrounds the carbonate shell forming a glassy barrier to further destabilization. An example of the use of shell material is given in the next chapter.

Another exotic temper material, added by the potter to temper the clay, is bone material. The bone fragments can be destroyed by heat, usually partially, and become a fusing agent. Bones are made of calcium (Ca) and phosphorus (P), and as is the case with shells, the calcium becomes a fusing agent upon the thermal destruction of the bone. However, in some early pottery, crushed bones were in such abundance that they were obviously used to temper the plasticity of the clay resource, not to become a fusing agent.

If the temper is not silicate in nature, straw or another organic material for example, it will be destroyed during the firing process without lasting effect on the stability of the ceramic body. It could be added intentionally to make the ware lighter. This may increase porosity but not necessarily permeability if the pores are not connected. It will however induce a reducing environment and keep the iron in a divalent, reduced state, thus promoting fusion of the clays present. Here, the temper is essentially a plasticity modifier, but could also bring cohesion and fire resistance as observed for ceramics tempered with sponge spicules of plant origin in the Amazonian Lowlands (Lozada Mendieta 2019). Vegetal tempers leave traces of reduced zones or voids which reflect their original volume and shape (see Fig. 5.6b in Chapter 5).

6.4.3
Thermal Properties of Oriented Clays

Clay minerals are, as has been mentioned many times thus far, sheet-like in shape (Chap 3, Sect. 3.2.2). This sheet structure is the result of a specific arrangement of atoms and chemical linkages in flat structures. The asymmetry in the clay mineral shape being a reflection of a specific internal structure gives physical properties which are not isotropic. The transmission of heat in a clay mineral particle is about six times greater along the sheet structure than across it. This is a very important property. When clay particles are aligned in a parallel fashion, they will resist heating across this structure. For example, the smoothed surface of a ceramic, made by wiping the surface when the clay is plastic or the polishing on semidry surface orients the particles. This gives a resistance to heat transfer to the surface.

In the production of ceramics, a certain amount of kneading and shaping is necessary, and this inevitably results in a partial orientation of the clay minerals, parallel to the surface of the pot. In order to overcome this effect, one can introduce temper, which is not only non-plastic but also has different thermal properties. Most temper is made of silicates, especially quartz, but also minerals, which have shapes that are not directed by a crystallographic cause. Hence, the transmission of heat in them is not oriented by the work of the potter. These minerals have a random orientation and in aggregate they transmit heat in all directions in the same manner, especially when many grains are present in random orientations. Thus, non-plastics tend to even out the heat flow in a pot and give it a better heating capacity, or at least one that is more homogeneous. Cooking pots should be full of temper materials beyond the needs of plasticity control.

The thermal properties of clay bodies are very important for cooking ware. If the ceramic is an insulator, it will take a long time for it to be heated but the heat will be maintained in the pot itself. If you want to make a stew, a slowly cooked dish, it is better to have a pot that heats slowly but keeps the heat in. This will allow a simmering of the contents. This is especially important for tough meats and grains. Maize is a good example, and pre-Hispanic ceramic production in the New World is partly linked to its culture. The structure and composition of a cooking utensil are important for their uses.

If clay minerals are highly aligned in the walls of a pot, the thermal conductivity of the ceramic is decreased. Heat is conducted six times more rapidly along the sheets in a clay structure than across them. The thermal conductivity of most silicate temper grains (quartz and feldspars) is similar to the sheet direction of thermal conductivity of a clay particle. Thus, if one adds silicate temper minerals to a clay-rich paste, it will tend to lower the thermal conductivity of the ceramic body wall. High orientation of clays will decrease thermal conductivity.

If many isolated pore spaces are present in the ceramic clay matrix the thermal conductivity is decreased greatly. Thus dense, low pore space and non-oriented fabric ceramic bodies will be good heat conductors. Hence, one should use clay-rich ceramics with highly oriented clay particles and pore spaces to give a slow simmering pot for making stew. If you want to bake a pie in a ceramic object, use a porcelain material where no clays remain. Figure 6.11 indicates these relations.

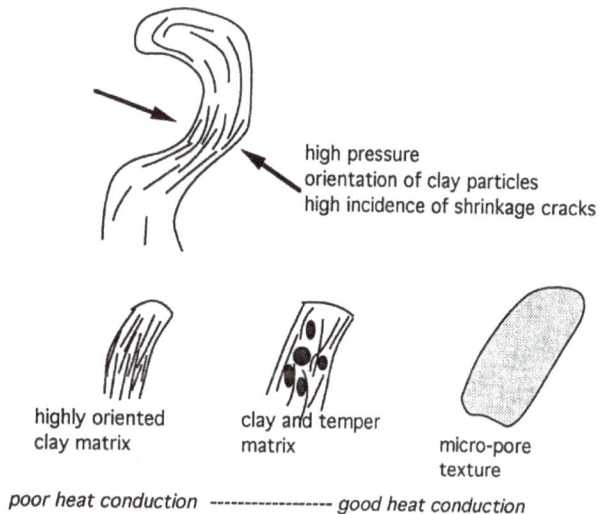

high pressure
orientation of clay particles
high incidence of shrinkage cracks

Fig. 6.11 Relations of texture of the clay particles, pores and temper grains as they affect heat conductivity.

highly oriented clay matrix

clay and temper matrix

micro-pore texture

poor heat conduction ------------------ good heat conduction

The poorest heat conductor will be a highly oriented clay matrix, with unconnected pores or shrinkage cracks. The best heat conductor will be a ceramic with a high density of highly connected pores in a micropore texture. In cooking, one wishes good heat conductivity at times and poor conductivity at other times, depending upon the type of cooking one is doing. Stews want to cook at low temperatures for long periods, here insulation or poor conductivity is wanted. Fast cooking wants good conductivity, connected pores or thin walls of the cooking recipient.

6.4.4
Hardness

The results of these manipulations give a ceramic of differing states of firing completion, or transformation. The problem is then to identify these different stages of transformation.

The different types of well-to poorly fired ceramics can be identified or classified by testing their hardness. *Hardness* is traditionally determined following the geologist's Mohs scale, where different minerals are compared among themselves by scratching one on the other.

These minerals are compared with such common materials as iron or steel and fingernails. Such tests are fraught with difficulty, not the least of which is the availability of different minerals in the scale of 10 hardnesses. Also, once a mineral is scratched, its use as a hardness indicator is greatly reduced because its surface is deformed and no longer has the initial properties of hardness. It is a good idea in theory, but not so in practice. Essentially, fingernails are a good test (hardness of 2.5). If you can scratch a ceramic surface with your fingernail, it is soft or poorly fired. If not, and if it is scratched by a knife blade point (hardness of 6), it is well fired. In between, copper gives a hardness of 3 and window glass is near 5. One can be scratched with the other, ceramic against the object or vice versa, in such comparisons. If one has trouble making an impression with an iron point, the object is probably in the porcelain or stoneware range of firing.

Other, more sophisticated, methods of determining hardness are available, following materials science protocol. The most common is to observe the dent that a metal or diamond point makes in an object when the point is forced into the surface of the object with a given effort. This method supposes a homogeneous, clean surface. If a ceramic material has a significant amount of temper, the difference between clay and temper can give a high variability to such tests. Clays and poorly fired clays are soft (below fingernail hardness) while quartz temper is hard (in the iron point category). Hence, ceramics are often not homogeneous enough for precise hardness measurements.

6.5
Oxidation-Reduction Effects

The firing program can be modified to change the oxidation of iron in the clay paste. Oxidation is important as is reduction. Normally, firing at high temperature creates an oxidizing environment. Clean burning wood or other fuels bring a high temperature in clean air. This changes the iron present into an oxidized state. Fe^{2+} becomes Fe^{3+}. However, if one uses poorly burning materials, some of the organic matter will not be completely burned and the iron will not be oxidized. Further, one often finds that the incompletely burned organic material is deposited on the surface of the ceramics. Earthenware used in cooking or kitchen is often intentionally blackened and the iron present is in the reduced, Fe^{2+}, form.

In clays which have an origin in soils, a major component in their overall composition is organic matter. Also, many clay-rich sedimentary rocks have a natural high organic content. Frequently, these sediments are called black shales, which are black (organic-rich) clay-rich sedimentary rocks. These materials, black shales and soils, will tend to have iron in a reduced state in the silicates and minor oxides. The color of the clay sources is initially grey to black. As firing in an oxidizing atmosphere is conducted to produce a ceramic, the clay body will normally become reddish or brown at its surface. The natural state of oxidized iron is one of hematite, red-brown in color. If the ceramic is fired for a period which has not transformed the paste to an oxidized state, the outer edges of the ceramic will be red to brown and the center grey or even black (we thus speak of incomplete oxidation).

Some clay sources are not organic-rich but are already oxidized. These will be little affected by firing in an oxidizing atmosphere, but they will, on the contrary, be blackened in a furnace with a reducing atmosphere. Here, the center will be red or tan while the outside will be black or grey. It is thus relatively easy to determine the types of atmospheric firing programs by looking at the succession of colors from the outside to the interior of a ceramic sherd.

6.5.1
Oxidation and Reduction Effects on Ceramic Raw and Fired Materials

The oxidation or reduction of elements (mainly iron) is effected by changing the oxidation state of iron molecules in the clay matrix. Oxidized iron is red or red orange in color which is common in clay mineral sources, and reduced iron is dark to back in color. This iron material is free iron oxide that is not incorporated in the silicate clay minerals in the clays and is therefore affected by the oxido-reduction atmosphere in the furnace. The presence of organic material in the clays materials tends to maintain the iron in a reduced state and basically black in color. However, the organic material present tends to give a dark to black color to clay sources also.

Oxidized free iron is essentially red in color and is called hematite. Reduced iron oxide is essentially black in color and is called wustite, or other names depending upon the structural arrangement of the oxygen and iron atom in the oxide when the clay source is modelled into a ceramic form the initial constituents of the clay source are present. The important aspect of iron oxidation state in clays as they are transformed into ceramics is the iron in a reduced form (black in color) which tends to combine with the free atoms on clays surfaces and form a chain or particles that solidify the structure. Oxidized iron (red color) does not have this function. Reduced iron tends to form more strong ceramics at lower temperatures than does oxidized iron.

Note that re-oxidation often occurs at the end of the firing, during cooling, when all fuel has been used up and more oxygen is available, or when opening the kiln (or firing structure). Thus, a carbon or dark core may be seen in a ceramic that has been reduced, while the sides are lighter. That core will be seen even if a true reduction is rarely achieved in so-called primitive firings, unless the potter manages to totally smother the fire to prevent re-oxidation at the end of the firing when the pots are cooling. Thus, a dark core does not necessarily imply the presence of (unburned) organic matter in the original clay. Many traditional or 'primitive' firings were done between 800-900 °C. By the time this temperature range is reached and depending upon the firing length the organic matter in the paste burned away. However, at that temperature, most if not all clays are still porous, so it is easy for carbon from the firing environment to penetrate the clay body and easy as well for it to dissipate in the cooling cycle, bringing about a lighter color, oxidized surface (personal communication, ceramist Andrée Valley 2017).

Experiments have also shown that the process of oxido-reduction-reoxidation can be mastered to adequately produce low-fired slipped and glazed ceramics in updraught kilns (see the work of Dawson and Kent 1999 for example).

Some of the configurations of firing oxidation-reduction effects are shown in Figure 6.12, with some sherd examples in Figures 6.13-14. Let us imagine a firing program which is simple; oxidation takes place on all surfaces exposed to the heated air. The clay paste is initially grey, having significant organic material present, but it contains iron. Oxidation will transform this material to a reddish clay when fired by burning the organic material and oxidizing the iron.

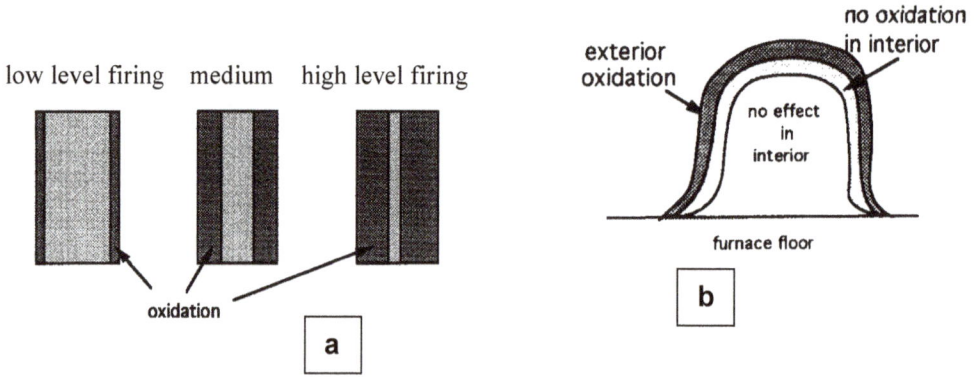

Fig. 6.12. Illustration of some of the oxidation-reduction effects in relation to firing. intensity. **(a)** Sequence of firing intensity (time-temperature) effects in an oxidizing furnace atmosphere, schema of a cut through a ceramic sherd. The longer the firing cycle, the thicker the oxidation zone produced. **(b)** Effect of position of a vessel in the kiln, cross-section showing the effect of isolating a portion of a ceramic from an oxidizing atmosphere during firing. Oxidation generally produces reddish or perhaps grey-buff ceramic color, depending upon the iron content of the paste. These changes can be seen with the naked eye, under the binocular microscope or in thin section under the petrographic microscope. Unoxidized paste can be grey or neutral to grey-black. The black colors of a ceramic generally indicate that the clay resource had a high content of organic matter, either as a soil or a sedimentary rock source. The organic matter is burned in an oxidizing furnace and the grey to black color is lost. The iron present gives a brown to red color. If no iron is present, the clay will become light-colored. If a ceramic form is placed mouth down as in **(b)** on the furnace floor, the inside remains un-oxidized during firing because hot air (oxygen) cannot reach it even though the temperature in the inside of the ceramic is the same as at its outside by thermal conduction of the furnace floor and the ceramic core.

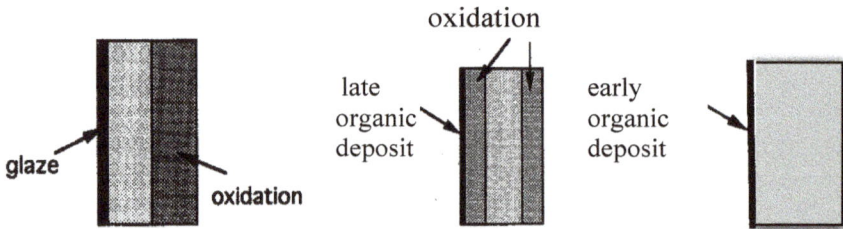

Fig. 6.13 Surface effects and oxidation state. A glaze layer on the surface of a ceramic protects the core from oxidation. Here, the glaze layer isolates the paste material from the outside furnace atmosphere and the paste is unoxidized while the unglazed side of the sherd is oxidized. The period of deposition of organic deposits can be identified by observing the core of the ceramic in section. An oxidized layer under the black layer of organic material indicates an initial oxidation during firing followed by a reducing atmosphere cycle at the end of the firing. The deposit of

organic matter on the surface comes from particles in the atmosphere of the furnace which are poorly burned organic materials (soot). A black layer on the surface of a fired ceramic which covers a reduced, or unoxidized core indicates a reduction atmosphere cycle of firing followed by an intense reducing cycle where organic material is deposited from the atmosphere on the ceramic. A reducing atmosphere in a furnace gives a blackened surface to a ceramic but if the paste material is black to begin with, it is not all that evident to identify the late deposition.

Normally, the oxidation proceeds from the outside to the inside of the ceramic, at medium temperatures (near 1000 °C). The longer the firing, the more the effect of oxidation should show. However, now let us suppose that the pot is placed with its top on the hearth floor of the furnace. The inside of the pot will not be exposed to the heated air and the oxidation will be either non-existent or retarded, creating an asymmetric pattern of oxidation in the wall of the pot. If an impermeable layer, such as a glaze, is placed on the surface of a ceramic, it will protect the paste from the atmosphere of the firing process (but not the heat that transforms the clays in the paste), and the side next to the glaze will not be oxidized but remain reduced. This gives an asymmetric disposition of the oxidized zones in a ceramic body (Fig. 6.14).

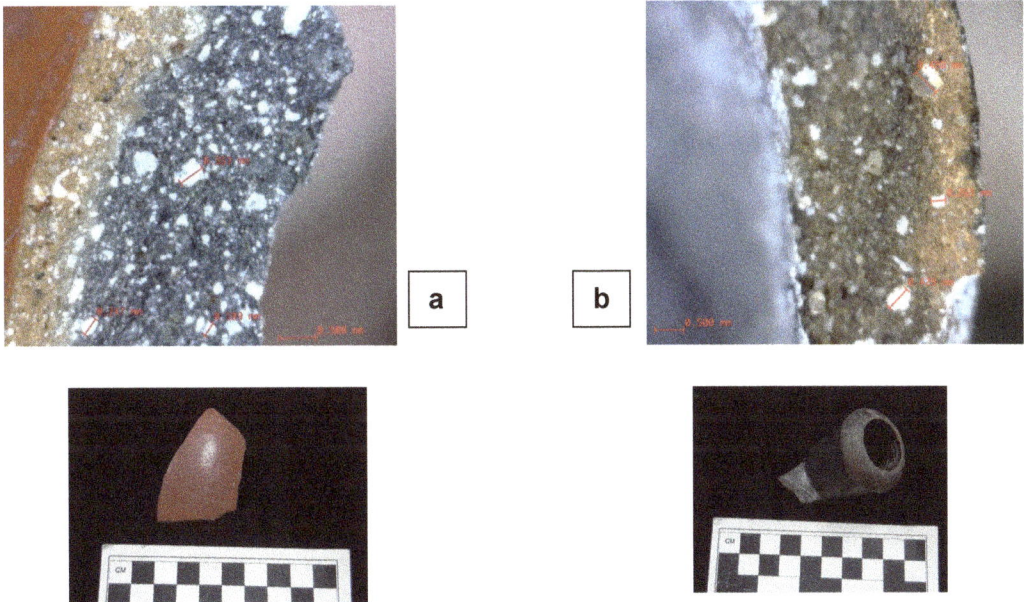

Fig. 6.14 Examples of oxidation-reduction in restricted vessels as seen in cross section with a digital microscope. (**a**) Red-slipped bottle KW66p, and (**b**) reduced bottle KW94p, cross section, Kuntur Wasi Peru, 1st mill B.C. Restricted oxygen access reduced the interior of the bottle, while the exterior was oxidized (**a**). In (**b**) the bottle was probably smudged to couteract the short re-oxidation at end of firing when more oxigen is available or when the vessel started to cool.

6.6
Oxidation-Reduction Cycles

Firing in a normal, free-burning kiln gives an oxidizing atmosphere, which changes the state of iron in the clay body. If the fire burns poorly, or if normal maximum temperature is not reached, the fuels do not give complete ignition and a certain portion of the fuel is in a state of incomplete combustion, creating a reducing atmosphere and blackening. This gives a reduced state to the iron in the clay body. The clays do not become reddish, they remain in their initial state of color even though the minerals change state as temperature and time increase during the firing process. A high degree of incomplete combustion gives a deposit of fuel gases and solids on the surface of the ceramics in the kiln. These are the reduced states of matter in the firing process. Clays containing organic material will not change color, and if a high state of reduction is attained, deposits of black, organic particles form on the surface of the pots. Obtaining a state of a reducing atmosphere is most likely done at the end of a firing cycle in that it is necessary to obtain this state by reducing the efficiency of the fuel combustion which gives a lower temperature and an inefficient burning process compared to the normal oxidizing fire. Hence, one would expect that a strong reduction cycle would come at the end of the firing cycle at lower temperatures. One would expect to see the black layers on the surface over a slightly or strongly oxidized inner layer in the clay body. However, many ceramics have most likely experienced a totally reducing environment, keeping the paste materials in a constant state of reduction. The grey-bodied ceramics commonly found in cooking ware have been fired in this manner. The end result is to enhance the melting characteristics of the iron in the clay portion of the ceramic, making it a more impermeable object. If you are going to cook, it is better to have a pot that does not leak.

Oxidation-reduction effects can be followed by Mössbauer analysis of the iron ions in the ceramic. This subject is briefly pursued further in Chapter 9.8 and 10.3.4.4 concerning the method and its application with a detailed example.

6.7
Mineral Reactions During Firing

The reactions during firing concerning water loss have already been commented upon in this Chapter, such as the loss of bound and crystalline water in the clay particle as well as the change of oxidation state of iron compounds (Section 6.4). These changes and others, like the formation of new minerals, are further detailed in this section. They are influenced by factors like the firing temperature, kiln atmosphere and paste composition.

Firing affects many mineral compounds and different reactions are observed as the temperature of firing rises. These various reactions include:

- **Water loss (dehydroxylation)**

The chemical water contained in the layers of the clay minerals escapes during firing. This water loss starts between 400 and 600 °C eventually leading to lattice distortion.

- **Loss of carbon and organic carbon compounds**

Some carbon and organic material may occur naturally in the clay resource used to make the pottery, or it can be added by the potter in the form of organic matter such as fibers, straw, plant ashes, etc. During firing, the carbon (whether natural or freed by the combustion of organic material) combines with oxygen and burns away as CO_2. The combustion of organic components acts as a reducing agent, taking away oxygen from the firing environment, paste material and immediate atmosphere surrounding the ceramic material. The amount of carbon loss and the extent to which the organic material is burned (oxidized) is dependent upon many variables, among which are the types of carbonaceous matter present and their original quantity before firing, the density of the clay, and, of course, the atmosphere in the firing environment, the firing temperature, and rate of heating and length of firing. In a reducing atmosphere, the reduction equation shows as: $C + 2O \rightarrow CO_2$. CO_2 replaces O in the clay body.

Carbon may also be deposited on the surface of the ceramic during cooking (use of the ceramic object) or when smudging occurs in a reducing atmosphere, thus complicating the assessment of its origin. Carbon loss from organic materials starts at around 200 °C carbon is largely removed at about 600 °C in low organic content pastes and 900 °C to 1000 °C in high organic content clays. If graphite is present in the paste (pure, well-crystallized carbon), it will usually not burn out (be oxidized by thermal reaction with the oxygen in the air) below 1200 °C (Rye 1981). This observation helps in distinguishing between graphite (mineral very rich in carbon) and carbon from organic material, which burns out a lower temperature. The length of the firing also plays a role, as a high carbonaceous paste needs more time to lose its charge than a carbon-poor paste. Thus, the presence of a carbonaceous core (a darker, grey-to-black zone in the center of a ceramic) may be interpreted as the result of insufficient firing under oxidizing conditions, or it might be an indication of the use of a paste of high organic content.

- **Iron oxidation state transformation**

Iron is found in two mineral types in most clay resource materials, either as oxide compound, i.e. as independent iron phases, or as silicates, combined with other elements such as silicon (Si) or aluminium (Al) and, of course, oxygen (O). Fe^{2+} in a clay matrix can be observed to be oxidized when heated in air beginning at 400°C in a firing cycle and be completely oxidized at 600 °C with the formation of hematite, Fe_2O_3, which gives a red color to the paste. In a reducing environment, the iron oxides are transformed into FeO (wustite) or Fe_3O_4 (magnetite) which are opaque black minerals and thus give a grey color in ceramics (see Sect. 6.1.1.4).

Hematite has been observed to form from the iron liberated from the clay lattice during dehydroxylation when the temperature reaches 800 °C (Gebhard et al. 1988, 1989). Iron-bearing alumino-silicates are observed when firing above 900 °C (Wagner et al. 1988, 1989). In a reducing atmosphere, dark iron-silicates form earlier, starting by around 500 °C.

– **Calcium carbonate transformation and lime forming (see Sect. 6.1.1.3)**

The decomposition of calcite into lime and carbon dioxide (CO_2) has been observed to start at about 620-700 °C and is completed by 850-900 °C. The differences in temperature are due to many factors, among them grain size (the smaller the grain size, the lower the temperature of destabilization) and firing time. The laboratory temperature of destabilization (large grains under a rapid heating regime) indicates a decomposition near 900 °C. Calcium carbonates (in shells, for example) disintegrate at around 900 °C. However, shell material is made of another mineral form of calcium carbonate. This is aragonite, which is normally unstable at surface conditions. Biological activity (the animal which made its shell) stabilizes this mineral.

– **Quartz inversions**

Quartz undergoes minor internal structural changes when submitted to certain temperatures. These changes occur around 573, 867-870 and 1250°C. There is an expansion of the volume (1-2 %) of the quartz grain during the first inversion, changing from alpha to beta quartz. The alpha- beta inversion is common in ceramic production in that most firing is done above 600 °C. The alpha-beta transition is reversible, and the volume expansion of the grain is largely insignificant as the clay matrix is under- going significant volume changes at the same time, allowing space for the quartz to expand (in coarse ceramics, at least). However, this inversion can produce shattering of the grain, which can be seen with optical methods, but this is not always observed.

The second and third changes are very slow and lead to the formation of new minerals: tridymite and then cristobalite. These reactions are highly dependent upon the length of firing time. They may not have time to form in many instances as many primitive firing cycles are short. Some do not last more than half an hour at maximum temperature.

– **Volatilization of some minor and trace elements**

Some elements may be lost when their boiling point is within the range of the firing cycle temperature. For primitive firing, usually below 1000°C the main elements concerned are arsenic, bromine, caesium, potassium, rubidium, sodium and zinc. The effect of volatilization is to reduce the concentration of the elements in the ceramic, which may cause problems in provenance studies. However, a study showed that no volatilization of these elements was observed, except for bromine (Cogswell et al. 1996).

– **Crystallization of new minerals**

This is the most important part of the effects of firing of ceramics apart from the amorphization of the clay minerals. Grim (1962, 1968) indicates the different minerals which one can expect to find forming from the common clay minerals. The minerals crystallized can be found in nature in rocks formed at high or very high temperatures (metamorphic or magmatic).

Montmorillonites form quartz (SiO_2) and at higher temperatures a high-temperature form, cristobalite. In laboratory experiments (more rapid heating programs than most kilns) this mineral is formed at temperatures above 1050 °C. Calcium feldspar, anorthite, is found also. Mullite ($Al_2Si_3O_5$) forms at 950 °C, while cordierite (a Mg, Al silicate) forms at temperature near or above 1200 °C.

Kaolinites form alumina (Al oxide) at 980 °C, mullite at 1000 °C, cristobalite at 1200 °C and glass above this temperature.

Illite forms spinel (MgAl silicate) at 1000 °C and mullite above 1200°C.

Many other minerals can crystallize when several clays react with iron oxide or a fusing agent such as calcium oxide. Wollastonite, a calcium silicate, is formed in such a way.

Here, one can see that the presence of a given newly crystallized mineral does not indicate the temperature of transformation in an absolute sense but the reaction series of a given clay mineral. Mullite is an example. One can also expect that a combination of clay minerals and fusing agents will change both the temperature at which the minerals will form and the time sequence of formation. Thus, mineral crystallization cannot give an absolute indication of maximum temperature of firing. For example, mullite has been observed to form from clay minerals and spinel between about 950 and 1275 °C (Rye 1981; Rice 2015).

–   **Vitrification begins between 700 and 950 °C**

It results in the destruction of the crystalline structure of the clay minerals. Temperature at which vitrification occurs is varying, depending upon flux agent, paste composition, and firing procedure. All of the above alterations may not be observed in a ceramic. Again, different variables of the manufacturing and firing process are factors. These transformations are important to identify, as they will inform us about the original firing temperature and kiln conditions. Researchers interested in these aspects look for these alterations using various techniques from optical microscopy to spectrometry, Mössbauer analysis and magnetic analysis (see Chap. 11).

6.8
Families of Ceramic Products

The different firing methods using materials composed of different pastes covered with different materials provide a potter with a great variety of production possibilities. Since potting methods have developed over long periods of time in very different cultures, the shapes and use of ceramics lead to still further variety. This is, of course the interest of ceramics in archaeology. However, in all such subjects of study, one attempts to simplify and categorize the different objects in order to organize one's thoughts in a coherent and internally referenced manner. It is always useful to speak a common language with people interested in the same subject. As a result of such a desire for order, ceramic products are placed into several large groups or categories. These groups combine the different facets of production, firing practice, paste composition and surface treatment.

Unfortunately, not only the production of ceramics but the study of ceramics has been done in different countries and in different languages. The development of the ceramic art and its study in each culture has also given a tradition of description which ignores in many instances experiences in other cultures. The big differences are perhaps between the Asian, American and European traditions of description and nomenclature. In Asia, porcelain has been the object of ceramic art from very early periods. Other materials are most often referenced from this state. Hence the different stages of achievement between crude earthen objects and the translucid thin-walled porcelain cup are chronicled with an eye to the end product. Starting materials in China (the components of the paste) were rapidly centerd on the resources of altered granites: kaolinite, feldspars and quartz. By contrast, the American archaeological experience has never gone as far as the porcelain stage in technological prowess, though results are often stunning. In European wares the full range of ceramic evolution has been produced, with true porcelain produced rather late but a long series of imitations have been chronicled in the majolique-faience productions of the Renaissance, and soft porcelains of later times gives a great variety of ceramic experiences.

Given this variety of ceramics and their study, some dissension is inevitable in even the major classifications of ceramics.

We propose here a modest unity of the different traditions based upon the ceramic-producing techniques. Our approach is from the materials point of view. The major categories are four: earthenware, pottery or faience, stoneware and porcelain. Depending upon the continent and tradition, the use of these four terms varies. We will use them here, not in an absolute sense, but in a comparative one. The most important thing to remember here is that a given firing practice can give different results depending upon the paste used. A fusing agent makes crude bricks very, very resistant to mechanical forces. That is their function. They may even appear to be more transformed than other tableware of the same period. The reverse is true. A given paste will be more or less transformed depending upon the temperature and time of heating during the firing cycle.

6.8.1
Earthenware

Earthenware should probably be most easily characterized by its "earthy" appearance and texture (to the touch). It can be plain (plain ware), that is uncoated, where the clay (usually brown, rose or earth-colored) shows through to casual inspection. In some earthenwares there is no effort to change the outward appearance of the ceramic. Earthenwares can also be polished and have a slip. If unslipped, the overall aspect of an earthenware object will be reddish due to the oxidation of clays or blackened by reducing conditions during firing. This ceramic type was and still is widely produced, notably in traditional ceramic productions. Earthenware is usually utilitarian. Some examples are the jugs and jars of Neolithic or prehistoric times. Others are the common ware for cooking and tableware in everyday circumstances, for example in Roman and mediaeval Europe. Bricks and tiles are earthenware. They were made according to the same traditions up until the end of the last century in European countries. These earthenwares or plain wares have been produced throughout the ages in most cultures at one time or another.

Earthenware produced in older cultures tends to contain large quantities of non-plastic, temper grain material, either natural or added in the form of sands and/or grog. Grog temper is especially common in bricks and other low-cost and aesthetically undemanding building materials. These earthenware materials are often coarse in surface aspect and rough to the touch, and more so when seen in thin section under the microscope. Clay and temper grains or other materials are often poorly mixed. Often one finds that primary pores are present.

Firing temperatures and length of firing cycles were usually not great in earthenware production. Addition of fusing agents such as lime (CaO) was often employed to rapidly produce the desired rough but usable object. Fusing agents hasten the destruction of the clays making a more durable material. This decreases the time-temperature equation of costs in the firing cycle. Earthenware objects then are identified by a medium range in transformation of the constituent clays to a destructured or amorphous state. They may have a surface, smoothed or slipped (a layer of clay-rich material). The range of transformation states of the clay materials is dependent on the use of fusing agents and the time-temperature cycle of firing.

6.8.2
Pottery, Terra Cotta, and Faience

Pottery. In some definitions, pottery is a general term for all clay-made vessels, usually utilitarian. Here, pottery is earthen- or plain ware. The body (paste) of the piece and temper sand grains are hidden by slips, paints and glazes on the surface. Pottery and faience are "up market" products. They are usually well fired but maintain a certain porosity. The clays are not completely destroyed but enough so to have formed a solid structure which is resistant to wear. When fusing agents such as calcium oxide (CaO) have been used to hasten the fusion effect, they have changed a large portion of the materials and welded them into a strong material. For some specialists, *terra cotta* seems to be favoured as a term covering much the same type of material as one would designate as pottery.

Pottery (terra cotta) is the most often studied material in Europe in the post medieval periods. Decorative plates and jugs as well as drinking vessels are common. The decorative elements are often the object of speculation, being used to identify origin and provenance. The surfaces can be of slip or glaze. The glaze can be opaque or transparent.

Faience. This term can be used to designate a pottery or terra cotta core covered by an opaque white surface often used as a mask to cover the color of the fired clays in the core of the ceramic. Typical opaque (white) glazed ceramic pottery is often called faience ware in European archaeological terms, developed from a base in Spain and more largely in the 15th century in Italy. It was a best-seller then and was imitated and transformed for centuries after. The principle is to use an opaque glaze as a substrate to be decorated. At times, the white undersurface is barely visible in more elaborate Italian objects, due to overpainting.

For modern potters, faience designates an earth type (or paste type), rather than a pottery type. Earthenware and faience can be polished and even have a slip. Plain ware has however no surface treatment. As one could expect, what we call pottery-faience and terra cotta cover a vast range of materials in a vast range of periods and cultures. The preference for one or another term seems to be one of present-day geographic use.

6.8.3
Stoneware

The stoneware paste is composed of much non-plastic material which has a low melting point, especially alkali (Na, K) feldspars. Here, the temper grains are the fusing agent and the clays are the resistant material. This is the case because the clays are either kaolinite or mica, both of which are resistant to heating (see Chap. 3). Both of these minerals tend to be of large grain size in the composition of stoneware, and hence fuse at higher temperatures. The clays then remain un-fused though their structure is destroyed by the heating. The firing of stoneware requires several days at temperatures above 1200 °C.

This material is usually white or cream-colored because it contains kaolinite, quartz and feldspars, none of which has an iron component which would give color to the ceramic product. The raw materials are then clean, they contain little iron oxide. Today's ceramics, plates and dishes are of this category. It is rare to find a commercial plate which, when broken, will show other than a white interior. Exceptions are, of course, hand-produced pots from artisans.

Stoneware is hard, has a low to non-existent macroporosity and is very resistant to breakage and wear. It was produced and common in China from an early date, about 2000 years ago, but more recent in Europe, becoming common in the 17th and 18th centuries, when jugs and crocks were common stoneware products.

6.8.4
Porcelain

This is the miracle product, of high quality and prestige in many cultures. It is the result of the complete fusion of both kaolinite clays and feldspar temper grains. Also most, if not all, of the quartz grains are melted. Curiously, the definition of porcelain usually used is not Chinese, where porcelain came from, but one of European origin. The standards set by the ceramists searching for the recipe for porcelain became so high that they eventually surpassed those of the inventors. Porcelain should be translucid, i.e. light should be seen through the edge of a porcelain piece.

In order to do this, the materials of the basic composition must melt. The temper grains fuse and melt the kaolinite; but in this process the high aluminium content of the kaolinite provokes the formation of a new mineral, mullite. This mineral grows in needle shapes, forming a mat of interlocked grains which are bridged by the melted areas. The identity of porcelain is its translucidity. Taken to the light, one can see light through its body. In order to do this, one needs a thin porcelain object, of course; but since porcelain is so strong, it can be made into very thin-walled objects. Hence, thin and delicate china cups and saucers have been found on the tables of the well-to-do over the past three centuries.

Porcelain manufacture requires high temperatures, above 1300 °C, and a long period of firing, many days. It is costly in energy and in the search for the right raw materials. Purity and chemical composition must be of a high standard. In terms of abrasion resistance faience and porcelain which includes objects of high quality and firing range are harder, copper- to iron-resistant, whereas the earthenware is usually softer.

As stated before, these definitions of the different types of ceramic productions must be taken rather liberally. In different parts of the world, the words porcelain, stoneware, pottery, earthenware, faience and so forth may have a different meaning. There seems to be no consensus. Our use of these groupings is not one designed to be of a strict categorization, but one that might be useful in describing the materials.

6.9
Summary

The information in this chapter relates to the physical processes involved in the firing cycle of ceramic production. Numerous variables are to be taken into account. The initial composition of the paste of the ceramic body is of greatest importance. It must be fired just right, not too much, not too little. The technique, firing structure and needs of the potter determine the length of time to be fired at a given firing temperature (in a kiln) or an integration of time and temperature in pit or bonfire firing methods.

It is evident that the better the control over the time-temperature variables, the better the product will be on the average. A stable furnace with consistent green or unfired ceramic objects in it will be much easier to master than a tricky bonfire subject to winds, the humidity of the fuel and the stacking density of the green pots to be fired.

The choice of the paste materials: grain size of the clays, proportions of the temper grains present, and so forth will determine the firing program in order to get the most out of the material and to avoid the production of cracks and fissures in the final product.

Surface treatments and applied decoration (slip, glazes or paints) will be done with materials which melt or transform at lower temperatures than the paste mass. They are meant to cover and flow over the surface of the ceramic and hence will need to be melted before the inner part of the ceramic changes its state during the firing process. If successive layers are needed or desired, successive firings will be used to make the desired effect. The application of glazes often necessitates successive firings. Some glaze melts at such a low temperature that it would flow off from the object in many cases if the firing were done in one step.

Other high-temperature ceramics such as stoneware or porcelain objects can be produced in two ways. Some have glazes which are applied on the green or air-dried material and the glaze melts as the clays are transformed in the core of the ceramic. This is the case for most ancient Chinese porcelain and stoneware. However, European porcelain is produced in a two-step process where the core is first fired to a given state of mineral transformation. The objects are then cooled and covered with a glazing material which is fired on the objects at a temperature lower than that of the first firing.

Further methods have been used to put a paint made of glaze-like materials on the surface of a high temperature glaze surface. These are called enamels. The enameling cycle is done at a temperature lower than that of the initial firing and the glazing cycles.

Firing cycles and decoration are dependent on mastery of the time-temperature equation. As one could expect, bonfire-fired objects can have decorative effects, such as paints, but not glazes or overglazes.

Bibliography

Acevedo VJ, López MA, Callegari A, Freire E, Halac EB, Polla G, and Reinoso M (2015) Estudio tecnológico de diseños "estilo aguada" realizados sobre fragmentos de cerámicos. In A Pifferetti and I Dosztal (eds), *Arqueometría argentina. Metodologías científicas aplicadas al estudio de los bienes culturales*. Aspha Ediciones, Buenos Aires, pp 109-125.

Bloomfield, L (2017) *Science for potters*. The American Ceramic Society, Ohio, pp 148

Cogswell, JW, Neff, H, and Glascock, MD (1996) The effect of firing temperature on the elemental characterization of pottery. *J Archaeol Sci* 23:283-287

Dawson, D, Kent, O (1999) Reducion fired low-temperature ceramics. *Post-Medieval Archaeology* 33: 164-178.

De la fuente GA, Martínez JM (2008). Estudiando pinturas en cerámicas "Aguada Portezuelo" (ca. 600-900 AD) del Noroeste Argentino: Nuevos aportes a través de una aproximación arqueométrica por microespectroscopia Raman (MSR). *Intersecciones en Antropología* 9: 173-186.

Dickinson, WR, Shutler, R Jr, Shortland, R, Burley, DV, and Dye, TS (1996) Sand tempers in indigenous Lapita and Lapitoid Polynesian plainware and imported protohistoric Fijian pottery of Ha'apai (Tonga) and the question of Lapita tradeware. *Archaeology in Oceania* 31: 87-98

Maritan, L., Ganzarolli, G., Antonelli, F., Rigo, M., Kapatza, A., Coletti, C., Mazzoli, C., Lazzarini, L., and Chavarria Arnau, A. (in prep). What kind of calcite? Disclosing the origin of calcite temper in ancient ceramics.

Gebhard, R, El-Hage, Y, Wagner, FE, and Wagner, U (1988/1989) Early Ceramics from Canapote. Colombia, studied by physical methods. *Paleoethnologica Buenos Aires* 5:17-34

Goffer, Z (1980) *Archaelogical chemistry*. Wiley, NY, pp 376

Grim, RE, (1962) *Applied clay mineralogy*. McGraw-Hill, NY, pp 403

Grim, RE, (1968) *Clay mineralogy*. McGraw-Hill, NY, pp 512

Jose-Yacaman, M, Rendon, L, Arenas, J and Serra Puche, MC, (1996) Maya blue paint: An ancient nanostructured material. *Science,* 273: 223-225

Levin, EM, Robbins, CR, and McMurdie, HF, (1964*) Phase diagrams for ceramists*. Amer Ceramics soc, Columbus, OH, pp 601

Rice, PM (2015) *Pottery analysis. A source book*, second edition. University of Chicago Press, Chicago, pp 592

Rye, OS (1981) *Pottery technology*, Manuals on archaeology 4· Taraxacum, Washington DC, pp 150

Some Reference Works and Studies on Ceramic Technology Applied in Archaeological Problems

Albero Santacreu, D (2015) *Materiality, techniques and society in pottery production: The technological study of archaeological ceramics through paste analysis*, De Gruyter, Warsaw/Berlin, pp 337

Brooks, WE, Piminchumo, V, Suárez, H, Jackson, JC, and McGeehin, JP (2008) Mineral pigments at Huaca Tacaynamo (Chan Chan, Peru), *Bull de l'Institut Français d'Études Andines* 37(3): 442-450.

Cau Ontiveros, MA, Day, PM, and Montana, G (2002) Secondary calcite in archaeological ceramics: Evaluation of alteration and contamination processes by thin section study, In: Kilikoglou, V, Hein A, and Maniatis, Y (eds), *Modern Trends in Scientific Studies on Ancient Ceramics* BAR International Series 1011, pp 9-18

De la Fuente, GA (2002) Diatomological analysis (provenance) application in archaeological ceramics: An exeprimental approach. In: Erzsébet Jerem and T Biro, K (eds), *Archaeometry 98 Proceedings of the 31st Symposium Budapest*, April 26-May 3 1998 vol II, Archaeolingua Central European series 1, BAR International Series 1043 (II), pp 501-511.

Druc, I (2000) Ceramic production in San Marcos Acteopan, Puebla, Mexico. *Ancient Mesoamerica* 11: 77-89.

Druc, I (2015) Charophytes in my plate: Ceramic production in Puemape, North Coast of Peru. In *Ceramic Analysis in the Andes* I. Druc (ed.), pp. 37-56. Deep University Press, WI.

Dumpe, B, and Stivrins, N (2015) Organic inclusions in Middle and Late Iron Age (5th-12th century) hand-built pottery in present-day Latvia, *J of Arch Sci* 57: 239-247 .

Eramo, G, and Maggetti, M (2013) Pottery kiln and drying oven from Aventicum (2nd century AD, Ct. Vaud, Switzerland): Raw materials and temperature distribution, *Applied Clay Sc* 82: 16-23)

Fabbri, B ed. (1995) *Fourth Euro-Ceramics Conference Proceedings*. European Ceramics Society Vol 14, Gruppo Editoriale Faenza

Goren, Y (2014) The operation of a portable petrographic thin-section laboratory for field studies, New York *Microscopical Society Newsletter*, Sept 2013 1-17 http://www.nyms.org

Heiman, RB (1989) Assessing the technology of ancient pottery: the use of ceramic phase diagrams. *Archaeomaterials,* 3:123 -148

Henderson, J ed. (1989) *Scientific analysis in archaeology*. Oxford Commission for Archaeology, Monograph 19, pp 312

Kamilli, D and Lamberg-Karlovsky, CC (1979) Petrographic and electron microprobe analysis of ceramics from Tepe Yahya, Iran. *Archaeometry* 21:47-60

Kingery, WD, and Vandiver, PB (1986) *Ceramic masterpieces; art, structure and technology*. Free Press, NY, pp 339

London, G (1981) Dung-tempered clay. *J Field Archaeol* 8(2):189-195.

Maritan, L, Nodari, L, Mazzoli, C, Milano, A, and Russo, U (2006) Influence of firing condition on ceramic products: Experimental study on clay rich in organic matter, *Applied Clay Sc* 31: 1-15.

Magetti, M, Galetti, G, Picon, M, and Wessicken, R (1981) Campanian pottery: the nature of the black coating. *Archaeometry* 23:199-207

Maggetti, M, and Schwab, H (1982) Iron Age fine pottery from Chatillon-s-Glane and the Heuneburg. *Archaeometry,* 24:21-36

Maggetti, M, Wesley, H, and Olin, J (1984) Provenance and technical studies of Mexican Majolica Using elemental and phase analysis. In: Lambert JB (ed), *Archaeological chemistry III*. Series 205 American Chemical Society, pp 151-191

Noble, JV (1988) The techniques of painted attic pottery. Thames and Hudson, London, pp 216

Ownby, MF, Giomi, E, and Williams, G (2017) Glazed ware from here and there: Petrographic analysis of the technological transfer of glazing knowledge, *J Arch Sci* 16: 626-626

Páez, MC, Giovannetti, MA, Arnosio, M (2013) Experimentation with ceramic pastes containing high amounts of pyroclastic materials: their relation to the manufacture of Inka vessels, *Archaeol Anthropol Sci* DOI 10.1007/s12520-013-0135-6

Philpotts, AR and Wilson, N (1994) Application of petrofabric and phase equilibria analysis to the study of a potsherd. *J Archaeol* Sci, 21: 607- 618

Pollard, EM, and Heron, C (1996) *Archaeological chemistry*. Royal Society Chemistry, Lethworth, pp 375

Quinn, P, and Day, P (2007) Ceramic micropalaeontology: the analysis of microfossils in ancient ceramics. *Jour of Micropalaeontology*, 26:159-168

Rasmussen, KL, De La Fuente, GA, Bond, AD, Korsholm Mathiesen, K, and Vera, SD (2012) Pottery firing temperatures: a new method for determining the firing temperature of ceramics and burnt clay. *J of Arch Sci* 39: 1705-1716.

Reid, KC. (1984) Fiber and ice: new evidence for the production and preservation of late archaic fiber-tempered pottery in the middle-latitude lowlands. *American Antiquity* 49(1):55-76.

Rhodes, D (1968) *Kilns: design, construction and operation*. Chilton Book Co, Radnor, USA, pp 251

Rice, PM (2015) *Pottery analysis. A source book*, second edition. University of Chicago Press, Chicago, pp 592

Rye, OS (1981) *Pottery technology*, Manuals on archaeology 4, Taraxacum, Washington DC, pp 150

Schiffer, MB ed (1978, 1979, 1980, 1981, 1982, 1983) *Advances in archaeological method and theory*. Academic Press, London

Schiffer, MB (1990) The influence of surface treatment on heating effectiveness of ceramic vessels. *J of Archaeol Sci* 17:373-381

Schiffer, MOM (1990) Technological change in water storage and cooking pots: some predictions from experiment. In: Kingery, WD ed, *Ceramics and civilization*, vol V, Am Chem Soc, Westerville, OH, pp 119-136

Shepard, AO (1963) Rio Grande glaze-paint pottery: a test of petrographic analysis. In: Matson, FR, ed. *Ceramics and man,* Aldine, Chicago, pp 62- 87

Trindade, MJ, Dias, MI, Coroado, J. Rocha, F (2009) Mineralogical transformations of calcareous rich clays with firing. A comparative study between calcite and dolomite rich clays from Algarve, Portugal, *Applied Clay Sc* 42: 345-355.

Vaz Pinto, I, Schiffer, MB, Smith, S, and Skibo, JM (1987) Effects of temper on ceramic abrasion resistance: a preliminary investigation. *Archaeomaterials* 1:119-134

Wallis, NJ, Cordell, AS, Newsom, LA (2011) Using hearths for temper: petrographic analysis of Middle Woodland charcoal-tempered pottery in Northeast Florida, *JAS* 38: 2914-2924.

7 The Making of Pots

The preceding chapters reviewed the geology of ceramic materials, clay groups, their composition and structure, as well as some of their properties, such as plasticity, water absorption capacity and weight loss upon drying. These properties are very important in pottery making. To some extent, they can be controlled by the potter, who modulates the preparation of the raw materials to obtain the right paste for his/her pots. This chapter deals essentially with the modulations introduced by the potter, their effects on the ceramic product, and with the potter's work as seen from the ceramic material.

As this chapter is oriented towards understanding the potter's art, strong emphasis is given to the paste, its properties in relation to the use or function of the pot, and the different steps in making a ceramic product. These steps can be identified through the identification of the temper grains to the building and finishing of the ware. They can be observed under a simple petrographic microscope (Chap. 10). Hence, the emphasis on paste texture and non-plastic grains, which are generally of larger size than the clay particles and therefore identifiable with this kind of microscope.

Before we plunge into the making of pots, it might be useful to clarify some terminology. The problem is the use of the word clay (see Chap. 1, Glossary). In the study of silicate minerals in geology and mineralogy, the word clay has the meaning of a specific family of minerals. Clay minerals are phyllosilicates, i.e. sheet silicates (Chap. 3), with a grain size of less than 2 µm in diameter. And a material composed of a majority of clay minerals is a clay (Fig. 2.12). These clays have a grain size inferior to 2µm. Hence, a clay can contain a certain proportion of other or non-clay minerals. In geological terms, a clay source is a natural occurrence where clays (a material with a high proportion of clay minerals) can be extracted. This clay material usually contains a certain percentage of non-clay minerals, of varying type and granulometry. For a potter (in current terminology), a clay is the material which serves as the base for making a ceramic, fine or coarse (with non-plastic grains that can be of different granulometry and type). Here, the term clay is related to the physical properties of plasticity.

Thus, clay mineral means 100% clay-sized material of a given crystallographic form. Clay is a material of high clay content (about 80%), clay resource is a material of "high" clay content, which is not specified. A potter's clay is a material with a high or certain plasticity when wetted and worked. The amount of clay minerals decreases in each definition. We will try to keep to the potter's definition when talking of potting, to the geological term for clay or clay resource when discussing the origin of clay materials, and to the term clay mineral when talking about the mineralogical material.

7.1
Temper and Tempering

A word must be said, again, about terminology concerning temper. In the archaeological literature, temper has been used extensively, not always with the same meaning. Sometimes it refers to all non-plastic grains in the paste. At other times, temper means the material added by the potter to the clay, be it another coarser clay or any other material. This definition is the one mostly used by archaeologists. It implies an act accomplished by the potter, hence giving an anthropologic dimension to temper.

However, a temper is, strictly speaking, a material that reduces the plasticity of the water-loving clay particles which, if taken alone, will form a sticky and fluid mass. In natural, geologic materials, the role of reducing plasticity is played by the non-clay mineral grains. These are described in Chapter 2, concerning the geologic origin of ceramic materials, and in Chapter 5, as the non-plastic portion of a clay source. The non-plastic grains do not associate much water with them in the paste, they do not attract or give up water during the paste-making process, nor do they lose water upon drying or firing. These non-clay grains, or non-plastics, can be called temper grains. These ideas of temper are based upon the opposition of clay vs. non-clay, plastic and non-plastic. Inclusion is also a term found in the archaeological literature. It usually refers to any non-plastic material in the paste, such as quartz grains, rock fragments, crushed shells or bones, plant remains, bits of charcoal, etc.

Non-plastic grains can occur in a clay source itself or be added by a potter in the process of assembling a paste. Or be accidental inclusions. This goes beyond the definition of non-plastics as temper. We propose to enlarge this definition to one more operational, and more anthropological in scope. Here, we include in the definition of temper all materials which when added to a given clay-rich water mixture decrease the fluidity or increase the viscosity of a paste. One given clay-rich material is tempered by another. Tempering is then the act of adding a temper material to the clay resource. In this sense, tempering is a voluntary act. It can be accomplished by adding different non-plastic materials, sand, straw, grog, or by adding another less plastic clay resource. Tempering can also play a role in the building and firing of a pot, yielding workability, enhancing strength, thermal resistance or porosity according to the temper added, as will be seen later in this chapter.

One of the reasons for reiterating these definitions is that potters at different times have used, and still use, a mixture of clays or clay sources to produce a workable paste. In this sense, a potter "tempers" a given clay with another in order to bring about the proper conditions for a given plasticity and firing. A potter can achieve the same thing by adding sand or fine grains of non-clay origin, or, on the contrary, by subtracting temper grains from a clay resource that is too sandy. If one keeps to the idea of a non-plastic material as a temper agent, compared to the clay, which is the plastic agent, the definitions can be kept at a minimum. However, when trying to identify the act of tempering, to distinguish the raw materials used when looking at a ceramic paste, we need to know what raw materials a potter can use.

7.2
Raw Materials

7.2.1
Clay Material

The level of purity and clay-composition control, characteristic of modern (19th-20th centuries) industrial production, is rarely seen in the ceramic productions found in archaeological contexts. Potters usually used the material at hand and adapted it to their needs. In fact, clay, the plastic material and binding agent of ceramic, is rarely found to be composed uniquely of clay minerals in nature. 'Impure' clays, thus, are very common in traditional and prehistoric wares. Various examples are shown in this chapter and in other chapters as well (Chapters 4 and 9) to give an idea of what a clay may contain besides the clay materials, and how useful it is to study them under the microscope.

Besides clay minerals, oxides and sulphides, clays include a certain quantity of non-plastic materials. These can be monocrystalline inclusions, rock fragments and/or organic components, all of varying grain sizes. Their composition varies according to the clay deposit's geological history. For example, primary clays due to rock weathering are coarser than sedimentary clays. They are unsorted, with fragments of the parent material as inclusions. They are often used in traditional ceramic production as well as for making building materials such as tiles, pipes or bricks. Sedimentary clays, on the contrary, are usually finer than primary clays, better sorted, often highly plastic, and will then require the addition of a tempering agent.

Thus, as starting materials one has fine to coarse clays (varying proportions of clay mineral and coarser grains) differing in their mineralogy, water absorption capacity and color when fired. Their plasticity, the way they react to the stress of drying and firing, makes them more or less suitable for pottery making. The type, quantity, granulometry and angularity of the inclusions help distinguish their different types and eventually their origins. This can be informative when studying ceramic production, the type of resource used, or the raw material's geological origin. For example, the clay illustrated in Figure 7.1a-b was collected on the slope of a high coastal valley, in the Andean piedmont, where the coastal Batholith (acid to intermediate intrusive rocks) outcrops, weathers, and provides much of the material that ends up as colluvium or eluvial deposits on the valley slopes and in river sand. It is yellow and gritty. This type of clay was used for traditional production.

Potters often refer to clays per their fired-color, such as red- or white-firing clays, which may be quite different from the raw material color. Then they also know that clays reach their best qualities at certain temperatures. Some will be good for the production of earthenware, others for stoneware, still others for high-fired products such as porcelain if mixed with other materials and following strict recipes. For example, if kaolin or ball clays are used (impure kaolin, kaolin+ball clay, or just ball clay) the potter needs to add some temper to improve workability and lower the firing temperature. Adding feldspars does some of the job. Then there is kaolin and kaolin. Primary kaolin (in-situ formation) is hard to work with and needs tempering. Secondary kaolin is transported and deposited away from the source, consequently it is less pure and more plastic than primary kaolin but still needs tempering. Ball clay which is a fine-grained secondary clay is very plastic, but it retracts a lot upon drying and thus also needs tempering. Its advantage is that it fires a light color (light grey, light buff) at lower temperature (c. 1300 °C) than kaolin clay which requires a high firing temperature of c. 1800 °C to mature to porcelain. Kaolin, however, can be fired at a lower temperature, which will yield more porous bodies (and one can add a flux to lower the firing temperature to reach the desired body). Pure kaolin will fire white whatever the temperature.

Addition of feldspar will act as flux to lower the firing temperature needed to reach a hard body. The Al-rich clays can be used for white earthenware bodies with quartz and K-feldspar as essential non-plastics, and addition of other silicates. White earthenware clays fire at lower temperature (between 950 and 1150 °C) than ball clays. To prepare these clays, potters follow long-trusted recipes. For example, Bloomfield states that "In the UK, white earthenware clay is composed of 25% China clay, 25% ball clay, 15% feldspar and 35% flint" (Bloomfield, 2017, p. 75).

There are also calcareous and marly clays which fire whitish, white-grey, pinkish, and have been extensively used around the Mediterranean (Albero Santacreu 2015).

The quality of the clay and the function of the pot will determine the measures the potter takes to improve his/her source material, such as adding temper or flux (fusing) agents, decantation, or fine grinding. For example, clays with sulphides produce scum on the surface of the pot and calcareous clays spall lime when fired. These clays will not yield high-quality ware unless special care is taken in preparing the paste (see below). Some clays are good for making cooking pots, for example clays with natural inclusions of low thermal expansion coefficient or refractory clays, able to withstand very high temperatures. Some can be used as they are by the potter, with no addition of temper, while others may be too lean and need to be mixed with finer clay. Coarse clays or clays with insufficient plasticity can be refined by decantation, levigation or another way of subtracting grains or separating grain fractions. Many clays, however, are too plastic (or fat) when mixed with water and require the addition of another material with a higher non-plastic content. This added material is then the temper.

Figures 7.1-7.2, 7.8, and several at the end of Chapter 4 (Fig. 4.9-4.14) give examples of different clays, collected as comparative materials around archaeological sites or at the house of potters who still work in a traditional way - that is, not using industrial clays and modern techniques. They illustrate how clays can vary in composition and granulometry of non-plastic materials, and color. Although the potters will choose their clays by 'feel' (grittiness, plasticity, cleanliness), taste, or performance, it is the composition and granulometry that will guide the archaeologist or the analyst when assessing provenance and technology.

a

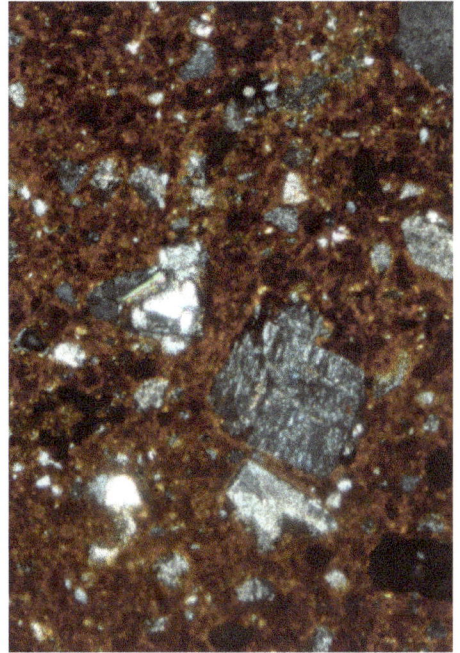

b

Fig. 7.1 (a) Primary clay, **(b)** thin section photomicrograph of the mineral composition of this material showing fine to coarse subangular to subround granitic fragments and derived crystals broken off from the lithoclasts, and occasional quartzites, sandstones, iron nodules. Fired clay tile HV41, Huarmey Valley. Peru. Granite fragment 0.4 mm long, xpl. (See Druc et al. 2020).

Fig. 7.2 Drying clays and clayey soils from three different sources mixed together to produce cooking pots in traditional production in Tarica, Callejon de Huaylas, Peruvian Andes.

7.2.2
Tempering Materials and Methods of Tempering

Tempering grains have been heretofore defined as being non-plastic elements in the clay-rich paste. These added materials can be simple, single-phase materials, or they can be complex substances made of clays and other types of grains. In non-industrial ceramics, the tempering materials are extremely diverse. They vary according to the resources at hand or those necessary to meet specific technical or functional needs, or even according to tradition. Three main categories of tempering material are found in ceramic pastes, alone or combined: mineral, vegetal and organic materials.

A mineral temper is defined, then, in a very wide sense, as any addition of mineral origin, such as clay, sand grains, and rock fragments of all classes. It is the most widely used temper and also the most difficult to identify in detail, in thin sections (optically) or by other methods. This temper material can be heated to render it more friable, it can be crushed, sieved, or separated from its original environment to be used as an additive to the clay base, which will render the paste less liquid when mixed with water. A few mineral tempers will be illustrated below with petrographic examples.

Organic materials include plants, carbonaceous materials, oxide compounds, biocarbonates, such as river and marine shells, and siliceous (SiO_2) foraminifer microorganisms (Fig. 7.3). Organic materials can occur in both the clay and the temper as impurities and be naturally present in clays in proportions reaching 17% (Johnson et al. 1988). When intentionally added to the clay, their abundance or nature will identify them as temper. They enhance certain paste properties such as impermeability or they act as flux (fusing agent) and coloring agents, as do iron compounds, frequently present in clays and soils. Plant ash (soda-rich) also can serve as flux.

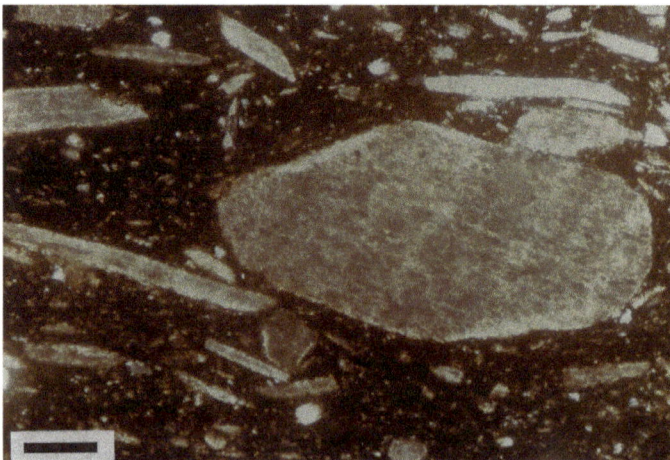

Fig. 7.3. Shell temper. The shell fragments are usually long and tubular grey plate-like inclusions. Exceptionnally, a large fragment is seen in the image center. The shell plates are oriented in the same direction, induced by the working of the paste. Few quartz (clear) grains are present. Aztalan age, Wisconsin, USA. ppl; *Bar*= 0.4 mm. Photo: J. Stoltman.

Vegetal tempers, a subclass of organic materials, are frequently found in building materials (bricks), but they are also used in ceramic production. These are plant fibers, straw or plant balls. Most vegetal tempers burn out during the firing of the ceramic, and their presence is deduced from the hollow imprints left. These form pores, enhance porosity and permeability if interconnected and open to the surface. The imprints on the surface of the ware or the voids in the paste are often characteristics, and may allow their identification. These voids usually differ from the pores due to the action of drying, forming and firing.

Carbonaceous materials usually burn out during firing, forming ferruginous nodes or spurs. If not totally oxidized during the firing of the pot, carbon will stay trapped inside the paste and form a black core. This is very often seen in archaeological and modern ceramics. Carbon somewhat enhances impermeability, as it blocks the circulation of humidity and the pores on the ceramic surface, when smudging is produced. In this process, carbon deposit is induced by the burning of organic-rich material in a reduced atmosphere (see smudging in Sect. 6.6). This is often seen in cooking pots. Many organic materials act as a flux (fusing agent), lowering the temperature needed for sintering (destruction of clays and melting of temper grains).

Calcium-rich phases of organic or mineral origin, such as calcite, shells and calcium phosphate as crushed bones, have also been used as temper. They yield high-quality wares when properly prepared and fired at lower temperatures. Careful preparation and firing are especially important with calcium-carbonate tempers because they may cause surface damage and internal cracks that may lead to breakage of the pot. These defects are linked to the decomposition of calcium carbonates during firing and recomposition with oxygen and water of the residual calcium, producing an increase in volume. An example is aragonite (a form of $CaCO_3$) a component of shells that alters to calcite at around 500 °C with a change in volume. In turn, the calcite decomposes around 620 °C changing into carbon dioxide and calcium oxide (CaO, lime). The carbon dioxide (a gas) tries to leave the ceramic and the calcium oxide is highly unstable. The latter may combine with the humidity of the surrounding atmosphere and significantly expand, causing lime spalling or breakage of the pot. In order to avoid this, shells or other carbonates have often been heated before being integrated into the paste as temper liberating the CO_2 gas before firing. Heating also makes crushing of the material easier. Besides pre-heating, the potter may finely grind his shell material to obtain very small inclusion grains that will not cause as much local expansion. A palliative technique is the addition of salt or salt water to the paste. Salt seems to impede the decomposition of calcium carbonate at its usual firing temperature, allowing the firing of the pot with less damage and preventing lime spalling. An instance reported is the addition of salt to a shell-tempered paste in the production of prehistoric Mississippian ware in prehistoric North America. Properly prepared, shell and calcium carbonates present coefficients of thermal expansion similar to that of the clay, thus offering a high resistance to thermal stress. They are frequent tempers in cooking pots. Calcium carbonate addition can be detected in elemental analysis by Ca concentrations higher than expected in regular ceramics. Ca-rich material can also be added as flux, and limestone was used as flux in early glazes from Chinese ceramics from Jingdezhen (Wu et al. 2020).

Many other tempers have been used to produce ceramics. For example, prehistoric and traditional ceramics have been frequently tempered with volcanic material, pumice, tuff, ash or volcanic glass (e.g. Figs. 7.4-7.5). Also, the addition of a coarser clayish soil is more common than thought. Often, the two materials differ only by the granulometry and proportion of the non-plastic grains they contain. Traditional potters in Japan today often use several clay sources to bring about a desired result in their pots. Not only plasticity but effects upon firing are very important to these potters, and hence they use a subtle mixture of clay sources to produce the desired result. There are also ceramics produced with only one material, without any addition or subtraction of non-plastic grains, as illustrated in Figure 7.6.

Fig. 7.4 (a) Mining area for temper for traditional production in Mangallpa, Peruvian Andes, and how it looks up close. (b) This is an eluvial deposit with pyroclastic material, consisting mainly of pumice, quartz, plagioclase, and biotite, with occasional hornblende and sandstone, and some clay as petrography shows in Fig. 7.5 (a) Mixed with a nearby clay, this temper is used to produce cooking pots (7.5b)

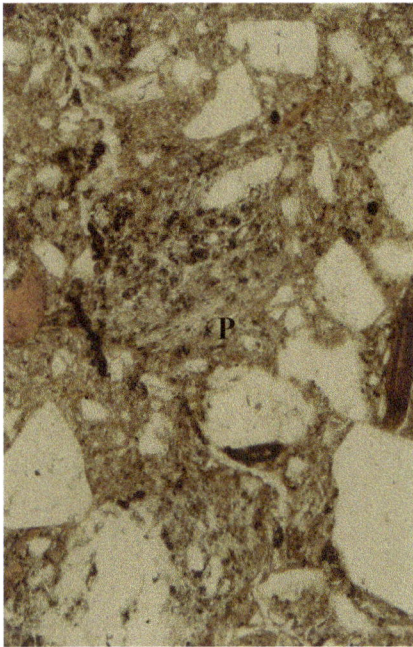

Fig. 7.5 (a) Ceramic paste with volcanic temper from the source shown in 7.4a. Large oval pumice (P) is 0.65mm long (ppl view). MM14a 40x, ppl and xpl..**(b)** Traditional cooking pot tempered with this material.

As mentioned earlier, the potter chooses the clay according to its firing color (in addition to all other properties). Here is a good example from Yi-Xin, China, a famous production town for teapots. All the (current) material comes from a very big hill close to the center of the town, with different veins of clay varying in color (Fig. 7.6).

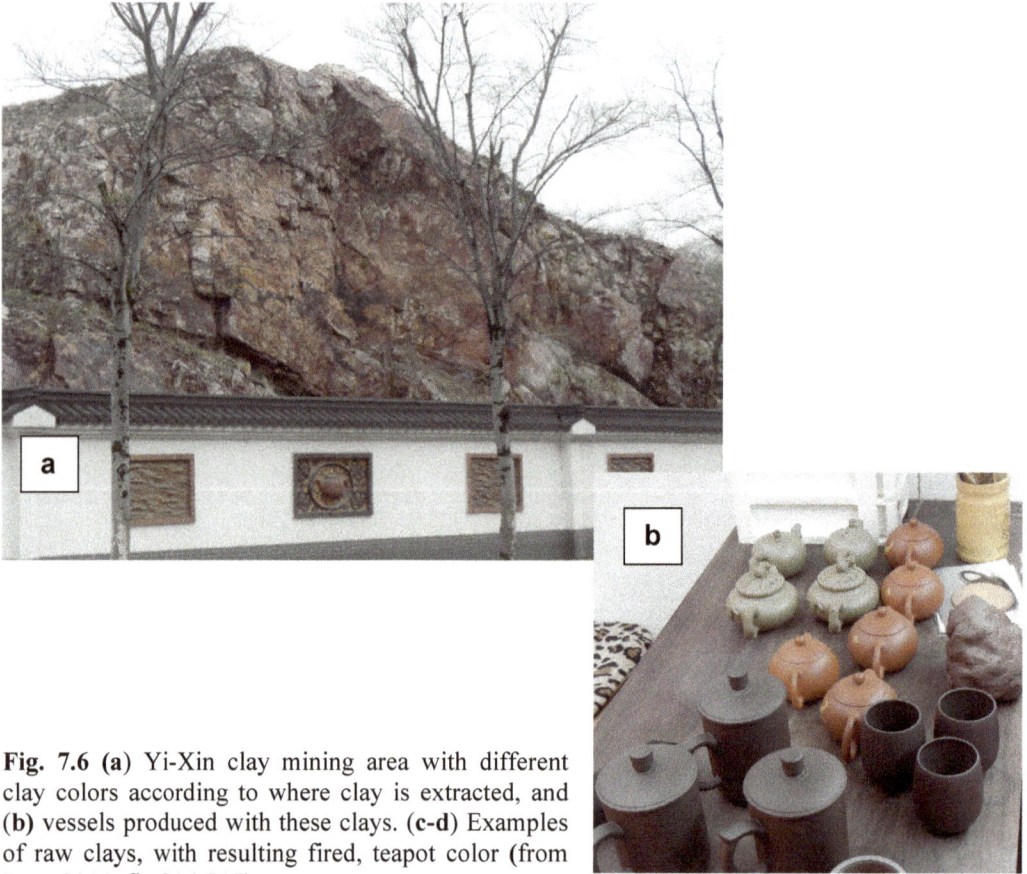

Fig. 7.6 (a) Yi-Xin clay mining area with different clay colors according to where clay is extracted, and **(b)** vessels produced with these clays. **(c-d)** Examples of raw clays, with resulting fired, teapot color (from Druc 2015, fig.216-217).

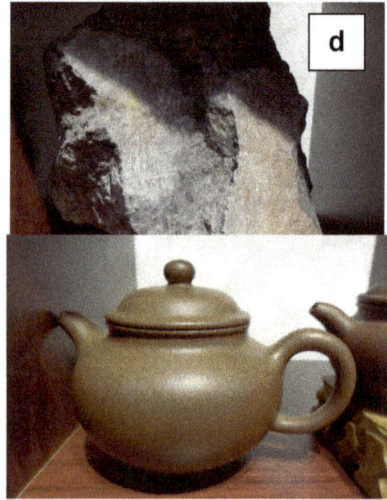

7.2.3
Tempering and Temper Identification

The petrographic example illustrated in Figure 7.7 (soil-tempered ceramic) raises the problem of temper identification when only ceramic fragments are available, as is commonly the case in archaeological settings. Many discussions are found in the literature about temper and its identification.

Depending on the temper used, identification may be straight forward, as with crushed shells or volcanic glass, or a very delicate matter, as when a second clay source is used. The addition of a second, coarser clay is a common occurrence in many pottery traditions. This process adds not only non-plastics but also clay particles to the paste. In the photomicrograph of Figure 7.7, the coarser felsic and quartzitic inclusions come from the temper, whereas the shale and mica fragments come from the clay. The small quartz and feldspars seen in the ceramic paste are found in both raw materials. Thus, a certain proportion of these small inclusions could be called natural, as they come from the clay. The rest of the inclusions are the added inclusions, as they come from the temper. In reality, such a fine distinction cannot be made, unless one has the raw materials to compare.

In this case of clay-tempered ceramic, the paste is composed of the clay materials of both components and of the inclusions contained in the clay and in the temper. This can be summarized in the following formulation:

$$x\mathrm{p} \ (\mathrm{I}1 + \mathrm{I}2) + y\mathrm{np} \ (\mathrm{I}1 + \mathrm{I}2) = z\mathrm{W},$$

where p stands for the plastic material, np for the non-plastics, I1 for the first ingredient (the clay), I2 for the second ingredient (the tempering agent) and W for the ceramic ware. $x, y,$ and z are the relative proportions of the ingredients in the paste. This equation shows the participation of both plastics and non-plastics in the clay and temper. It further explains the difficulty in determining which inclusion pertains to which clay source ingredient. This problem is seen in sourcing raw materials and in chemical studies. Chemical studies, by neutron activation or X-ray fluorescence, measure elemental concentrations in bulk samples. The fragments and clay minerals have been reduced to powder and the analysis reflects total composition. Thus, the different clay sources, which can contain clay minerals as well as non-plastic materials, will all be analyzed together. For this reason, the distinction between one clay and the other is near impossible, besides identifying the different temper grains in the paste (for temper identification, see Chap. 8).

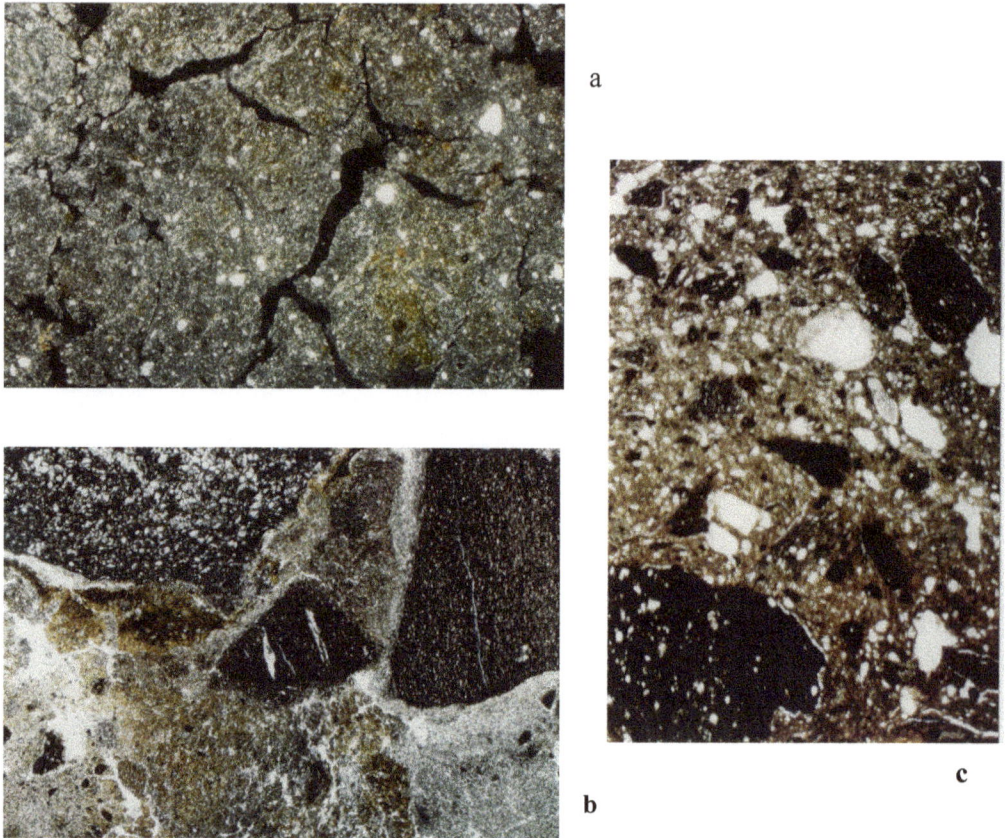

Fig. 7.7. Photomicrographs show the materials used in the production of an Andean cooking pot: (**a**) is a yellow soil (4x, xpl), (**b**) is the unprepared black clay (4x, xpl); (**c**) shows the resulting ceramic paste (4x, ppl). The two different soils were ground in the dry state, then mixed with water. The ware is coil-made and fired on the ground. The black clay is coarse with large shale and mica fragments that will be partly removed after sieving, and small quartz crystals. The yellow soil (ground) acts as temper by virtue of its higher quartz and feldspar content, although it has clay minerals in it. Traditional production, Marcara, Peru.

Another example from the same region is developed in more details in Chapter 9, section 9.5.

In general, the different materials used to form a paste (the clay and the temper) have each their own grain distribution, hence the bimodality expected when mixed, and the idea that bimodal grain distribution in a ceramic paste point to the presence of temper, but not always. There are examples, where some silty clays show the natural presence of medium to large granitic fragments while the fine grain fraction is nearly null. A ceramic produced with such a clay (without adding anything) will show a bimodal grain distribution (Druc et al. 2017). Thanks to the sampling of comparative clay materials, it was possible to avoid misinterpretation of the ceramic data, and to think that coarse granitic sand had been added to the clay base. Sampling local raw materials (soils and clay deposits), as well as knowing what is (or was) the regional ceramic tradition, greatly helps in interpreting the mineral and chemical data.

Natural sediments may show a plurimodal distribution as there could be several modes present, leading to a continuous grain distribution curve (Fig. 7.8a). In an ideal bimodal distribution, the finer fractions are attributed to the clay, and the coarser to the temper (Fig. 7.8 b). In practice, however, the curve is rarely this well defined, and the curves of grain distribution in raw clay and in the temper partly or completely overlap (Fig. 7.8 c), according to the raw materials used. This is schematized in Figure 7.8. Such distribution can be obtained from modal analysis or from granulometric counting and surface measures done with image analysis programs, such as described in Chapter 8. The total area represented by all the inclusions in the surface area analyzed offers a quantitative measure of the percentage of non-plastics vis-a-vis the clay matrix. It is not a measure of the material added by the potter to temper the plasticity of his or her clay.

There have also been attempts to separate clay particles from non-plastics in order to compare the fired clay to the raw material; but without distinguishing added from natural inclusions, the comparisons are bound to give approximate results. There are, of course, exceptions where non-plastic inclusions are few, unimportant, or their influence attenuated. Hence, the usefulness of petrographic studies in identifying the inclusions in the paste.

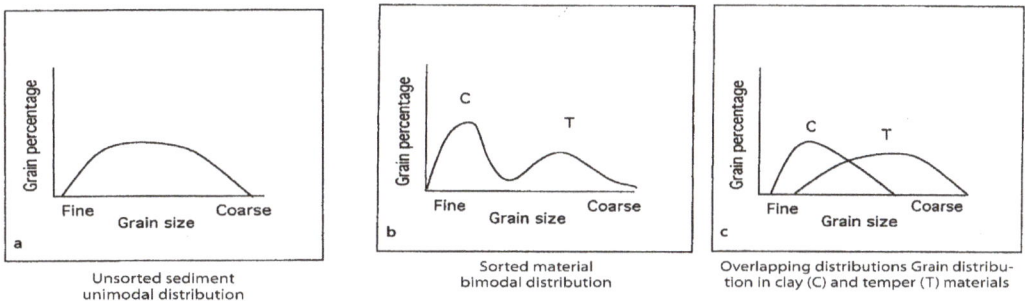

Fig. 7.8. Schematic grain distribution in (**a**) unsorted sediment; (**b**) sorted material used in ceramic production, ideal bimodal distribution; (**c**) observed grain distribution in clay and temper, overlapping curves.

Tempering a ware with volcanic ash or pyroclasts is also quite frequent. Volcanic material resists heating cycles well and when pumice constitutes a good part of the temper, lightness of the material could have played a role in its selection to produce lighter pots. The type, form and size of the fragments, and their alteration give clues as to the possible origin of the material. When the plagioclase, hornblende and biotite fragments appear 'fresh' and in particular the pumice and glass shards are unaltered this suggests that the acquisition area of the material was close to the original source (i.e. no water transport of the material, little weathering). While the quartz and plagioclase in pyroclastic material can be very angular the tuff and pumice fragments are often oval to round. This is due to their softness and not to grain transportation. Pumice and glass shards weather rapidly (e.g. devitrified glass appears brownish and could recrystalize into microcrystals of quartz). This type of temper is of course found near volcanic areas, but ash deposits can cover extensive areas, as they are easily carried far by winds.

On the coast, the sands and deposits available to potters are frequently of mix mineral composition, the grains are often rounded (but individual minerals can also be angular, in particular quartz) and carbonates are much more present, as illustrated in Figure 7.9. However, beach sands are not always of mix compositions. The study by Dickinson et al. (1996) of ancient Lapita ceramics in Polynesia presents an excellent 'counter' example. After sampling sands for comparative purposes from different island beaches, it was clear that each island had a particular volcanic signature, which enabled the researchers to show that Lapita vessels were mostly locally made and did not travel as far as originally thought, being exchanged mainly amongst island clusters.

Fig. 7.9. Here, the grains are weathered, worn out, with rounded edges, and rock fragments are of heterogeneous composition, reflecting the geology of the area (volcanic and sedimentary sources). A few carbonates and spicules of marine shell origin can be seen to occur randomly in the paste. One elongated biocarbonate is present in the center of the photograph. The whole temper assemblage suggests the use of coastal sand as temper where elements of marine origin are mixed with rounded, transported mineral grains. Black cooking pot, 1st mill. BC, Ancon, Peru. Thin section, xpl; *Bar* = 0.3 mm.

One must stress that paste analysis is linked to archaeological questions related to much larger issues than just temper identification, but this is one of the important steps. The constraints affecting ceramic production (environmental, cultural, social, political, economic, etc), the organization of production, how far ceramics were distributed, the relationships between artisans and patrons, and of course groups' identity are often objectives orienting this type of research. Ethnographic studies coupled to archaeometric ones are great to help us realize the variability of materials, ceramic productions choices, and recipes that can exist. Nadia Cantin and Anne Mayor (2018), for example, combine petrography and the study of fashioning techniques to understand production constraints and identify cultural groups. The study shows how different potters' communities and diffferent ethnic groups, living in the same area (Southeast Senegal) use different paste recipes, mixing various raw materials (e.g. straw, grog and mineral), but also, in one case, different chaînes opératoires according to the wares to produce -so in this case one cannot equate a chaine opératoire to a cultural group, nor should we necessarily associate a paste composition to a particular community of potters. As the authors observed with petrography, even when the raw materials involved are known, it is not always easy to recognize that two mineral materials (clayey soils, sands, rocks), have been used or that they have been crushed. Alicia Espinosa (2020) is also using this multi-disciplinary approach (surface analysis and petrography) to recognize cultural identities within the same valley and site. The following sections in this chapter should certainly be read with this larger picture in mind.

7.3
Making a Pot: Physical and Chemical Reactions

The last section was largely concerned with raw materials. Making a pot requires more than adding tempering material to clay, however. The manufacturing process includes many steps (i.e. forming, drying and firing), during which physical and chemical reactions occur. These reactions are important to understand, as they explain some of the measures the potter takes to control them and the quality of the final product. Three points are discussed in relation to manufacture and to the physical and chemical reactions influencing ceramic production: plasticity and the role of temper, shrinkage during drying, and non-plastic expansion during firing. Preparation and forming techniques, surface coatings, and firing of the ware will be discussed later.

7.3.1
Needs as a Function of the Object

The potter knows, intuitively or by experience, what is the right paste for his/her pots. This knowledge can be deduced by looking at the properties of the pot and the raw material used. The basic physical characteristics of the ceramic paste are its plasticity, which enables the potter to form the desired pot; its ability to retain the given shape once dry; and its hardness when fired. From the plastic phase to the hardness of the final product, the ceramic material goes through different stages of structural and chemical change due to the drying, firing and cooling of the pot. The "visible" reactions are mainly retraction and expansion processes tightly bound to the quality of the paste, the type and quantity of the raw materials, the particle size, the amount of water present, the preparation of the paste and the forming techniques used.

7.3.1.1
Plasticity and the Role of Temper and Non-Plastics

Plasticity is obtained by mixing the clay with water. It enables the potter to model the paste by pressure. As discussed in the last chapter, the level of plasticity relates to the water absorption capacity of the clay, which varies from one clay type to another. A simple test that potters often do to see if a clay is suitable for pottery making is to roll and make a circle to see how the clay will dry (Fig. 7.10). Next step is to fire it to see how it resists temperature and what color it takes. As clay is often too plastic, or sticky, when wet, the addition of an adjuvant is required to make it "lean" (compare the French term *dégraissant* or the German *Magerung* for temper). The role of temper, as added material, is thus to give workability to the paste (usually by lessening its plasticity introducing a less plastic material) and to strengthen the body. It can also be added to enhance specific properties of the ware, such as porosity, hardness or thermal stress resistance. This role is assured by the presence of the non-plastics, which are often coarser or more abundant in the added temper material than in the clay base. Non-plastics help the pot withstand the drying and firing phases of the manufacturing process. They open the paste, managing space for shrinkage and expansion, allowing water and air to circulate. As non-plastics reduce the difference in shrinkage rates between the core and the ceramic surface, the chance of breakage is also reduced.

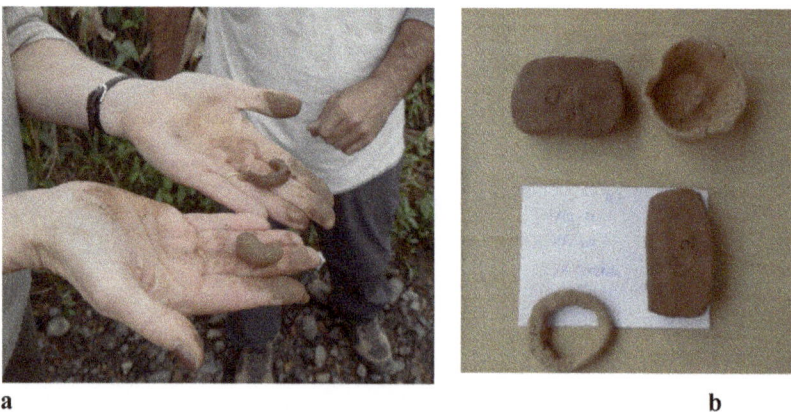

a   b

Fig. 7.10. Plasticity and drying tests. When looking for suitable clays, beside tasting it, the potter can roll a sample to see its plasticity (**a**), if it will crack upon drying, and do some firing test (**b**).

7.3.1.2
Drying and Shrinkage

Shrinkage is caused by the evaporation of the water added to the clay to make it plastic (the mechanical water). If the paste has not been well kneaded, water evaporation can cause defects, such as small voids and cracks. Drying can also accentuate the defects caused by bending or compressing material in the building process (Fig. 7.11). In addition, certain clays lose more water upon drying and will retract more producing thin long voids, or differences in drying rates between clay and non-plastics may cause excessive shrinkage and paste retraction around grains.

Water evaporation takes place during the drying of the pot, usually at room temperature, through capillary action from the interior to the exterior of the pot. When the excess water is gone, that is, when the water film surrounding the clay particles has evaporated, drawing the particles close to each other, the surface is said to be leather-hard. It is still not totally dry, therefore not too hard, but strong enough for the polishing and the application of decorative designs without deforming the pot. Surface decorations, like incisions, engraving and punctuation, are often applied at this stage. The work will leave clean marks, as opposed to the chipping and scratches left when decoration is done on a surface that is completely dry or, on the contrary, the clay ridges when incision is done on a wet surface. These features can be observed macroscopically or under a microscope.

The type of clay, number and size of inclusions, and pores will all influence shrinkage. A coarse clay with large pores and non-plastic inclusions has a low water content and good capillary action. It dries quickly, with less shrinkage than a very fine clay. The fine clay, because of the water films around the many clay particles, and because of its fine capillaries, shrinks considerably and needs more time to dry-unless tempered.

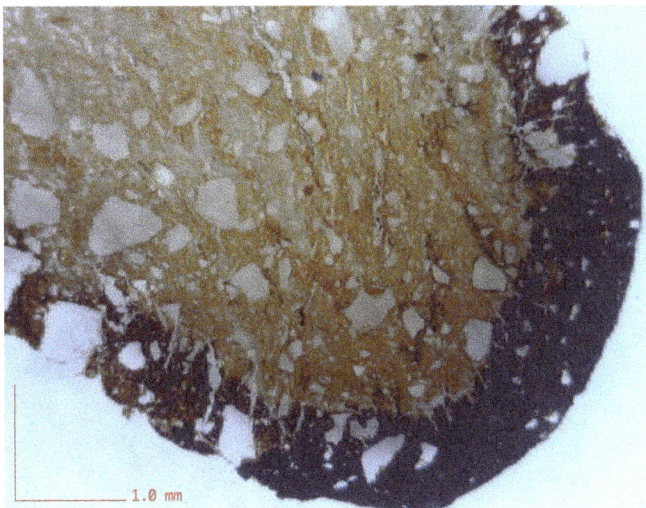

Fig. 7.11. Multiple planar voids perpendicular to the rim and in the clay body. They may have initiated during drying but firing of the ware accentuated the process. The cracks form thin, irregular (white here) voids. Photomicrograph of ceramic thin section, ppl view.

The orientation of particles also plays a role in water evaporation. This is induced by the forming and the finishing techniques (wiping, smoothing and polishing of the surface) in wet or half-dry state. Oriented parallel to the wall, as seen in Figure 7.12, the particles slow down the process, opposing a barrier to evaporation. Large connecting pores (Chap. 6, section 6.4) and non-plastic grains randomly arranged (in a deflocculated state) or perpendicular to the wall accelerate water evacuation. Orientation also plays a role in heat propagation and thermal stress resistance (see below). Orientation can be determined by the forming technique. This is often seen in wheel-made pottery, but it is also found in paddle-and-anvil or coiling techniques (see below). The use of different forming techniques on the same pot, resulting in varying particle alignment or water content, can induce differential shrinkage. While particle orientation can strengthen, it can also weaken a ware. Thus, breakage is easier along cleavage planes, as in the case of a paste of schistose texture with many mica fragments forming parallel planes of weakness.

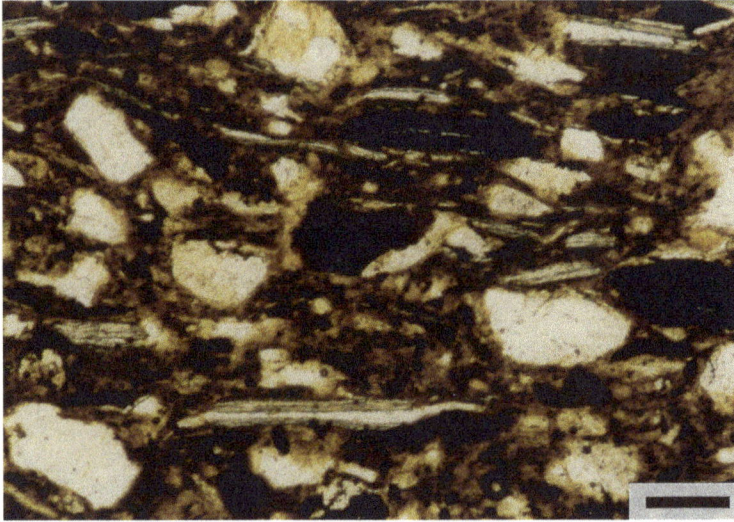

Fig. 7.12. In this figure, the long white mica flakes are oriented in the same direction, parallel to the ceramic wall. This orientation could result from the use of a technique similar to paddle-and-anvil to finish building the walls, where applied pressure gives the strong particle orientation seen here. Peruvian jar, 1st millennium B.C. xpl; *Bar* = 0.18 mm

7.3.1.3
Material Expansion

No further significant water loss is expected after the drying stage, as it is only the interlayer water in the clay particles that evaporates during firing. Here, another process is taking place: the expansion of non-plastic grains while kiln temperature rises. This can cause cracks or surface defects. Again, the potter can reduce the risks, by using the right materials or paste preparation. He can also add agents that influence the temperature and chemical reactions at which some non-plastics expand. Certain minerals, like quartz or calcium carbonates, expand when heated above critical temperatures or change their chemical state. Quartz undergoes a first inversion (from quartz alpha to quartz beta) around 573 °C and calcium carbonates expand above 750°C. The quartz inversion is unimportant if quartz is not too abundant in the paste, as expansion occurs while there is still some water loss, allowing space for the expansion.

The defects due to the use of calcium carbonates (see Chap. 6, Sect. 6.1.1.3, 6.7) may be somewhat counteracted by very finely crushing of the ingredients or preheating them. If overheated, or if the paste is not porous enough to allow the volume intake and the rapid escape of gases, damage may result to the ware. The expansion rates of the inclusions and of the paste must then coincide for the vessel not to crack or show strains. Low expansion rates also offer better thermal stress resistance. This is the reason why mica, with a thermal expansion close to that of clays (because they have the same mineralogical structure), is preferred to quartz when producing cooking ware, which must endure rapid heating and cooling throughout its life. Other minerals with low thermal expansion coefficients include grog (crushed ceramic), calcite, crushed shell, plagioclase feldspar, pyroxene, amphibole and basic rocks.

7.3.1.4
Grain Angularity

Grain forms influence drying and the strength of the ware, depending on the bonds they offer with the clay matrix. Angularity provides a better bond between grains, which are well inserted in the clay matrix. Angular grains, however, can cause breaks at junctures. Spherical grains allow better drying because of air circulation around them, but provide a weaker bond. In public works construction for roads, dams and so forth, the role of angular grains versus rounded ones is well known. Angular grains tend not to slide by one another and as such they give a greater mechanical strength to the clay-grain mixture. Angular sands give greater shearing strength to the material. In terms of potting, an angular temper will give a stronger, more rigid body with less tempering agent than one with rounded grains.

Also, grain angularity can yield interesting information on production process and origin of materials (see also Chap 8, Sect. 8.3.4.2).

7.3.2
Paste as Related to Function, Form and Manufacturing Requirements

Economics play an important role in ceramic production, as the potter generally uses the source material at hand. The function or quality of a pot may thus be influenced by the local geology and the potter's knowledge. A coarse paste, fairly porous, with poorly oriented particles, and low-firing, is more defect-resistant. It is a multipurpose paste often found in prehistoric wares. It is seen not only in cooking ware and jars, but also in fine bowls or plates (good looks are not always a warranty of fine paste).

Regarding the kind of pot produced in relation to the paste and manufacturing technique, no firm rules exist. Physical laws may apply, such as the fluidity of the paste if using a wheel or the necessity of building thick walls for large pots. At one extreme, there is the case of potters producing all their wares, whether for serving, cooking, or storing liquid, with the same paste and manufacturing technique, as many traditional pottery communities do. At the other extreme, paste composition can vary according to the pot and the parts of the pot (bottom, neck, lower-upper body), as in the case recorded by DeBoer and Lathrap (1979) among Shipibo-Conibo potters in Peruvian Amazonia.

Most potters, however, slightly modulate their paste to produce the desired pot. For this, they can use the same mixture of clay, temper and water, but they will grind it differently (from poor grinding to fine) according to the pot to produce (a coarse cooking pot or a fine bottle), or they will add more of one ingredient, usually temper, for larger vessels. They can also subtract the coarse non-plastic grains in the clay (which refines the clay) and re-add this coarse material into the clay base (see Chap. 8). There are also cases of careful selection or preparation of the materials to obtain a particular paste for the production of a specific kind of ware. A good example is the adjunction of bone ash in English bone china ware, to give the ceramics their translucent appearance. Another is the addition of salt to calcic or shell-tempered paste in order to reduce lime spalling.

Another factor influencing production is the size of the object to be produced, which will determine the amount of paste to be used and the wall thickness of the pot. For example, a small bowl will require less paste and thinner walls than a big jar. In turn, wall thickness influences the rate of water evaporation and thermal stress; and these are determinative of the function of the ware. Strong wares are needed for food preparation; strong but light for transport, thick and thermal shock-resistant for cooking; heavy and stable for storage; fairly porous for liquid containers to allow evaporation to keep water cool; impermeable for drinking, and so on.

The type of paste used in relation to the function of the pot can be seen in thin sections, on the SEM, or with X-radiography. With bulk chemical analysis, this relationship can be deduced from the ceramic classification in a dendrogram or scatter plot obtained with statistical analysis. The correlation of chemical data with formal categories may show compositional tendencies related to the production of a certain kind of ware. For example, all the cooking wares are grouped together, whereas bottles and decorated bowls are in another cluster. This result can be interpreted as an indication of different provenance for coarse ware versus fine ware, but it can also indicate differences in paste preparation.

7.3.3
Needs as a Function of Use of the Object

It has been seen that certain raw materials and manufacturing processes affect the properties of the paste. This section will discuss the use-related properties of ceramics. A whole field of research has developed in materials science for the study of these properties and to understand the relation between raw materials and the functional uses of ceramics. A number of authors and teams working in this field are given at the end of the chapter (see also Chap. 10).

Ceramics fall into different functional categories: cooking vessels, serving dishes, liquid containers. These ceramics require specific physical and technical properties, such as thermal stress resistance, durability, strength, and impermeability. These use-related properties are influenced by the nature and texture of the paste, such as the size of the inclusions, the particle orientation, angularity, and preparation and firing conditions.

7.3.3.1
Durability and Breakage Resistance; Strength and Hardness

Durability and breakage resistance are related to the strength and the hardness of the ware. A pot is usually expected to last long and to with- stand mechanical stress, such as hitting, scraping, pounding or dropping (when processing food or during transport). These properties are influenced by the paste composition and texture, wall thickness, the shape of the ware, the way it was manufactured and fired.

Strength is a notion in relation with the whole vessel, whereas hardness is rather used to describe the ceramic surface. Strength is enhanced by the size of the non-plastics. A small-grained paste has more strength than a coarser one, because of the expansion problems linked to the presence of coarse grains. Porosity also plays a role in strength and breakage resistance, as pores may stop crack propagation and allow material expansion. However, when pores are too large, strength and resistance to breakage are reduced. Strength is also enhanced by a good bond between inclusions, which is favored by grain angularity, and small inclusions randomly arranged. These may stop crack propagation. Finally, firing may largely induce strength, as sintering and melting strengthen the bonds between the clay and the inclusions.

Hardness is generally measured with the Mohs' scale used in geology (see Chap. 9.6.2). It is enhanced by surface treatment, such as polishing, which hardens the surface, orients the particles and seals the pores. In the laboratory, strength is measured by applying mechanical stress to a surface, and looking to see how much force it can take before breaking. Laboratory studies have shown that a high resistance to breakage by mechanical stress is best obtained with fine platy particles. Their orientation parallel to the surface opposes cracks propagation through the body. They also yield more elasticity to the paste. A strong orientation of platy particles may induce weakness along cleavage planes, however. Use of pyroclastic and pumice material has also been shown to allow weight reduction and higher resistance to breakage (Páez et al. 2013).

Breakage resistance may be enhanced by wall thickness and the shape of the pot. In general, rounded shapes are stronger than angular ones. This is explained by the fact that angles are stress traps (mechanical or thermal). Wall thickness should also be kept constant, to avoid differential drying and firing, which would weaken the ware. Other weak spots are handles and other additions to the main body, prone to break away during handling. Breakage may also occur at juncture points of the manufacture, between two coils or between body and neck, for example. In the ceramic corpus recovered in archaeological context, handles, necks, and rims are often found alone.

One should not forget that what we see in a ceramic may not relate to the choice of raw material, provenance or technology of production. Wear, internal fractures, surface concretions (notably salt and calcium), iron deposits, may give clues about vessel usage or the environment in which the ware or ceramic fragment laid dormant during hundreds of years.

7.3.3.2
Porosity, Density, Permeability, Impermeability

Porosity is an expression of the volume of pores present in a ceramic body. It is influenced by particle size, grain distribution, inclusion shape, manufacture technique and firing. Porosity plays a role in thermal stress resistance, air and water circulation, allowing material expansion and the escape of gases. Porosity can be estimated by physical methods (i.e. immersion, water absorption or by image analysis).

Density is related to true porosity. Its measure takes into account all the pores, closed and open. On the contrary, apparent porosity is a measure of water absorption, where only open pores are considered (they are the only ones that can absorb water). It is measured by looking at weight differences between dry and water-impregnated samples.

Permeability is a slightly different concept as it expresses the transfer of air or a liquid across the ceramic walls. Besides compositional characteristics, it is influenced by viscosity, air pressure, temperature, and wall thickness.

Impermeability (no transfer) is strongly influenced by firing temperature and surface treatment. High temperature allows sintering and smelting. Sintering and vitrification enhance vessel strength and impermeability. The vitreous phase is stronger than the crystalline phase. At this stage, pores are sealed or disappear with shrinkage. This process reduces porosity and increases impermeability. Different surface treatment, such as polishing, application of slip and glaze, also induce impermeability, mainly by sealing the open pores on the ceramic surface.

7.3.3.3
Thermal Stress Resistance and Thermal Conductivity

Thermal stress resistance is a required property for cooking ware, and for all pots in general. A cooking pot is submitted to thermal stress, that is, to rapid heating and cooling, and to repetitive heating cycles. Thermal stress may cause breakage. This is due to volume differences, originated by material expansion during firing. Materials with low expansion rates when heated, or with rates similar to the clay, warrant a good resistance to thermal stress. Mismatch creates thermal stress and cracks.

Thermal stress resistance is a function of the inclusion type, size, distribution and shape. It is also related to porosity. Fine inclusions reduce material expansion risks, whereas coarse inclusions open the paste, allowing the escape of gases, hence inducing thermal stress resistance. When clay particles and platy inclusions are oriented perpendicular to crack propagation (parallel to the walls), thermal stress resistance is stronger. How- ever, this reduces heat propagation.

All these modulations are and were done by the potters to produce the desired ware. For example, cooking pots need such properties as thermal stress resistance and heat conductivity. The latter is enhanced by the presence of interconnected pores, large and open, which allow the circulation of air and the escape of gases. Thus, a coarse and fairly porous paste is preferred for the production of cooking pots. On the contrary, porosity opposes heat propagation, when pores are small and not connected. This yields wares with high thermal insulation. Wall thickness also influences heat propagation, thus, thin walls offer less resistance to heat propagation. Besides, walls of varying thickness and angular forms induce local concentration of thermal stress.

7.4
Material Preparation

The raw materials used by traditional potters usually require treatment before use. The clay is often acquired in chunks excavated from a pit, a field or a quarry. It is easier to transport dry clay than a wet, plastic material. These dry chunks of clay must be crushed, ground, and sieved to obtain a finer material, taking away the coarser fractions and impurities. The process is tiring, as the clay can be very hard when dry. The crushing is done on a millstone or a stampmill (mechanized or not). Also clay clods can be spread over a hard surface and crushed with a heavy stone, a hammer or a similar tool. The clay can be further ground on a handmill and passed through a screen or a sieve. In other cases, clay can be soaked in water in a vat or pit to allow material settling. The clay particles and fine material will stay in suspension and settle above the coarser material, allowing their separation. The process is similar to natural sedimentary settling and is called decantation. The finer material is removed and can be sieved; or else the clay is allowed to dry and the potter has only to collect the fine powder obtained through this process. A more sophisticated process that also involves settling uses a slightly inclined surface along which the liquid clay runs, leaving the coarser material behind. This process is called levigation. Examples of material preparation for clays or tempers are illustrated below (Fig. 7.13-7.14).

The temper is also subjected to prior treatment before mixing with the clay. Preparation will vary according to the nature of the temper. Mineral temper is usually crushed in the same way as clay, on a stamp mill or with a hard tool. It is then ground and sieved to remove the coarser fractions and unwanted inclusions. Decantation and levigation are not used unless the temper is another coarser clay that needs refinement. Some materials, such as shells or rocks, may be heated to facilitate their crushing. In the case of shells, heating also prevents further material expansion that would be harmful to the ware during the firing process. This technique, discussed in previous sections, was often used in prehistoric times. Other material, such as tree bark, can be reduced to ashes before blending with the clay. Chaff, straw, cotton balls and the like can be chopped, while grog (old ceramic fragments, wasters, and fired clay) and bones are ground and sieved to remove unwanted coarse inclusions.

a   b

Fig. 7.13 (a) Drying clay lumps in front of a potter's house, Cajamarca, Peru. **(b)** Decantation tanks for the clay, Mina Clavero, Argentina. The tanks can be used for levigation as well.

a

b

Fig. 7.14 (a-c)   c

d

Fig. 7.14 (a) Sieving mineral temper at the extraction place **(b)** and crushing clay **(c)** material, Sorkun, Turkey; **(d)** Clay-mixer, PhuLang, Vietnam. This clay is very plastic and brought by boat or truck, and only needs to be homogenized. It is bicolor, with strikes of grey and pink. The electric-powered machine eliminates the tiring work of kneading by foot.

The same material can also be used for obtaining the clay component and the temper additive. The coarser and finer inclusions of the clay are separated by water immersion, as in the decantation or levigation process. The fine sediment is collected for use as the clay, whereas the coarser material is removed and used as temper.

Once the clay and tempers have been prepared, they are mixed together with water and kneaded until a well-homogenized paste is obtained. The kneading process is very important, since, if properly done, it allows the complete blending of raw materials. If the paste is not well-homogenized, air pockets will be present in the fired product. They may induce breakage during firing or body weakness and attest to incomplete kneading. The paste is kneaded by foot or by hand on a flat surface on which material such as sand or ash may have been spread to prevent sticking. The paste is used right away or allowed to rest for some time. In Peru and Mexico, many traditional potters prepare their paste the day before and keep it in a cool place, covered by a textile or plastic sheet to prevent the paste from losing its humidity and plasticity. In some places, the paste (or just the clay) is allowed to "age" for weeks or months. Ageing allows the decomposition of organic materials, the growth of bacteria, chemical exchange, and water penetration. Ageing can be helped by the addition of organic material such as acids like vinegar or stale wine, as reported by Rye (1981). These materials yield more plasticity and a better paste to work with.

Finally, two points should be stressed relative to the results of material crushing. First, the fragmentation planes of the crushed rocks may be different, yielding grains of differing angularity. Thus, a rock of granite type with well-formed crystals, angular or subangular, will fragment along the crystal junctures when crushed, breaking down into smaller angular fragments. In contrast, rocks with a fine matrix, softer, with no individual phenocrysts, like tuffs or shales, will yield round or elongated fragments even if they are crushed. Hence, the angular criteria for determining if a raw material was crushed before use must be carefully considered in relation to the type of material present.

The second point to consider in relation to the end result of material preparation is the number and type of the individual inclusions in the paste. The crushing frees up inclusions of larger rock fragments now found in the clay matrix, adding to the original clay inclusions. For example, quartz grains and large biotites found loose in the matrix may come from the crushing of a quartz-mica schist temper. If so, modal analysis will show higher quartz and mica content in the clay matrix of the paste than if the analysis were done on the original raw material.

All the above preparation procedures alter the characteristics of the original materials used to produce ceramics. This is an important point to remember in ceramic analysis, when searching for provenance or studying production techniques. This is still important when raw materials are compared to archaeological ceramics to prove their local or non-local production. Also, one should remember that chemical modifications may be introduced by the soluble salts and trace elements contained in the water added to the clay-temper mix, elements that may obscure the comparison between raw materials and ceramics in chemical studies.

7.5
Forming Techniques

Several forming techniques can be used to produce a pot. The primary forming techniques are hand-forming (pinching, coiling, slab building, molding, casting, and wheel-throwing). Secondary forming techniques are used to finish the building process. Among these, the beating and paddle-and-anvil techniques are very common. The forming techniques reflect the technological advances in ceramic production known to the society that produced the ware. Existence of the technology, however, is not a guarantee of its use. Potters may produce handmade pots in one village, while the wheel-throwing technique is used in the next. The choice may be economic or linked to tradition or aesthetics. Several techniques can be used on the same vessel, as seen in Vietnam (Fig. 7.15, see Druc 2016 for more details).

Pinching, a hand-forming technique, can be used to produce a vessel in whole or in part. The vessel is formed by finger pressure, squeezing the clay out of a ball and thinning the walls through compression. No special particle orientation is observed with this technique. This forming technique can be used for constructing the base of pots that will later be formed with the coiling technique. Pinching is also used to form the rim of vessels built with other techniques. One example is provided by Mexican potters, who finish the coil-made rim of a bowl by pinching it to achieve special undulating effects. In this instance, pinching serves as a finishing technique.

A technique akin to pinching is the *drawing* technique, where the vessel is formed by drawing up a claylump, after opening it with the fingers or the fist. As for the pinching technique, particle orientation is random, but finger marks may be seen on the vessel interior if not obliterated by the finishing technique.

A very common hand-forming technique is *coiling* (Fig 15a). A coil is a lump of paste rolled between the hands or on a flat surface so as to make a long rope. The rolling action induces the particles to orient in a loosely circular way, parallel to the long axis of the coil. This may be recognized in a fresh cut perpendicular to a coil, where inclusions can be seen sticking out vertically from the paste. While not easily seen in thin sections, particle alignment roughly parallel to the coils shows up on X-radiographs. However, the finishing technique (paddling and polishing) can blur these features, reorienting the particles that are close to the surface.

Coils are placed one on top of the other, in rows, to construct the walls of the pot. Coil thickness and length vary. They depend on the diameter of the pot, its wall thickness, the potter's experience, and paste composition. Usually, one row consists of one coil. The number of coils can also vary, as a medium-sized pot may use three large coils, for example, or six small ones. The coils are joined together by hand pressure and water to seal the joins. This introduces a fragility factor, and pots break preferentially along the coil juncture lines.

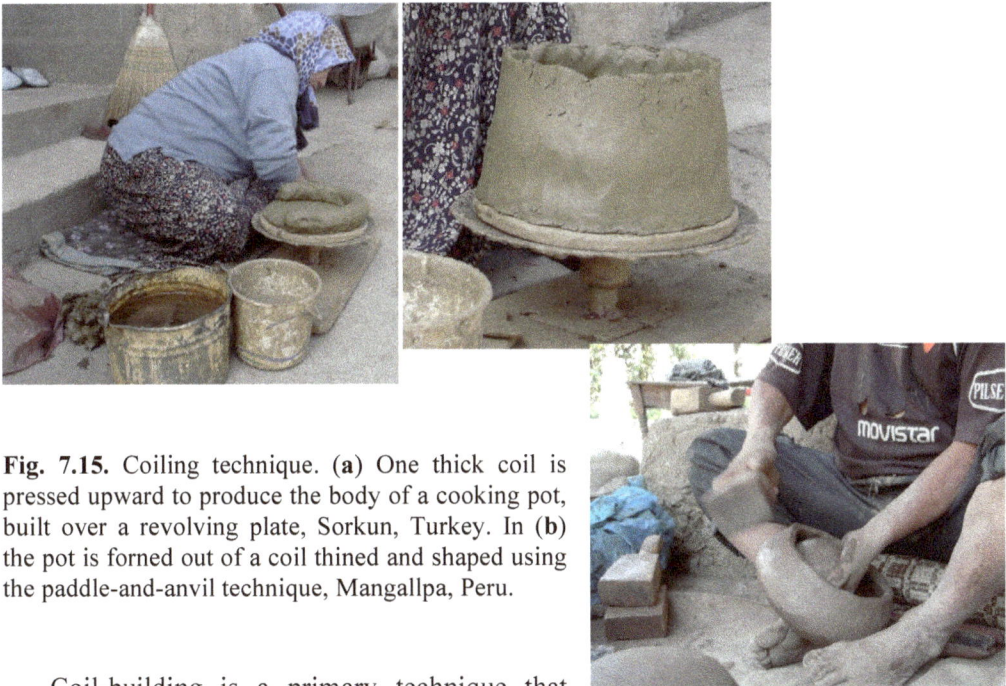

Fig. 7.15. Coiling technique. (**a**) One thick coil is pressed upward to produce the body of a cooking pot, built over a revolving plate, Sorkun, Turkey. In (**b**) the pot is forned out of a coil thined and shaped using the paddle-and-anvil technique, Mangallpa, Peru.

Coil-building is a primary technique that may be followed by the paddle and anvil technique (Fig. 7.15b). The paddle can be a small wooden bat, flat and smooth, while the anvil is a round rock or clay tool used to counteract the paddle strokes and hinder deformation. Some modern potters of the Peruvian Andes use a mushroom-shaped clay anvil, the "stem" being the anvil's handle to hold. This technique is used to strengthen coil-made walls or lift them up. It hardens the surface, compacting the paste and forcing the inclusions into it. It also plays a functional role, as material compaction induces impermeability, because particles are oriented parallel to the walls by the repeated paddle strokes, slowing down water penetration and evaporation. Paddling may thus be part of the building process or a finishing technique. Coils can also be turned, as seen in PhuLang, Vietnam (Fig. 7.16).

Fig. 7.16. Coils are prepared to be turned on the wheel, Phulang, Vietnam. Three coils are sufficient to build a jar. (See video documentary: PhuLang, Ceramic production in North Vietnam, Druc 2015, https://vimeo.com/134255181).

Several other handmade techniques are described from ethnographic contexts by Cantin and Mayor (2018), Gosselain (1992, 2000), Livingstone Smith (2000, 2007), Roux (2011, 2019) to cite a few researchers who have investigated the diversity of manufacturing techniques, the traces left by these, the chaîne opératoire, and patterns that one could look for when studying archaeological ceramics.

Necks, rims and handles can often be made out of coils, even if the walls are not formed with the coiling technique. The base, however, is usually a clayslab or claycake, molded over an existing base form or shaped by hand. The claycake is obtained by flattening a claylump (like dough for a pie or a pizza). The base can also be formed out of a claylump, opening it at its center with the thumb or the fist. The *molding technique* can be used to produce the base and the walls. There, the claycake is applied over (i.e. an upside down pot) or inside a mold and pressed down with the hand (Fig.17). In this technique, particle orientation is heterogeneous, except some alignment close to the surface if a paddle is used.

Another hand-forming technique is the *slab* technique, where units are welded together (Fig. 7.18). Clay slabs are flattened to a desired size, usually in rectangular form, and the rim surfaces are wetted and joined. This is used to produce preferentially angular objectss, but it also yields rounded pottery. Some very large jars are produced in this way. Here again, particle orientation will not be strong. X-ray radiography can be used to investigate the building technique used and distinguish between slabs and coils, as slab dimensions and row junctures show up on the radiographs.

Fig. 7.17 Molding technique. A large round slab is pressed inside a bivalve mold, and the two halves are then joined. Workshop of Manual Ocas Heras and Felicita Aquino Minchan, Cajamarca Peru.

a

b

Fig. 7.18 (a) Slab preparation (Phulang, Vietnam). Pile of clay in the back with rectangular form to produce slabs. **(b)** The slabs are put together to produce caskets (drying in front). A long coil is flattened (on board at right) to hold the seams in place from the interior. (See Druc 2016).

Casting is a technique which calls for the paste to be cast into a form or mould. This technique requires the use of a fluid paste, in which clay is the dominant component, with little or no tempering material. The ingredients are finely ground, as this augments plasticity and reduces the risks of material expansion during firing. This technique tends to be associated with mass production and reflects a level of specialization, organization and development in ceramic production rarely found in domestic production and early prehistoric societies. Thus, the type of technique determines the type of paste to be used and its composition.

The *wheel-thrown* technique requires a fine and plastic paste. The pottery wheel was not known to all prehistoric societies: it is a technological advance linked to the invention of the wheel. However, there are other types of wheel besides the true wheel with continuing rotation. The best known is the tournette, which is manually rotated. It can be a simple slab or a large piece of broken pottery set over a support that enables its rotation when propelled by the potter's hand, the potter's foot or a helper. The speed of rotation is irregular and less significant than with a true pottery wheel. Hence, it is less suited to mass production. As well, the paste requires less fluidity. The forming principle is the same, however. A lump of clay is set over the turntable or thrown onto the rotating wheel and modelled to form the base and walls. The uplifting and thinning of the walls are controlled by hand pressure. This process leaves regular horizontal or upwards-moving marks, in spiral, such as striations and undulations. These are influenced by the wheel's speed and axis of rotation and finger pressure. They are best seen on the inside of the pot, as the outside is frequently subjected to a surface finish of some kind, often meant to obliterate the building marks. The base of wheel-thrown pots is flat and usually detached from the wheel surface with a string. This leaves characteristic cut-off marks on the bottom of the pot, such as a straight line across the bottom surface, dragged inclusions marks or spiral patterns. Particle orientation is also typical for a wheel-thrown pot, as the wheel rotation and hand pressure orient the particles parallel to the wall in a homogeneous way.

In the *wheel-thrown* technique, paste composition is very important not only homogeneity, plasticity and fineness of inclusions, which require a well-prepared paste, but also the drying property, which plays a crucial role. If the pot dries too quickly during the process of manufacture, the forming is more difficult, and cracking and damage may occur during the drying phase. The drying rate is influenced by the clay's capacity for water absorption and evaporation, the temperature, and the rotation speed that prompts evaporation. As seen in earlier chapters, water absorption capacity and evaporation rate differ from one clay to another. Thus, some clays are not as well suited to wheel-throwing as others, even if the potter is careful to keep the pot surface wet.

These descriptions allow us to understand how a forming technique can influence the way particles are oriented, how it is dependent on the com- position of the paste, and what kind of external and internal features are left as clues for the researcher when investigating ceramic production. Different techniques, such as wheel-throwing, coiling or beating, are associated with different sets of features. Some external features are the presence of striations and undulating ridges on wheel-thrown pottery. Internal features are porosity, and the orientation of grains and voids, as explained above in relation to the different techniques. As seen earlier, in their center, coil-built wares show a preferential particle orientation parallel to the walls. Towards the exterior of the coil, inclusions are unevenly oriented. This is enforced by the finishing step, as the polishing process often pushes the grains below the surface. The mode of fracture also follows the coils at the juncture of rows. In contrast, wheel-thrown wares present an internal homogeneous and fluid structure.

It should, however, be noted that particle orientation is best seen by observing elongated inclusions such as micas (see Fig. 7.13). If inclusions have rounded shapes, if they are few and too small, their orientation will be difficult to assess. The material used and paste preparation, including paste homogeneity, may blur this particle criterion. The building features can be further obliterated by the finishing steps of the manufacturing process. The finishing techniques include wiping (cleaning or sweeping the surface while the pot is still wet, with the hand, a cloth, grass), scraping (to take away excess paste and surface irregularities), and polishing. Slips and glazes are surface treatments (see Chapter 5, section 5.3).

All these treatments modify the external aspect of the ware's surface. Furthermore, some of the building features can be misleading, as different techniques may yield similar features. For example, the ridges left by the wheel process may appear on superficial examination like the ridges of the coils. Flattened coils may resemble slabs. Void and particle orientation are even more difficult to assign to a particular technique, especially if the potter is experienced. Besides, different forming techniques can be used on the same pot. For example, the base may be formed on a mould, while the walls are made of coils lifted up and thinned by the-paddle-and anvil technique; or a pot may be started on the wheel and finished by hand; or the main body may be modeled while the shoulder and rim are coil-made; or a tournette may be used instead of a true wheel, or clay slabs instead of coils. Depending on the fragment analyzed, different conclusions can be drawn. For sounder studies of forming techniques, therefore, it is best to examine different fragments of the vessel if the whole pot is not available. X-ray radiography and petrography are successful analysis techniques for this purpose.

7.6
Surface Coatings as Related to the Function of the Ware

Of concern here are the coatings applied to a pot, such as slips, graphite coating and glazes. Coatings have both a functional and a decorative role (Fig. 7.19-7.23). They hide surface marks left by the manufacture process and present a nice visual effect. They are also used in relation to the function of the pot, mainly to enhance impermeability or ease cleaning, in earthenware production (impermeability is also reached with high firing as for stoneware and porcelain). Cooking pots require a rather rough surface for handling (the same as for containers). Hence, they usually lack coating, except smudging. Smudging is the result of carbon deposit on the surface of the ware, produced when organic-rich materials are used during the final stages of firing in a reduced atmosphere. The surface is blackened, carbon is deposited on the surface and penetrates the open pores of the ware, which enhances impermeability. Bowls are often coated with a slip, for visual effect, and to prevent food from sticking or penetrating into the porous paste of the ceramic body. Many pots are slipped only on their upper part, like water jars, or on the surface exposed to the view. Other coatings are used, such as organic coatings to reduce porosity (grease, resin, plant juice, and so on). Unfortunately, they rarely last on archaeological ceramics, exposed to weathering and post-depositional alteration.

Fig. 7.19 Mica-rich slip for coating cooking wares. The wares are pre-heated close to a small wood fire for better adherance of the slip. Sorgun, Turkey (see Druc, 2016 and video documentary "Women potters of Sorkun" on Vimeo https://vimeo.com/35526032).

Slip, paints and glazes require specific technical knowledge, control and firing technologies. They attest to a certain level of development in ceramic production and pyrotechnology. Slip and paints can be fired below 1000°C, whereas most often glazes require higher temperatures and longer firing times, not reached by open fires. The next figures illustrate slips and glazes as seen under the petrographic microscope.

Fig. 7.20 Red slip (iron oxides) covering the volcanic coarse body of a bowl decorated with cross-hatch pattern of polished lines, Kuntur Wasi, Peru, ID12. Thin section photographed with a portable microscope DinoLite and back light stage for work in transmitted light mode. Grains and slip width calculated with the image analysis software of the microscope. Slip is 0.076 mm thick. ppl.

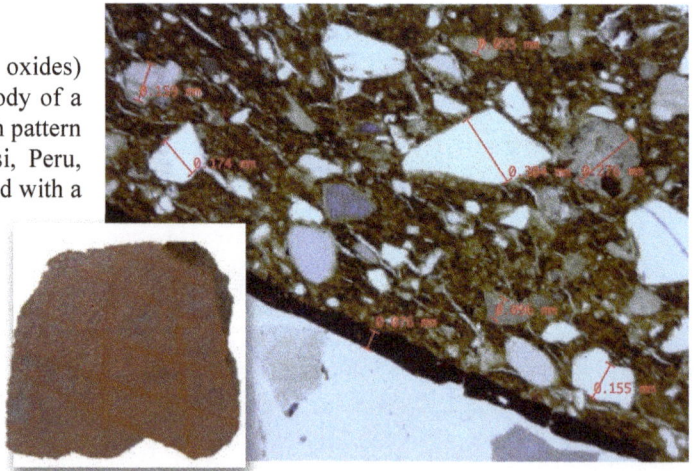

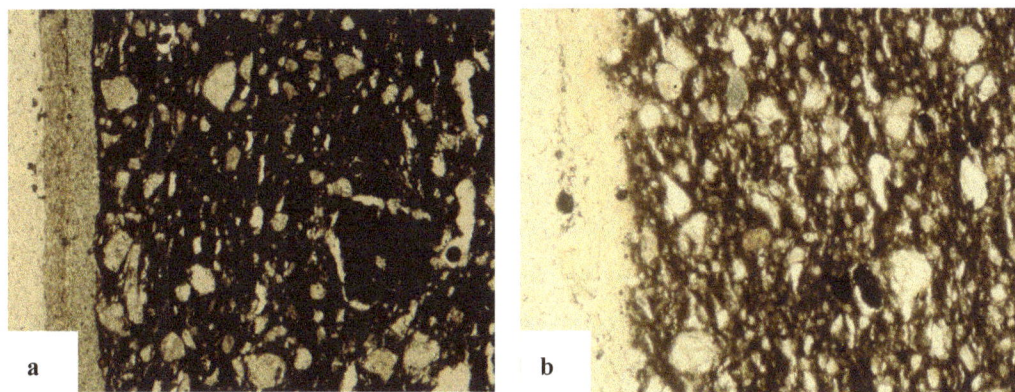

Fig. 7.21 (a) Slip over stoneware jar. 40x, ppl. and **(b)** Glaze over stoneware body, 40x, ppl, Phulang, Vietnam. The glaze in ppl is transparent, while opaque in xpl, as the glassy material is amorphous and does not permit the transmitted light of the microscope to go through (not shown here).

Fig. 7.22 Slipped stacked cooking pots and green-glazed jar in the producing village of Phulang, Vietnam (see Druc 2016).

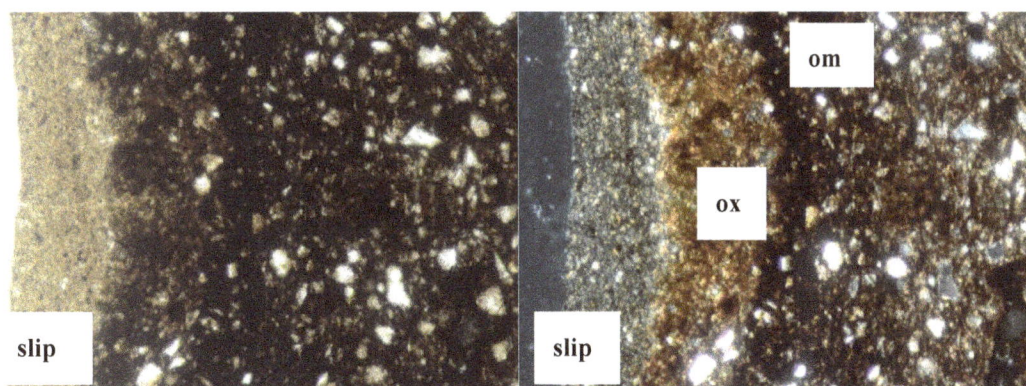

Fig. 7.23 Ceramic coated with a white clay slip to hide the brown clay body. Late prehistoric Longshan period white ware, Liangchengzhen site, Shandong province, China. Slip is 0.125 mm thick, 10x, ppl. The xpl view (right) shows that the surface just under the slip is oxidized (ox), while the body is incompletely oxidized and organic matter (om) is still trapped inside.

7.7
Firing and Furnaces

We saw earlier that paste composition, grain size, material expansion rate and porosity play a role in the firing phase. This section expands further on the firing process itself, on the influence of time, temperature and firing atmosphere, and presents the main furnace types found in traditional ceramic production. All these variables influence the end result, that is, the vessel produced, its color, its durability and even its impermeability. For example, according to the firing atmosphere, temperature and time of exposure, the decomposition of oxides and carbonates in the paste will vary, yielding ceramics with different characteristics. The potter can also modulate the firing to achieve special effects, functional or aesthetic. One example is smudging. Smoke is produced by feeding the fire with special fuel (i.e. wet grass), and prevented from escaping, oxygen access is blocked. In an open fire, this may be done by completely covering the pots with dirt. In a kiln, all air access is closed. The atmosphere is reduced and the carbon compounds (solids) are deposited on the surface of the ware. If this carbon is deposited before cooling, the black residue will stay. It can also be polished, sealing the surface pores and yielding a nice dark impermeable ware.

By controlling the firing atmosphere, the potter can modulate the color of the pots. A reduced atmosphere, even after an initial oxidation phase, will produce a black or dark ware, whereas firing in an oxidation atmosphere will yield red or light wares, providing the clays used are red-firing clays. Remember that oxidation is produced by an excess of oxygen, while reduction is caused by the absence of oxygen, inducing the formation of carbon deposits. The cooling period is also variable, but it should allow time for the pots to adjust to temperature differences, to prevent breakage. The type of pots fired, their wall thickness and porosity influence the rapidity for cooling down, in relation to thermal gradient.

Firing atmosphere and temperature are largely dependent on the type of firing and fuel. Some materials fire rapidly and reach high temperatures, such as straw, while others are low-firing, such as cow-dung. Different types of wood burn differently and can produce hot air more or less laden with organic materials, which give various amounts of oxidation or reduction to ceramics being fired.

In general, the combustion of wood produces three effects:

1) **hot coals** which emit heat in the form of radiation (red color) which can be absorbed by the kiln walls. Temperature increases as the amount of heat is increased. The radiation produces oxidation in ceramics accomplished when heating in a normal atmosphere.

2) **hot air** that moves upward and outward of the kiln. This air is charged with organic materials which are produced by the burning wood. They produce a chemically reducing atmosphere which varies depending upon their composition and water content of the burning wood.

3) **charcoal** left when the wood has been ignited and the combustible elements have been released into the air and the carbon residue remains in the form of charcoal which radiates heat.

There are three main firing types: open fires, semi open structures (pit kilns, open kilns), and closed structures (closed kilns). Many variations of these three basic types exist, adapted to each ceramic production tradition, culture and economic situations (see also Chap. 6, Sect. 6.2).

These different firing and furnace types will be now briefly presented, as their performances and limitations influence the production and quality of the products fired.

7.7.1
Open Fires

Open fires or bonfires have no physical or permanent structures (Fig. 7.24). The surface may have been slightly excavated to produce a shallow pit or simply cleared from unwanted stones, branches and other material. The pots rest on a fuel bed, are intermixed with fuel, or are covered with the fueling material (tree branches, leaves, straw, animal dung, and other vegetal or organic materials). The heap formed by the pots and the fuel material is then put on fire. The maximum temperature reached is mainly a function of the type of fuel used and winds exposure. These types of fire can reach temperatures of 800 °C but rarely exceed 1000 °C. The time of firing depends on the ceramic material, and the fire temperature. We know of potters firing their ware close to 8 h, while others have it finished in 2 h.

There is not much control over firing atmosphere. Potters may cover the structure (pots and fuel) with dirt or large broken pottery sherds, to favour reducing conditions. However, access to oxygen is difficult to restrict in this kind of open fire and oxidation can happen when fuel has burned out and during the cooling period. The way pots are disposed, close to the flames, upside-down or not, or in contact with each other, influences air circulation and firing effects or defects, as seen in Figure 7.25a. In open fires, the ware loss may be high. Some may break due to thermal stress, others will have smoke or fire marks because of direct contact with the flames when fired in open fires or pit kilns. Many traditional and prehistoric ceramic production were fired that way.

Fig. 7.24 Open fire in the Andes, traditional production. The pots are stacked above a wood frame, and covered with straw, while broken pots hold the structure. See Druc 2012, Ceramic production in Tarica, Peru, https://vimeo.com/42790326 or Druc 2006, Andean Potters of Conchucos, Peru, https://vimeo.com/35529198

Fig. 7.25 Effects of firing on the exterior of a ceramic. Modern cooking pots produced in the Andes. The *dark spots* on the cooking pot (**a**) are smoke marks produced by contact with the flame during the firing cycle. The white paint used as decoration is kaolin. To the right (**b**), the dark thick patches are carbon deposit, resulting from numerous and long cooking cycles.

7.7.2
Pit-Kilns, Semiclosed Structures, Open Kilns

Pit-kilns, semi-closed structures and open kilns all refer to types of furnaces incompletely closed, sometimes with walls, but with no roof. Various structures are known. They can be semicircular, circular, oval, pear-shaped. The lower part is built or excavated. It may or may not have a combustion chamber. The kiln is often closed, once loaded, with old broken ceramic pieces, or metal sheets. It can be covered with branches, sawdust or dung. Higher temperatures and better atmosphere control can be obtained than in open fires. Open kilns can produce stonewares and glazed vessels, but not porcelain, for which higher and better control of the temperature and firing atmosphere are needed.

Bonfire and pit kilns are dominated by the contact of hot air charged with organic material, that can be of variable carbon content depending upon the type of wood used and its dryness. The hot air and radiation transfer of heat from the wood are dominated by organically charged air that has a chemical reducing capacity. When the wood has burned (production of organic molecules in the air) heating continues by radiation from the wood coals which transfer heat by radiation. Thus late stages of pit firing can be of an oxidative nature.

Open kilns are permanent structures without a roof, as illustrated in the example below (Fig. 7.26) of a traditional Mexican potter producing utilitarian glazed ware. The production of glazed ware usually requires two firing cycles. The first firing produces the 'biscuit', on which the glaze material is applied. The pot is bathed in a tin of lead oxide mixed with water. Once dry, the pot is fired a second time. The kiln used in this case is a circular semi-open structure, made out of bricks, with walls about 1 m high.

Fig. 7.26 Modern kiln in San Marcos Acteopan, Puebla, Mexico. The fuel is loaded in a chamber below the sole, which leaves the hot air go through and reach the pots. The pots are covered with broken vessels and aluminum sheets. Here the kiln is still loaded with pots cooling down. Atmosphere is oxidizing.

The firing box is just below the chamber where the pots are piled up. The chambers are separated by three spaced vaults. A small aperture at the bottom allows fueling and air alimentation. One hundred and eighty medium-sized round pots, sun-dried, are disposed upside down. This allows a better weight repartition and protects open-mouth pots from receiving too much heat and collapsing. Once loaded, the top of the structure is loosely covered, with broken sherds. The furnace atmosphere is thus more or less oxidizing, and the pots produced are of light red color. This first cycle aims at producing regular fired pots, to harden their surface before applying the coating to be glazed. The first firing period takes about 2 h, and pots are allowed to cool down slowly. Firing temperature is below 1000 °C.

The potters use any kind of wood, readily at disposal, except for guava wood which is more expensive and reserved for the second firing. This wood burns more slowly, with less flame. It allows sintering and adherence of the glaze to the ceramic body during the second firing. The second firing takes 2 1/2 to 3 hours. Pots are then promptly pulled away from the fire, with a long pole to fish them out. This is done to prevent the vessels from sticking to each other, before the glaze has hardened. The result of this quick withdrawal, in the microscopic point of view of the ceramic analyst, is a clear border between the glaze and the body.

7.7.3
Closed Kilns

Kilns are closed structures with a firing chamber (or firebox) in front, below or beside the chamber where the pots are placed (Fig. 7.27). In sophisticated kilns, the flames and heat are not in direct contact with the pots. The hot air is diffused through vents into the chamber. The air circulates, escaping through a chimney. Temperatures can reach 1300°C. Atmosphere is controlled, as well as heating rate. The heating and cooling periods are much longer than in open fires and open kilns, and the whole firing process can take days.

In an updraft kiln, the combustible is below the ceramic materials which are positioned on a permeable flooring (sole). Hot air moves upward heating the ceramics and escapes from the top of the kiln from a restricted number of holes in the roof. A kiln creates to a large extent a chemically reducing atmosphere and is fuel efficient. The heating and ceramic transformation is largely accomplished by the transfer of heat by radiation processes and hot gases from the burning wood. In a closed kiln, the air from the burning wood is laden with organic material and it is chemically reducing (Fig. 7.28). The effect of chemical reduction is lessened if openings allow air to get in. The result is one of variable oxidation-reduction effects, in part depending upon the fuel and its release of carbonaceous material into the air. Re-oxidation also may happen during cooling if oxygen is made available.

Kilns allow for mass production. Because of the higher temperatures reached in closed kilns, and good control of the firing atmosphere, better quality ware such as stoneware and porcelain can be produced. Kilns hallmark advanced technology in ceramic making but require an economy allowing their construction and maintenance. They are often shared by several workshops. The archaeological example of the kilns in Aventicum, Switzerland, illustrates the advances of firing technology used during Roman times. Two types of kilns were found: drying ovens used for drying the ceramics up to temperatures of 950-1050 °C, and kilns that reached temperatures up to 1050-1200 °C to fire the vessels. At that time and cultural/geographical nexus, Ca-rich clays were used for ceramic production. Ca-poor clays were prefered as construction binder for the kilns as they are more refractory (see Eramo and Maggetti 2013).

Fig. 7.27 Kiln, Ban Muang Kung, Thailand, firing chamber in front, and fired jars on side (oxidizing atmosphere).

Fig. 7.28 Reduction kiln, modern production, Shandong province, China. The kiln is in the process of being unloaded.

7.7.4
Paste Type and Surface Color Related to the Type of Firing

The firing process must allow the transformation of the original clay structure, up to the point of losing its plasticity. An unfired clay body, or incompletely fired one, could return to its plastic state. Upon firing, the paste changes into a crystalline phase or, when fired at very high temperatures, into a vitreous phase. Nearly all clay types are good for low firing, that is below about 800 °C, but they can be fired up to 1000 °C. These temperatures are rarely exceeded in open fires. For most prehistoric and traditional ceramic production of earthenware, in open fires, clay refractoriness was or is not a concern.

However, higher temperatures may cause the body to bloat, fuse, or deform unless the clay used can withstand high temperatures. This problem arises with the use of semi-closed furnaces and kilns, where higher temperatures are achieved. High temperatures allow the sintering process, when clay starts fusing and a vitreous phase may form. This is a characteristic of stoneware and porcelain. For example, stoneware production requires temperatures between 1150 and 1300 °C. Glazing can be done at lower temperatures, but often these are above 1000 °C. Thus, the production of stoneware and porcelain requires the use of high-temperature furnaces and a material withstanding these high temperatures. Also, too rapid heating or cooling may cause serious defects in the pot, unless the material used can withstand the rapid temperature changes. Thus, both heating and cooling periods are very important. Often, (at least in 'modern' ceramic production) a first slow firing is done to obtain a biscuit ware to remove all chemically bonded water in the clay (above 350 °C), up to dehydration (500°C) and up to circa 900 °C where the product is strong, sintering might have started, but the ware is still porous. Once cooled, glazes can be applied, which then requires a second firing to adhere to (and melt onto) the ceramic surface. Temperatures at which you bisque fire and fire the glaze depends on the temperature at which the clay and glaze will mature (be dense, not porous, strong). The heating rate of the bisque fire is also very important, at least at the beginning.

The way in which a pot is treated before firing, the raw material used, paste composition, inclusion size, wall thickness, porosity and material compaction will influence the rate at which a ware can be heated, fired and cooled. A dense paste, where the particles are compacted with few voids and pores to allow hot air and water to circulate and escape will require a slow heating phase. Size and orientation of non-plastics and pores will also ease or hinder air circulation through the body. The same is observed for pots with a slip closing the outside pores. If this is not respected, material expansion is too rapid, the air and water cannot escape quickly enough. These stresses cause cracking, leading sometimes to destruction of the pot. Wall thickness influences this process. The stress is caused by the resistance to heat propagation while crossing from wall to wall and temperature differences between the inside and outside of the ceramic body (thermal gradient). The thinner the walls, the less stress goes through the body. As seen before (Sect. 7.3.3.1), uneven wall thickness may also cause breakage, as the thermal stress will concentrate on weak points and not diffuse uniformly. This is also related to the form of the pot, where thermal diffusion is not the same for round or angular wares. For example, thermal stress will be higher at juncture points (corners, carinated shoulders). Round wares, for this purpose, are well resistant to temperature changes during heating and cooling, and good pots for open firing. It is not surprising to find this form preferentially among cooking pots.

Finaly, a clay may change color between its raw and fired state, and the surface color can be modulated according to the firing athmosphere: red (reddish, pink, clear) when oxidizing or black when reducing (see Chapt. 6, section 6.5.1). Final surface color can also be modified during the last part of the firing or just after, if you smudge the vessel (e.g. putting it in hot ashes). Figure 7.29 shows two cases using the same clay.

Fig. 7.29 Ban Muang Kung, Thailand (**a**) clay; (**b**) oxidized (red) and reduced (black) small vessels. Reduction is achieved by burying the (red) pots in hot ashes, then polishing the surface (with gloves to handle the still hot pots).

The preceding observations show that kiln firing, because of the higher temperature reached, will require, the use of a clay resistant to high heat, and a slow heating and cooling cycle. A well-homogenized paste, with small inclusions, a fairly porous body, thin walls and rounded shape, ensures the production of well-fired pots.

7.8
Summary

This chapter introduces the potter's work in relation to paste's properties and physical reactions while making a pot. To obtain the desired paste to reach specific functional or decorative requirements, the potter modulates the raw materials, works the paste and monitors the whole production process. He or she adapts the material to his or her needs. The potter's expertise has important consequences in the success of the product. Bad kneading, wrong temper proportions, inadequate forming technique or firing may cause serious defects in the ware. The potter's actions, as well as the materials used enhance the properties for which a pot is often acquired: durability, impermeability, thermal resistance or aesthetics. The way in which non-plastics are oriented, their angularity, granulometry, the drying and expansion rates of the raw materials, the form of the pot and the firing process all play a role. This chapter aims at understanding this role, which can often be observed under the petrographic microscope or measured by simple materials sciences techniques.

Bibliography: Ceramic Production

Arnold, DE (2005) Linking society with the compositional analyses of pottery: A model from comparative ethnography. In Livingstone Smith, A, Bosquet, D, and Martineau, R (eds). *Pottery Manufacturing Processes: Reconstitution and Interpretation*. BAR International Series 1349. Archaeopress, Oxford, pp 15-21

Bloomfield, L (2019) *Science for potters*. The American Ceramic Society, Ohio, pp 148

Cantin, N, and Mayor, A (2018) Ethno-archaeometry in eastern Senegal: The connections between raw materials and finished ceramic products. *J of Arch Sci Reports* 21: 1181-1190

Dickinson, WR, Shutler, R Jr, Shortland, R, Burley, DV, and Dye, TS (1996) Sand tempers in indigenous Lapita and Lapitoid Polynesian plainware and imported protohistoric Fijian pottery of Ha'apai (Tonga) and the question of Lapita tradeware. *Arch. Oceania* 31: 87-98

Druc, I (2016) *Traditional Potters. From the Andes to Vietnam*. Deep University Press, WI, pp. 137

Druc I, Bertolino, S, Valley, A, Inokuchi, K, Rumiche, F, and Fournelle, J (2019) The Rojo Grafitado case: Production of an early fine-ware style in the Andes. *Boletín de Arqueología, PUCP*, Advances in ceramic and pigment analysis in archaeology, Part 1, 26: 49-64.

Eramo, G, and Maggetti, M (2013) Pottery kiln and drying oven from Aventicum (2nd c. AD, Ct. Vaud, Switzerland): Raw materials and temperature distribution, *Appl Clay Sc* 82: 16-23

Espinosa, A (2020). Filiations culturelles et contacts entre les groupes sociaux de la côte nord du Pérou à la Période Intermédiaire Ancienne (200 av.–600 apr. J.-C.): Étude des traditions techniques de la production céramique Viru-Gallinazo. PhD Thesis, Université de Paris I, Fr.

Gosselain, O (1992) Technology and style: Potters and pottery among Bafia of Cameroon. *Man* 27(3): 559-586.

Gosselain, O (2000) Materializing identities: An African perspective. *J of Archaeological Method and Theory* 7(3):187-217.

Livingstone Smith, A (2000) Processing clay for pottery in Northern Cameroon: Social and technical requirements. *Archaeometry* 42:(21-42).

Livingstone Smith, A (2007) *Chaîne opératoire de la poterie. Références ethnographiques, analyses et reconstitution.* Tervuren: Musée Royal de l'Afrique centrale.

Lozada Mendieta, N (2019) Tecnologías cerámicas y redes de intercambio precoloniales en el Medio Orinoco: Aplicación de análisis petrográficos y de fluorescencia de rayos x para el estudio de sitios arqueológicos multicomponentes en los rápidos de átures (Venezuela). Congreso Latino Americano de Arqueometria, Bogota, Colombia. June 7, 2019.

Noble, JV (1988) *The techniques of painted Attic pottery.* Thames and Hudson, London, pp 216

Orton, C, Tyers, P, and Vince, A (1993) *Pottery in archaeology.* Cambridge University Press, Cambridge, pp 269

Rice, PM (2015) *Pottery analysis. A source book*, second edition. University of Chicago Press, Chicago, pp 592

Roux, V (2011) Anthropological interpretation of ceramic assemblages: foundations and implementations of technological analysis. In: Scarcella, S (ed). *Archaeological Ceramics: A Review of Current Research*, BAR International S 2193. Archaeopress, Oxford, pp. 80-88

Roux, V (2017) Smoothing and clay coating: reference collections for interpreting southern Levant Chalcolithic finishing techniques and surface treatments. *The Arkeotek Journal*, n°2. The Arkeotekjournal.org.

Roux, V (2019) *Ceramics and society. A technological approach to archaeological assemblages* (in collaboration with Courty, MA), Springer Nature, Switzerland, pp 329

Rye, OS (1981) *Pottery technology.* Manuals on archaeology 4, Taraxacum, Wash DC, pp 150

Schiffer, MB, Skibo, JM, Boelke, TC, Neupert, MA, and Aronson, M (1994). New perspectives on experimental archaeology: surface treatments and thermal response of the clay cooking pot. *American Antiquity* 59: 197-217

Shepard, AO (1965) *Ceramics for the archaeologist.* Carnegie Institution of Washington Publication 609, pp 414

Sinopoli, CM (1991) *Approaches to archaeological ceramics.* Plenum Press, London, pp 237

von Dassow, S (2009) *Low-firing and burnishing. Ceramics Handbook.* A & C Black, London and The American Ceramic Society, Westerville, OH

Wilmsen, Edwin N., Griffiths, A, Thebe, P, David Killick, D, and Molatlhegi, G (2016). Moijabana rocks and Pilikwe potters: acceleration of clay formation by potters employing simple mechanical means. *Ethnoarchaeology* 8:137-157.

Wu, J, Ma, H, Wood, N, Zhang, M, Qian, W, Wu, J, and Zheng, N (2020) Early development of Jingdezhen ceramic glazes. *Archaeometry* 62(3): 550-562

8 Optical Observation of Ceramics

The study of ceramics from the inside is a very important aspect of ceramic analysis. It is, in fact, the basic subject of this book. What one can call, perhaps incorrectly, ceramic petrography is basically the study of the different constituent elements, clay matrix and non-clay grains in the ceramic body. This study involves the proportions of clays to non-plastic minerals, the size distribution of the non-plastics and their composition (mineralogy), as well as the spatial distribution of the elements, which is called texture. However, surface details such as slips, glazes and paints can also be observed optically at different scales. Three tools are available at present to do this work: computer scanners working in the submillimetric scale, binocular microscopes, where one can see objects to 0.05 mm in diameter, and a petrographic microscope, which allows one to identify objects to 0.005 mm in size.

8.1
Methods: How Can One See a Ceramic Sherd?

The different methods discussed below give information on different scales; some of it is overlapping, but often each tool gives a different set of observations.

8.1.1
Computer Scanner

The most simple and readily available tool for treating surfaces is the computer scanner. Most common instruments can determine sub-millimetric objects (0.01 mm in diameter). All that is necessary is to have a cross-cut of a ceramic object. Normally, diamond or resistant abrasive wheel cutters can be used to obtain such a surface. The surface should be somewhat polished to take away the bumps and irregularities as well as putting the temper grains in visual contrast with the clay matrix of the core of the ceramic. For the observation, one poses the object on the scanning surface, one scans and then uses the magnifying capacity of the machine to enlarge the sector of the table scanned which contains the object. Trial and error can give the operator the limits of the magnifying capacity of the machine. Normally, it is useful when the ceramic has a width of several millimeters (more than 4 mm or so) and the grains are on the order of 0.01 mm diameter. The scanner is extremely useful and can be compared with a binocular microscope with a camera attached.

Figure 8.1 shows the temper grains in a Chinese stoneware where the small grains shown (simplified in a binary image) are on the order of 100 µ or 0.01 mm diameter.

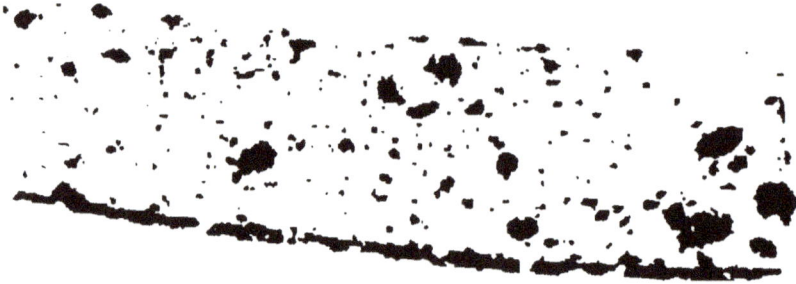

Fig. 8.1. Temper grains in a Chinese stoneware sherd. The image was obtained using a computer scanner by direct application of the sherd to the scanner. The width of the sherd is 5 mm and the smallest grains (enhanced by using a binary black or white imaging process) are on the order of 0.01 mm in diameter. The dark edge of the sherd shows a deeply colored slip surface.

8.1.2
Binocular Microscope

As with a scanner, this instrument sees the surface of an object. This surface can be natural, i.e. that given by the finished ceramic product, or can be induced by a saw cut. Observation by binocular microscope gives the special relations of the different parts of the ceramic in essentially one- dimensional space. It is possible to observe the succession of operations on the surface of the object such as paints, slips or glazes. In cross-section one can tell the larger temper materials from the clay matrix and one can have a very good idea of the proportions of each. This is especially true if a sharp, polished cross-cut has been made on the sample. Here, when a clean aspect of the interior is available, one can obtain a very good idea of the structure of the ceramic object.

8.1.3
Portable Digital USB Microscope

Since about 10 years, a new generation of microscopes is used in the study of archaeological materials. These are portable, the size of a cigar, with a USB connection to display the image on any computer, and now even on your cell phone or tablet. They work in reflective light, that is the object is illuminated from above, while in a petrographic microscope, the light is transmitted from below the sample, which needs to be in the form of a thin section, or extra thin. These portable microscopes work like a binocular works, and basically gives the same information, but image analysis programs can readily be used to process the pictures taken, allowing size measurements, calculate percentages of inclusions (in the case of ceramic pastes), identify basic mineral and textural categories. They are portable (small and light), meaning you can bring them to the field, across borders without much questioning. You can examine hundreds of samples, establish paste categories and then choose the best samples for further analysis, because you <u>need</u> further analysis. This will not allow detailed mineral analysis, which is essential for provenance studies, but some very good information can be obtained regarding technology -size and amount of grains for example are most often related to the way the raw materials have been used or treated. This also allows you to constitute a reference digital library of all (or most) of your corpus of study. Finally, many ceramic pieces (and fragments) cannot be taken out of the archaeological site or the museum where they are displayed and cannot be 'thin sectioned' for petrographic analysis or reduced to powder for chemical study. Portable devices palliate this problem, and most do not damage the piece to analyze (at least not in a visible way). PXRF, micro-Raman, and portable digital microscope accomplish just that.

The surface you examine needs to be clean from surface deposits, so if you want to analyze the paste, you may chip a very small part of the sample with a player for example. This is called a cross-section. However, the surface does not need to be extra flat and better not cut with a saw otherwise you will produce marks and erase technological information. The latest digital microscopes combine different images with different depth of field to give you the best image possible, with very good resolution (and publishable). An analysis protocol and identification Atlas are given in Druc (2015). This kind of microscope is also an excellent tool for examination of the ceramic surface, paint or manufacturing marks. Additional portable accessories can convert this microscope into a small petrographic microscope to bring to the field, for example by adding a backlight stage that will illuminate the sample from below (Figs. 8.2-8.3).

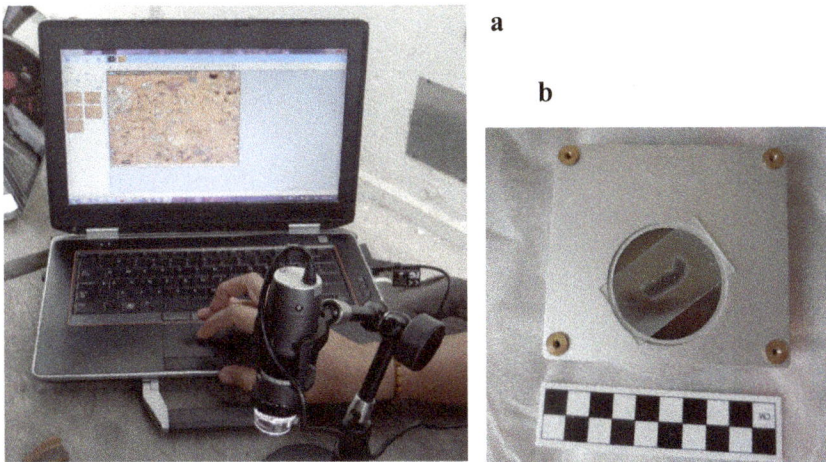

Fig. 8.2 (a) Portable digital microscope of DinoLite trademark, and image of a ceramic paste displayed on a laptop. **(b)** Backlight stage polarizer (DinoLite BL-ZW1) used with the digital microscope, to conduct transmitted light microscopy (petrography) for the analysis of thin sections (one is seen on top of the platform).

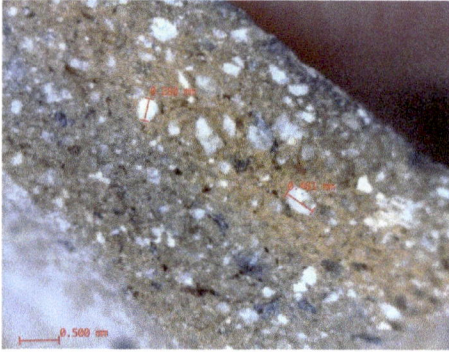

a

b

c

Fig. 8.3. Examples of ceramic paste seen in reflective light (**a-b**) with a portable digital microscope, and in transmitted light in thin section petrography (**c**). CP89 Bowl, Kuntur Wasi, 1st mill B.C. 55x DinoLite, Petrography: 40x xpl.

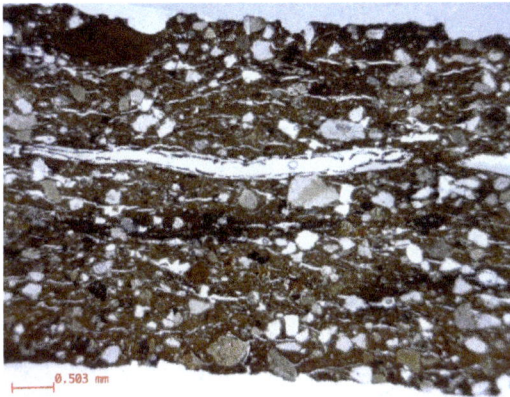

a

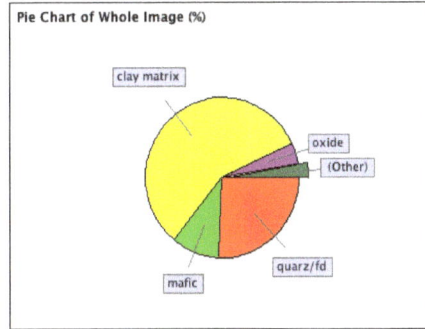

b

c

Fig. 8.4. Examples of image analysis with the program JMicrovision, point counting analysis,. You can use (**a**) a ceramic thin section (portable microscope + transmitted light platform as illustrated in Fig. 8.2b); or (**b**) a ceramic cross section (DinoLite portable microscope, reflective light, 80x). You decide the number of points to be counted, if on a random basis or not, and categories or classes of interest (**c**). Each 'jump' the program moves to a different place and you must identify each grain (where the point is) so it is categorized in one of your chosen classes. (**a**) sample PUCA16 (**b**) sample ID10. Other analyses can be performed, such as granulometric analysis; in that case the program would measure the length of the grains (for example), and can group the results by size range (fine-, medium-, coarse sand, etc).

a   b

c

Fig 8.5 Detail of the pigments on the surface of a bowl 55x (**a**), and cross section (**b**) of the body (volcanic paste) 50x, RA19. A Raman study coupled to pXRF was then conducted on this fragment. Photos taken with a DinoLite portable microscope (and regular digital camera for the corresponding ceramic fragment, **c**). See also, Druc et al. 2020.

8.1.4
Petrographic Microscope and Thin Section Preparation

  The great advantage in using the optical petrographic microscope (Fig. 8.6) is the ability to identify the minerals according to their optical properties. The shape of the grain, cleavages and the color of the minerals are useful identification criteria. Phase identification is very important for a distinction of different sources of materials (Chap. 4), as well as identifying special characteristics which might be determinative of a ceramic production or provenance. The presence of a specific mineral or group of minerals might well be used to distinguish a group of pots from another even though the origin of these minerals cannot be given with precision.

  In some cases, the microscope can be coupled to a video camera linked to a computer. The result is the same as when using a scanner, except that a thin section is used and the magnification is higher. The image captured by the computer can be treated and analyzed with image analysis programs. Such a procedure is illustrated further on in Chapter 9.

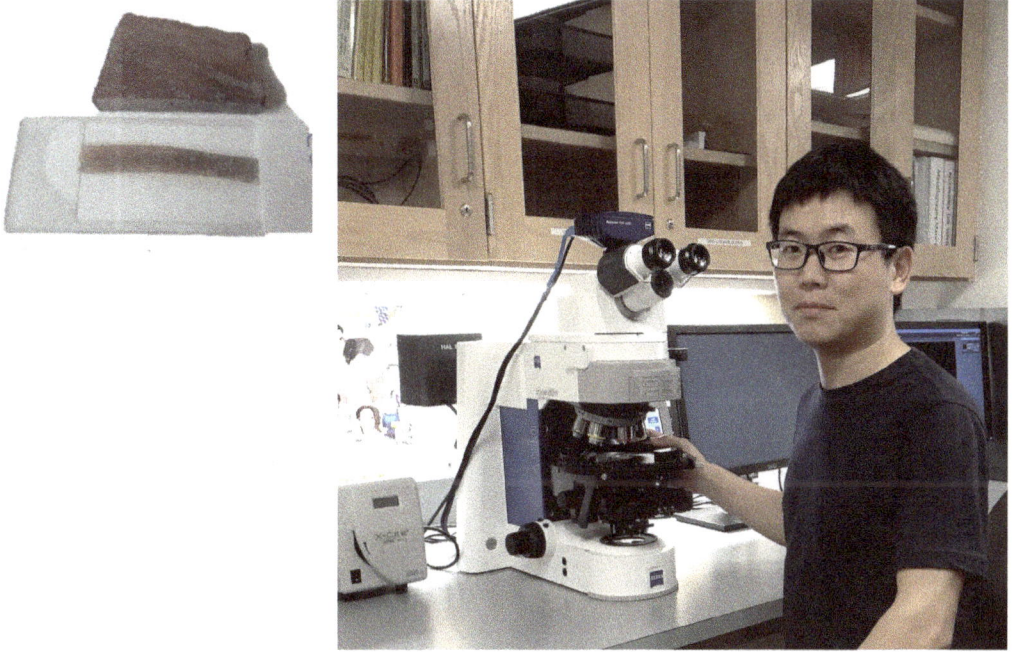

Fig. 8.6. Petrographic microscope with a camera inserted into the top ocular of a trinocular microscope and linked to a computer screen. A small digital ocular can be inserted into one of the oculars if only working with a binocular. Above left: ceramic fragment and thin section. Dr Qingyu Lu at the microscope, Shandong University, Ceramics Laboratory, China.

In order to conduct petrographic analysis, one needs to prepare a thin section, a slab of ceramic 0.03 mm (30 microns) thick glued to a glass slide. This is made out of a thin slice taken from the ceramic fragment or piece to analyze. The cut usually includes the core, interior and exterior sides of the vessel, to have a good view of not only the mineralogy but of the texture and some aspects of ceramic technology. Preparation of the sample is usually left to a specialized technician, as it is a rather difficult process to be done well and requires particular machines (e.g. Fig. 8.7). However, it is possible to equip oneself (or a group) with a small saw or razor blade, hot plate, lap grinder, polishing wheel to make one's own petrographic thin sections of ceramic objects. It takes a large amount of dedication, but the rewards are great in the long run and it costs less once you reimburse the cost of the equipment. Note that commercial labs to prepare thin sections can ask between $17 (for precut, unstained, uncovered, samples) to $60 or more if you want a polished section that you could use for SEM-EDX analysis for example. Yuri Goren (2014) published a good article on how to set up a thin section laboratory in the field. There are preparation directives and advice online, mostly by and for geologists, but good for archaeologists too.

For ceramic materials, unlike rocks, it is usually necessary to consolidate the pre-cut sample before cutting it furthrer and polishing it in order to secure the grains in the paste. If not, they might be torn away during the preparation process. Epoxy resin or a similar polymer can be used for this matter and to glue the ceramic slab to the glass slide. A transmitted light microscope (aka petrographic microscope) is also used to verify the thickness of the thin section. Correct thickness is reached when quartz grains, the most common non-clay minerals in ceramics, are clear and white in cross-polarized light. Yellow quartz grains indicate that the thin section is still too thick.

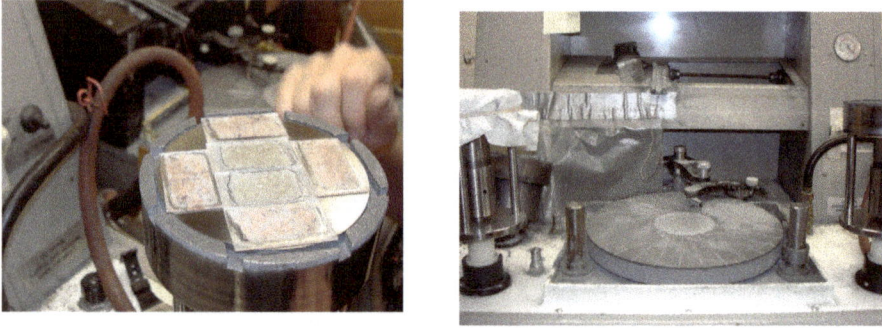

Fig. 8.7. Polishing wheels for thin section preparation.

8.1.4.1
To Count or Not to Count: Granulometric and Modal Analyses

One of the advantages of petrographic analysis is to be able to identify the components of the paste (non-plastic grains, clay vs. inclusion domains) and paste texture (granulometry, grain angularity and orientation within the clay matrix, voids, paste color, etc). These components are linked to the type of raw material(s) used (and are also useful for provenance studies), but also to the processes they have been subjected to by the potter when preparing the paste, which is indicative of ceramic technology and production. However, a thin section or any other sample taken for analysis whatever method is used, is but a very small portion of the whole vessel. We may ask ourselves how representative of the whole is our sample. In some instances, especially when the paste is coarse, this small window (a 'rectangle' of 1 x 3 cm or so) into the composition of a ceramic may not show all the diversity of sizes and types of minerals and rock fragments present in the paste. Some researchers may submit more than one thin section (or powder for other analysis) taken at different places of the vessel (the neck, body, bottom) to see if differences exist, if time, budget and sampling availability allow it.

Once we determine that our sample is fairly representative of the whole (prior examination with a regular optical microscope on cross-sections can help), then we need to evaluate how frequent are the paste components we see: percentages of grain sizes, of types of minerals and rock fragments, of pores, bioclasts, clay matrix vs. inclusions, etc. This can be done with the aid of visual tables like those proposed by Matthew et al. 1991 or by counting a certain number of grains (or other inclusions). The number of grains counted can differ according to the composition of your sample, but generally, to allow for more accurate percentages, a minimum of 300 to 500 counts should be counted. Several counting methods are known, borrowed from geology: area count, grid count, ribbon count, or point count. In these, you count and classify all the inclusions, grains, or features you encounter be it along a transect, within a grid or at specific intervals (say every 0.3 mm, according to the size of the inclusions in the paste). For the grid count, you need to superpose a thin grid on your thin section or determine a specific area you will count. Prudence Rice (2015) gives a good summary of these methods. See also Stoltman (1991) for point counting methodology and example. For point counting, a mechanical stage (Fig. 8.8) or an automatic stepping stage (with a keyboard) are helpful to perform quantitative and modal analysis, equipped with stepwise increments to advance on your thin section along a transect and repertory the size and type of grains encountered on your way.

These methods (visual estimates or counting) help you determine granulometry and modes of your inclusions. Granulometric analysis can be performed also using a digital microscope or computer scanning methods coupled to image analysis softwares as shown in this chapter. A modal analysis, however, is really best performed with petrography and will give you excellent results for provenance analysis. Modal analysis is one where you repertory both the size (granulometry) and the type of inclusions you encounter, the type of minerals and rock fragments, size and type of pores, etc. It gives you a very detailed account of paste composition. Automated counting stages are available in many geological departments and laboratories allowing faster analysis (see Quinn 2013: 109-110 for illustration), while mechanical stages (Fig. 8.8) can be bought online for less than US$50.

A new automated method is now available, called QEMSCAN which combines petrography and SEM-EDX (see Chapter 10.4.2).

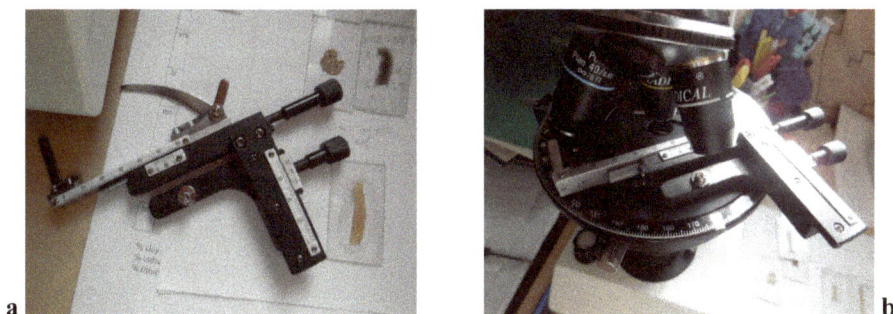

a   b

Fig. 8.8. (a) Mechanical stage used for point counting analysis, **(b)** mounted on the microscope stage. See a demonstration in Druc 2009, Ceramic petrography, an interview with Jim Stoltman, https://vimeo.com/35922518. Stolman popularized the point counting technique for ceramic analysis (Stoltman 1989, 1991).

In terms of the type of analysis one can perform to evaluate the frequency or percentages of the minerals and rock fragments present in a thin section (which later leads to provenance studies), a few methods exist. The Gazzi-Dickinson method of point counting, is known for William R. Dickinson who devised it, originally used for calculating the percentage of minerals composing sedimentary rocks such as sandstones. This serves for provenance studies and evaluating the deposition or transportation history of the sediment under study. In this method sand-size grains are counted as individual grains even if they are part of a rock fragment. In the Folk method, the sand-size crystals within a rock fragment are considered part of that fragment and counted as a lithic. In modal analysis, you would have to identify that lithic to classify it correctly, be it just volcanic or igneous intrusive, or more precise if you specify which volcanic or intrusive that grain is part of.

Then one can go on with statistical analysis, comparing groups of samples based on point counting data, and/or build ternary diagram showing sample distribution and groups such as in Figure 8.9.

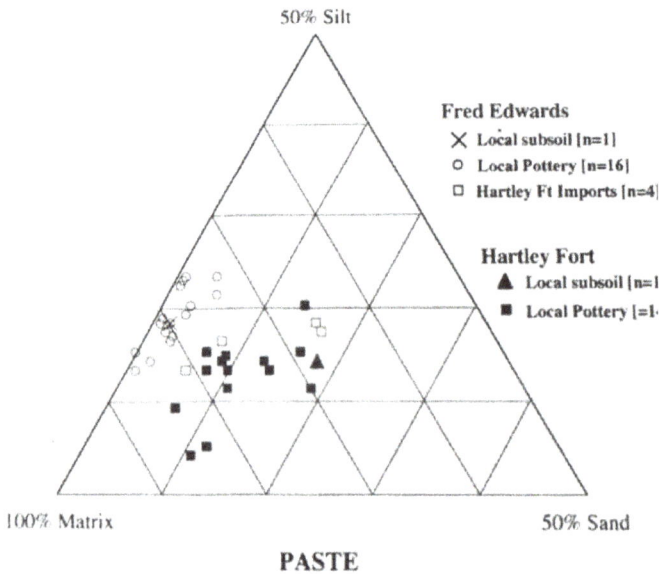

Fig. 8.9 Ternary diagram, to show the distribution of the samples according to their composition (or other characteristics). Taken with permission from Jim Stoltman's study of Woodland vessels, from Mississippian-contact sites, dating to c. AD 900-1150, central USA (see also Stoltman 1991).

8.1.5
Reflected Light Microscopy

Much of the ceramic petrographic studies with optical microscopy involve looking at thin sections in transmitted light with a petrographic microscope. Most of the photomicrographs shown in this book were taken that way. It allows the identification of the rock fragments and minerals in the paste. However, not all minerals in a thin section are 'visible' in transmitted light, several metals and oxides stay opaque as the light cannot go through them. The use of reflected light (when the beam of light comes from above the sample, not underneath) reveals another picture and allows identification of those minerals. This type of lighting can be connected to a regular petrographic microscope. In that case, the researcher changes the path of the light, selecting transmitted from beneath the sample, or reflected with the light source coming from above. To conduct reflective light microscopy, the thin section must NOT be covered (no glass, no plastic glued on top of the ceramic sample) and must be well polished. David Killick (University of Arizona) strongly advocates a more global technique, where the sample is examined in transmitted and reflective light, while Rob Ixer coins the term 'Total Petrography' when combining transmitted and reflected light observation with whole rock geochemistry (Ixer 2014). Figure 8.10 gives an example of what can be seen in reflected light microscopy. Goethite, pyrite, hematite, magnetite, etc. are typically found in ceramic thin sections, but without reflective light, they cannot be identified and are only classified as opaque minerals. It is a technique much used for the analysis of ancient metallurgy, but it is also useful for ceramic. For example, see Killick and Miller 2014 for a case study, Marshal et al. 2004 for an Atlas of Ore Minerals or go online for the Virtual Atlas of Ore Minerals: http://atlas-of-ore-minerals.com/ix_t_0.htm.

Mining, stone work, exploitation of ore minerals, iron, pigments, metal working can be activities influencing ceramic production be it because of firing technology, similar mining areas, or reuse of by-products for temper (from marble grinding - Maritan et al, in prep), pigments (hematite and cinnabar - Prieto et al. 2016, 2019) or combustible (charcoal fuel - Goldstein and Shimada, 2007). Ceramic materials and products may also be used for metal working (e.g. crucibles, ceramic blowing pipes for metal production) or as saggars or canisters for firing and glazing steatite beads (Kenoyer and Miller, 2007; Li, 2007). Co-crafting activities imply proximity, associations, or connections between the respective productions.

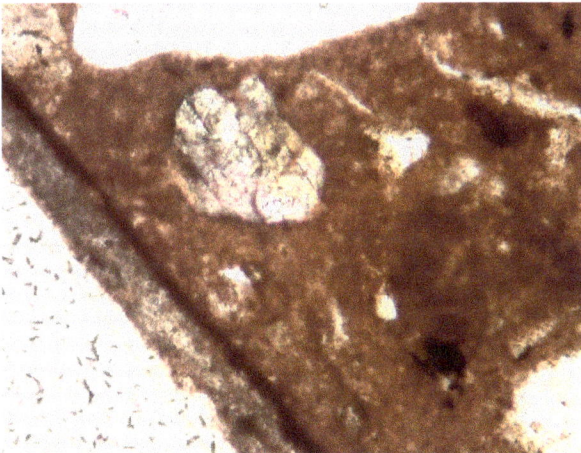

Fig. 8.10 Hematite slip underneath a paint application is visible in this Middle Horizon ceramic fragment from the site of Huarmey, Peru.

8.1.6
Scanning Electron Microscope

The scanning electron microscope can also be used as an optical tool to look at ceramics, pigments, paints or paste, although sophisticated and expensive in use. The energy source is an electron beam and not natural light. The image is one of electrons, those which are given off either by rebound or secondary emission. And it is in black and white (unless transformed by the program). The object analyzed can be a whole fragment (providing it is not too big) or a thin section. In this case, the surface must be high polished to give the best picture. This microscope makes it possible to reach much higher magnification than with a simple microscope, well below 0.001 µm, and to perform chemical analysis. It is the possibility of chemical analysis (if coupled to an Energy Dispersive X-ray Spectrometer (EDS) for example) and surface imaging and visual examination of specific grains or areas that is great with this type of instrument. This is discussed further in Chapter 9, with an example of SEM analysis (section 9.6) and in Chapter 10 (section 10.3.5.1).

8.2
Characteristics Observed: What Can One See in a Ceramic Sherd?

The different parts of a ceramic, exterior and interior, can be observed at different scales, scanner and binocular microscope and then petrographic microscope. The observations of the surface components and those in the interior of the object can be made easily, once it has been "opened" to view via a cross-cut into it. They are as follows.

8.2.1
Slip, Glaze or Paint

One can see the succession of the surface treatments by just looking at the external and internal surface of a ceramic section under a binocular microscope. A Song pottery sherd from south central China can be given as an example. Figure 7.8 gives two views of the sherd, one its surface and the other an idealized view of a clean cut from left to right across the sherd perpendicular to its outer surface. In the second view, one can see the succession of layers of glaze. A dark brown, opaque glaze material was applied to the surface of the ceramic which masks the light grey color of the clay-temper body. This glaze is covered with a transparent glaze material. There are thus essentially two layers of materials which cover much of this ceramic.

However, in looking at the surface of the sherd, in the decorated area, one sees that the dark glaze undercoat is not applied to the entire surface. A design of prune flowers and branches has been left free of opaque coating. One sees equally that the prune flowers have been in some cases entirely covered with transparent glaze, and in other cases only partially. This technique gives different shadings to the flowers which are rendered in different shades of grey-pink depending upon whether the ceramic body was protected from oxidation (under the transparent glaze), and thus grey, or subjected to the furnace atmosphere, becoming pink from oxidation of a small amount of iron in the paste (essentially in the clays). The flowers have been decorated with an iron oxide-rich paint, which defines their stamens. This paint material is oxidized when it has been applied on the surface without a glaze covering giving an iridescent quality. When applied on the pot surface and subsequently glazed over, the paint is, of course, not oxidized and it remains black without an iridescent lustre.

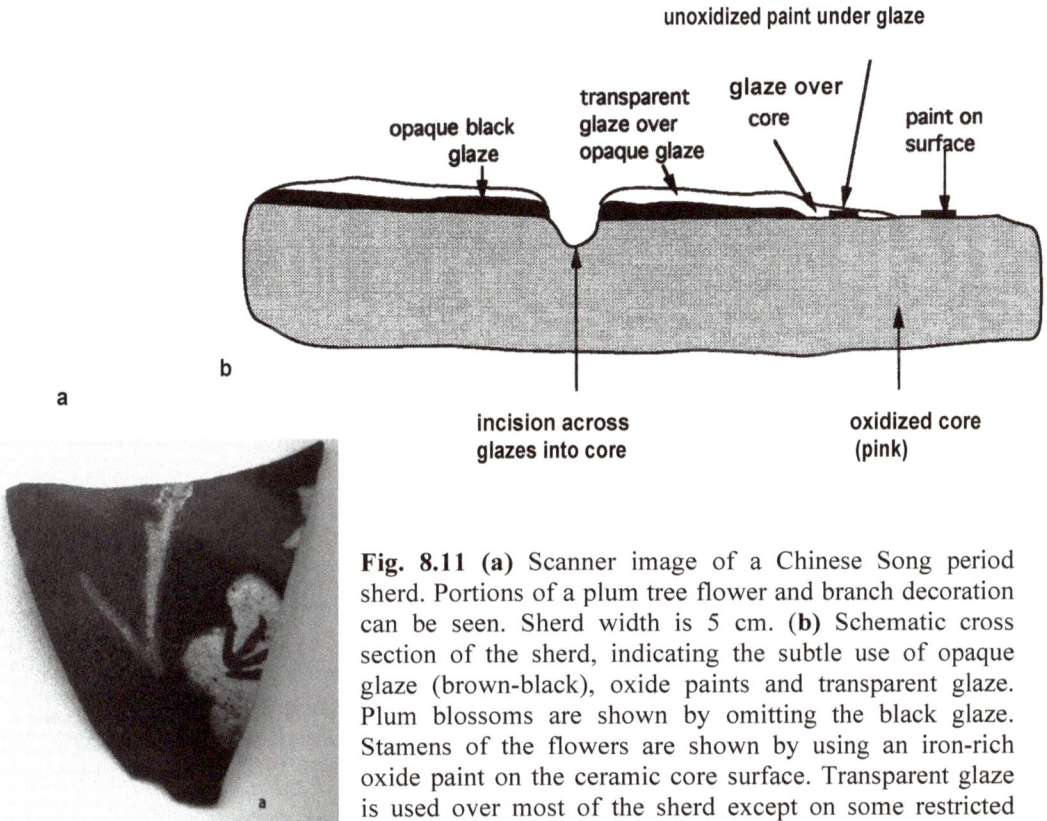

Fig. 8.11 (a) Scanner image of a Chinese Song period sherd. Portions of a plum tree flower and branch decoration can be seen. Sherd width is 5 cm. **(b)** Schematic cross section of the sherd, indicating the subtle use of opaque glaze (brown-black), oxide paints and transparent glaze. Plum blossoms are shown by omitting the black glaze. Stamens of the flowers are shown by using an iron-rich oxide paint on the ceramic core surface. Transparent glaze is used over most of the sherd except on some restricted areas of the flowers. Direct exposure to the oxidizing atmosphere of the furnace gives a slight pink color to portions of the flowers. The stems or branches of the tree are made by scratching through the opaque glaze and into the ceramic core. The transparent glaze has flowed over parts of the scratch, giving an irregular oxidized-unoxidized coloring.

In some areas of the sherd surface, a deep incision has been made into the body of the object, cutting through the transparent and opaque brown glaze layers. These scratches are now partially filled by flow of the transparent glaze which occurred during firing. The incisions give a greater relief to the surface, a more distinct outline to the incision and a subtle nuance of orange to brown on the edge of the incision where the glaze and paint flowed on the edge of the cut. These nuances are not present in the flower decoration.

Thus, one can see and describe the rather complicated decorative techniques of such a ceramic by using a simple binocular microscope and a saw-cut edge of the ceramic object. In general, one can see slip layers and glazes with the binocular microscope in cross-section. Thin paints cannot generally be detected by analytical means in most cases in cross-section if the paint layer is too thin (< 5 μm) and must be observed on the surface to which they were applied. Hence, both interior and surface aspects are important in observing paints, slips and glazes.

8.2.2
Temper Grains and Clays

Temper grains can be observed in thin section or on a cut, polished surface at different scales of observation depending upon the methods employed. As mentioned above, a simple computer-driven scanner can be used to distinguish temper-clay relations of many samples, those where the non-clay, temper grains are above 0.01mm in diameter. Here, it is necessary to have a smoothed, cut surface for the observation and hence a saw adapted to this work. If the samples are big enough, one can use a simple scanner hooked up to a computer. All that is necessary is to position the sherd on the glass of the scanner. However, it is wise to put a transparent sheet of plastic on the glass before the operation in order to protect the glass from scratching. Figure 8.11 shows the use of such a now commonly available system. The magnification printed from the scanner is two times natural size, which is sufficient for an acceptable analysis. The figure shows a cross- section, the surface of a cut Roman amphora handle found in Southern France. The shape is that of a slightly flattened rolled strip of clay paste which gives a slightly oblong shape in section. In the image showing different shades of grey, one can observe spots in the surface, some darker than the paste (clay-rich part) and some lighter than the paste. This wine amphora came from the Roman comagmatic volcanic area of Italy, where grapes were grown on the slopes of volcanic hills. Hence, the majority of the sand grains in the paste are of volcanic origin, and being so, are dark in color. They are mostly pyroxenes and olivines, which contain significant amounts of iron and are dark in color as a result. The white spots are shells, indicating that the source of the sands was somewhere near the sea, perhaps tens of kilometers. Of course, this is compatible with an origin of the ceramic production just south of Rome or north of Naples.

An interesting trick to use with a computer is to select portions of an object by selecting their different intensities of shading. For example, one can intensify the selection to take only the light shell grains (Fig. 8.11 b) or just the dark pyroxene and olivine mineral grains (Fig. 8.11 c). Here, one has direct access to a proportion of the image which has dark temper grains and light ones. One can then make a grain-size analysis, as we will see below (Sect. 8.3.3).

Depending upon the magnification and size of the object, one can do the same work as indicated in Figure 8.9. If a microscope is used, a photograph would have to be made or an image could be captured with a video and computer system attached to the microscope. However, a scan of a photomicrograph works as well as a direct image.

A good discussion about digital image analysis, how to perform it and its accuracy is proposed by Livingood and Cordell (2017). The authors compare point counting to the results of digital and image analysis. One of their conclusion is that the performance of such a technique really depends about the complexity of the sample and the objectives of the study.

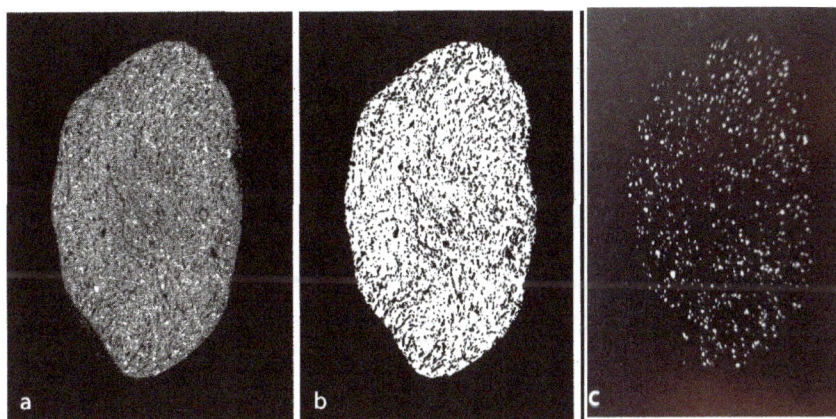

Fig. 8.11 Direct scanner image of Roman "black sand" (see Chap. 8.1) amphora cut across the handle fragment. The image was made directly from a cut surface using a computer scanner system. The image shows 2 cm in greatest dimension. (**a**) Initial image shown in shades of grey. One can see that temper grains are both white and black. (**b**) Enhanced dark (black sand) temper grains and shrinkage cracks are figured. The mineral grains are composed of pyroxenes and a few olivine grains plus one or two yellow garnet grains. c Enhanced white areas show white shell temper grains.

8.2.3
Temper Grains and Size Distribution

Temper grains, either naturally included in the clay resource, or added by the potter to temper the plasticity of the clay resource, can range from almost pebble to silt size (several mm to 0.01mm). The scale of observation necessary to identify these grains will vary, depending upon the object. In general, it is wise to try to study up to the smallest grain size, that just above the clay particle size. In order to do this, it is often necessary to use a petrographic microscope. Here, the thickness of the rock sample is 0.03 mm, which allows one to identify yet smaller grains at the surface of the thin section. Petrographic microscopes and thin-section preparation allow one to see the full range of non-plastic grains. Clay grains have a 0.002 mm diameter, as you remember.

Using the petrographic microscope, one can identify the grains according to their mineralogy, giving an idea of provenance or perhaps particularity of a sample or production type; and, of course, one can count the grains, and thus estimate the total amount of temper (non-plastic) material present in the sample. Further, one can estimate the grain-sizes and the relative amounts of the different-sized grains, i.e. the grain size distribution. This can be done manually or with image analysis programs.

Figure 8.12 is intended to illustrate some of the potentials of this method of analysis. The initial image is one that has been treated from a photomicrograph taken with a petrographic microscope.

Fig. 8.12. (a) Reproduction of a photomicrograph of a thin section of a common ware sherd from Saint Marcel, Central France. A thin section is a two-dimensional representation of a three-dimensional object where the sand grains are cut at less than maximum diameters. The photomicrograph was scanned and treated to enhanced temper grains of quartz and feldspar which are initially light in color but shown in black here. Note the large differences in the sizes of the temper grains. *Bar* = 0.3 mm

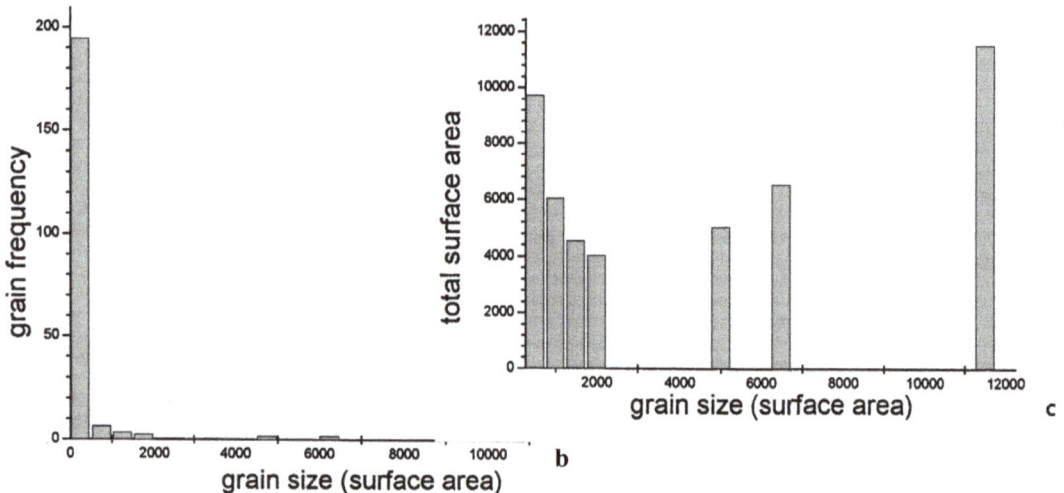

Fig. 8.12. (b) Histogram of grain frequency by relative sizes. Most grains are small, few are large. **(c)** Importance of the different categories of grain sizes is shown by their total surface area in the thin section. Here, each grain-size category (surface area) is multiplied by the number of grains in the category, shown on the Y-axis as total surface area. One can

estimate the relative importance as to the volume of the sherd occupied by large and small grains. The many smaller grains just equal the few larger grains in surface area.

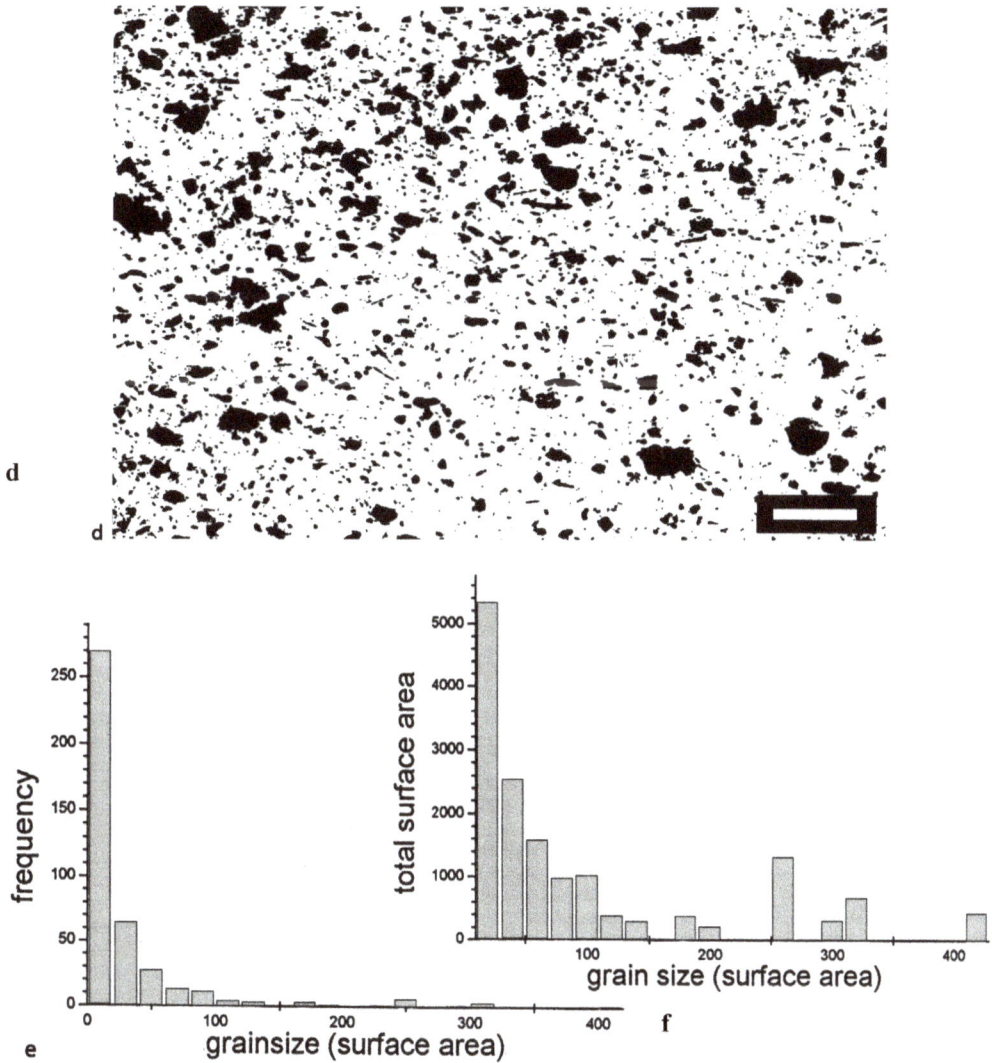

Fig. 8.12. (d) Grains from another thin section from the same site in Central France where the temper grains are smaller and show a more even distribution of grain size and abundance. *Bar* = 0.3 mm. € The frequency diagram showing the number of grains of different sizes is similar to that in the example of a. (f) The total surface occupied by the grains in the different size categories is shown. Here, the abundance of the smaller grains is reflected by a large surface area compared to the larger grains. The contrast between this figure and c is due to the smaller size of the larger grains in the second example.

In Figure 8.12, the temper grains, for the most part originally white, have been shown in black. Under the microscope the different elements are more easily distinguished by their differences in color. Most often, the clay matrix is brown, orange to yellow in color and the temper materials are white (transparent) brown, green, or other easily remarked colors. One rapidly has a general idea and can give a general description of the matrix and temper grains. Such terms as coarse, fine- grained, irregular come to mind. However, they are not very quantitative. These terms should be demonstrated by a numerical analysis. The analysis is available, or can be effected, using simple means. In our case, we use a program which isolates the grains, determines their surface, perimeters, length over width parameters which can, in turn, be classed into a frequency plot. Subsequent treatment (establishing histograms of frequency) can be done using spread sheet methods. We will use only the grain size frequency possibility in this demonstration.

The temper grains, which are light in color, are selected from the initial scanned image (in shades of grey) and given a white code under the scanner transformation process. Non-white material is coded as black. This gives a black-white picture of binary composition (two components). For use in the computer program the white grains are given a black coloring, which is the convention of the program to identify the object to be analyzed. This white-black conversion or inversion is easily done with the image analysis programs. Such a conversion is standard practice in such systems. Thus, the white grains are given a black coloring to be treated by the analysis program.

The last step is to give numerically the relative abundance of grains having a given size or apparent diameter. This step is arbitrary in that the definition of the classes is at the choice of the operator. It is not really important which values are used, but one should try to give the same sizes to the categories of samples which are apt to be compared one to the other. This aids in visual interpretations. If one changes scale on each sample, the different analyses cannot be easily compared.

Figure 8.12b shows the distribution of grain sizes. Here, it is evident that the most frequent grains are in the smaller size categories. In this case, there is a greater abundance of grains of greater sizes. However, in Fig. 8.12a one can see that there are a large number of small grains and a few very large ones which appear to dominate the photograph. One can assume that the smaller grains are associated with the clay source and the larger grains have most likely been added as needed according to the physical properties of the clays and the needs of the potter to make the ceramic.

Whatever the reasons for the grain-size distributions, these distributions can be used as a signature of the pottery-making process of a given ceramic object. This signature can be compared to other objects in the same production or to objects coming from other sites. Conversely, one can use the grain-size distributions to identify elements coming from other sources. In some cases, one can identify the grains as being exotic to the day sources, or as coming from the same source. In any event, a numerical comparison is especially important.

One aspect of the grain-size analysis can be done in a slightly different manner. In each category there are a certain number of grains but, in a way, the total abundance of this material is more important. That is to say, the mass of grains of a given size is more important than their number. It is evident that a larger number of small grains is needed to give a tempering effect than a smaller number of large grains. Thus, if one wishes to indicate the tempering effect of different grain-size categories, one should multiply the number of grains in a category by the size of the category. This gives an idea of the mass of the material in the smaller sizes compared to the large ones. For example, a large grain of one millimeter in diameter gives a non-plastic character that hundreds of 0.01 mm grains would. The effect is not really the same, but the comparison is not uninteresting. This comparison is easily done by multiplication in the spread sheet. Figure 8.12c shows this relation. The large grains take on a very great importance. One can call this analysis the grain abundance. Another example, but with surface area counts is given in Chapter 9.

A second example is shown from the same site in Central France. Here, the sample is different, the temper grains are much less coarse and the size appears more regular (Fig. 8.12d). The same treatment of the scanned and analyzed data for the grains gives different histograms. The frequency-grain size (surface area in the section) is similar to that of the first example above. However, the analysis of total surface area (Fig. 8.12f) for each size category shows a histogram which is similar to that of the frequency diagram (Fig. 8.12e). This is strikingly different from the first example, where a few grains dominated the surface area distribution. By comparing the frequency and surface area diagrams, one can gain a better idea of the amount or mass of temper grains in each size category.

In such analyses of grain frequency and abundance there is one point which should always be kept in mind; this is the peculiarity of the thin section. In fact, a thin section is a slice of three-dimensional space. The very thinness of the thin section of a ceramic surface gives a two-dimensional view of what is, in fact, a three-dimensional problem. The problem is simple. Take a sphere, and cut in a slice, looking at the surface cut. A maximum diameter is a low probability cut. There will be only one, in fact. All other cuts will give a diameter inferior to that of the grain. Hence, most of the grains seen in thin section are only a small slice of the true grain size. Thus, any grain cut at random will be seen as smaller than it really is. Further, a series of grains of the same diameter will give a spread of different diameters. This is indicated in Fig. 8.13. If we consider an unusual ceramic with only one grain size for its temper grains, the effect of cutting a two-dimensional slice into a three-dimensional object will be shown on a frequency diagram as in Fig. 8.13b. The single category is made artificially into several of the same abundance. If the sample has temper grains of two grain-sizes, the effect can be illustrated as in Fig. 8.13c. The apparent frequency distribution is flattened and widened.

The use of a two-dimensional cut into a three-dimensional assemblage of objects is one of a loss of precision. If one is aware of this, it is all right. Using the same method on all samples spreads the error onto all of the objects to the same extent. However, one should be aware that the two- dimensional cut is only an approximation of the three-dimensional reality. Visual measurement diagrams for grain abundance and shape are given in Matthew et al. (1991).

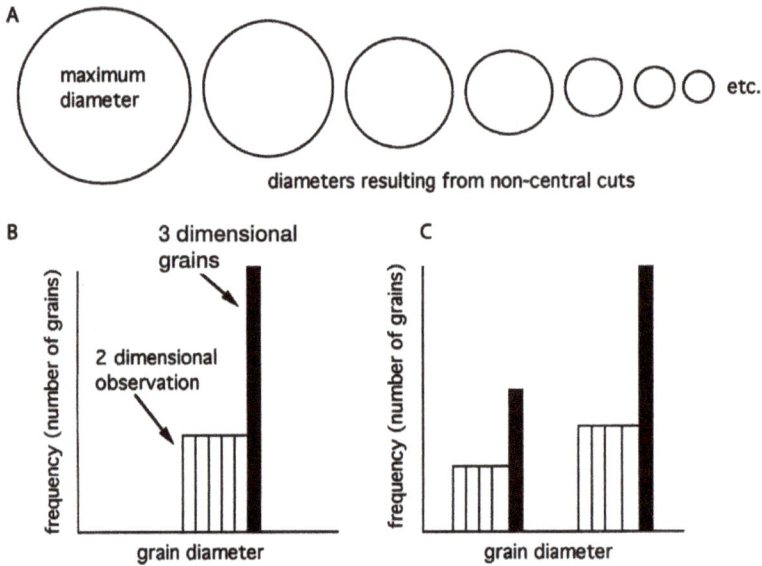

Fig. 8.13. Demonstration of the effect of cutting grains of a given size at different positions in their volume. (**a**) Given a maximum diameter, the range of sizes from cuts in different parts of the grains are less than the true diameter (maximum diameter) of the sphere. (**b**) Distribution of grain sizes in a frequency diagram when a two- dimensional cut is made in a ceramic when a temper material has a single grain size. **c.** Frequency diagram of a ceramic with two temper grain sizes as represented in two-dimensional cuts

8.2.4
Grain Shapes

8.2.4.1
Crystal Shapes

The shape of temper grains can give some immediate information as to the type of mineral grains present. Micas are usually elongated in thin sections, because the shape of the crystal is a sheet structure, similar to a clay mineral (Chap. 3, Sect. 3.2.2). A cross-cut of a sheet gives in most instances a thin, linear shape. In Figure 8.14 the long mica plate can be immediately identified as a biotite, iron-magnesium micas. In the case of their being white or colorless, they would be muscovites (see Chap. 2, Sect. 2.3.2.1.5). Some amphiboles present long, lath shapes. These minerals are usually green in color, and hence are not to be confused with micas. Otherwise, mineral grains tend to be more equant in shape, but weathering can round them up or yield irregular shapes as for feldspars. Quartz can also be irregular, equant or round according to the source. The clear grains shown in Figure 8.14 are quartz and plagioclase feldspars, of more equant shapes. If crystal grains are broken they can show very different shapes.

Fig. 8.14. Crystal shapes as seen in a ceramic thin section (4, xpl): biotite mica (elongated, platty), plagioclase (equant, 2-color grey twinns), and quartz. Here, the quartz crystal (bright white) is coming from a volcanic rock and shows irregular faces.

8.2.4.2
Angularity

Angularity, texture and general aspect of the mineral inclusions (eroded or not; rounded or angular) can also reveal the presence of different components. It yields information on the origin of the material and whether the material was acquired close to its source or not. In principle, the more a grain has been transported, the rounder it is (see Chap. 2). A frequently used scale of relative angularity is shown in Fig. 8.15. The grains are classified as angular, subangular, subrounded, and rounded. More precise measures exist now with quantitative image analysis programs, where the sphericity of inclusions is calculated, on a scale from 0 (maximum angularity) to 1 (maximum sphericity). In looking at grain angularity and alteration, one must remember that the fragmentation planes may differ from one mineral class to the other. There are "hard" rocks or minerals that give angular fragments when crushed (quartzitic rocks, plutons, coarse-grained volcanic rocks) and "soft" rocks giving more rounded grains (tuffs, schists, shales, argillaceous material, siltstones). Also, some rocks and minerals alter more quickly than others when exposed to weathering or chemical alteration, for instance feldspars and mafic minerals (see Chap. 2.3).

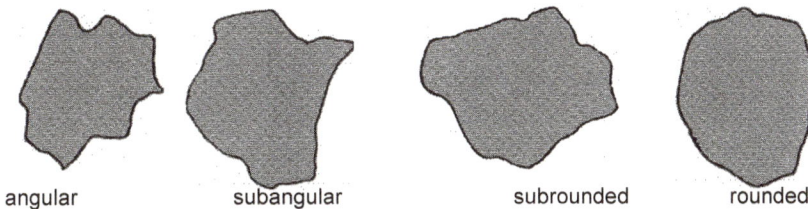

angular   subangular   subrounded   rounded

Fig. 8.15. Angularity chart as a visual method of assessing grain angularity in ceramics.

Some Examples. Figure 8.16a shows grains which are very rounded in shape. These are beach sands, found in Roman amphorae from the Italian peninsula. Their dark colors, in reality greens and browns, and the strong cleavage planes in the grains identify them as pyroxenes with occasional olivine grains. The rounded shape of the beach sand grains indicates water transport which rounded their edges. The grains in Fig. 8.14b are sharply different, with irregular-shaped feldspar and quartz grains found in a Neolithic pottery from north-central France. Here water transport was probably not important in the formation of the temper grain shapes. The temper grains are naturally included in the loess deposits found along river valleys and on adjacent plateaus.

a

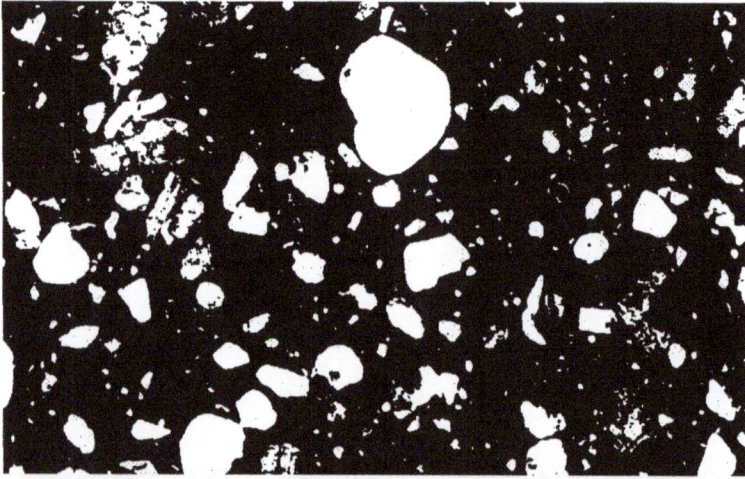

b

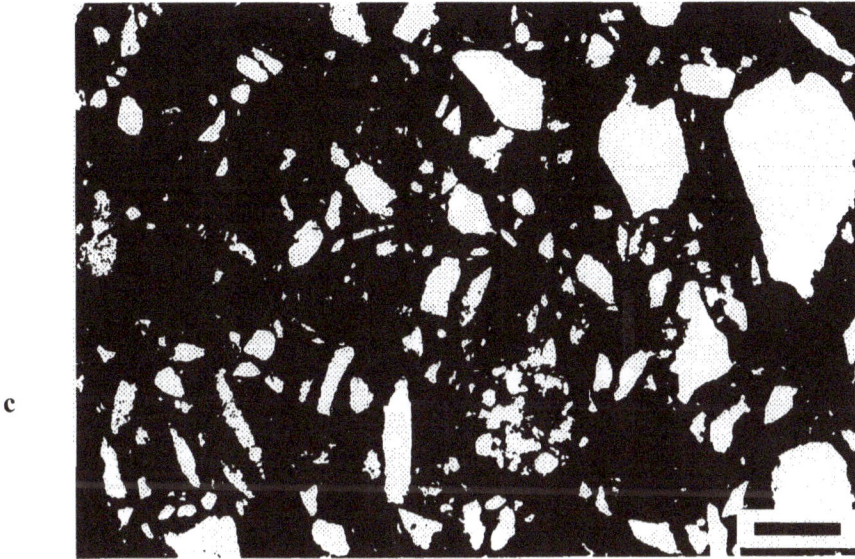

c

Fig. 8.16. Grain shapes in several ceramics. (**a**) Black and white representation of a photomicrograph of amphora temper grains (shown in white). Many of the larger grains are very rounded, indicating that they are found in stream sands. A strong bimodal distribution of sizes indicates the existence of finer temper grains which probably originated in the clay resource. They have a more irregular shape (are more angular) than the rounded larger grains. Bar = 0.3 mm. (**b**) Black and white representation of a photograph of a thin section showing sharp and angular quartz grains (shown in white) found in a prehistoric ceramic sherd found in north-central France. These sand and silt size grains are a natural component of the clay source; they are so abundant that they are a major component of the clay material. The resource is a loess deposit which is the result of wind and stream action along the front of continental glaciers which covered northern Europe. *Bar* = 0.3 mm. (**c**) Black and white representation of a photomicrograph of a thin section of a ceramic from Turkey (16th century) where ground and broken quartz (shown in *white)* has been used as the matrix base. Note the highly irregular shapes and the great variety of sizes. Photo length = 3 mm.

An example of irregular angular grains is seen in Fig. 8.16c where a thin section of a ceramic from the Middle East, 16th century, Iznik Turkey shows a high content of temper grains which are almost all angular. The ceramic is made of quartzic materials and calcium oxide plus some clays. The temper grains are, in fact, crushed quartz grains which show an unusual sharpness. These materials were produced by grinding larger quartz materials. The grains are a majority in the sample. Calcium oxide is added to the highly rigid paste. The fusion of the quartz by calcium oxide during firing is the method used to form the ceramic material. Almost no clay is present and one would expect that the paste for the ceramic material would not be very plastic. However, moulding methods allow the use of a very rigid paste in this process. The firing produces a very hard, stable material which approaches porcelain quality.

8.2.4.3
Size Distribution of Temper Grains

The overall size distribution of temper grains is important also. It can be used to describe a production method, and perhaps the origins of the different components.

In Figure 8.17 we have displayed three different types of grain sizes in sherds coming from the same archaeological site. This is a Gallo-Roman site in central France (Saint Marcel, Creuse) where a large number of common ware pottery samples were found. The clay material is very much the same, as is the mineralogical composition of the tempering grains: quartz, potassium and plagioclase feldspars, biotite and muscovite micas. The clay matrix contains significant and varying proportion of temper grains of small size. However, other larger temper grains are present in different proportions.

In Figure 8.17a the temper grains are small and evenly distributed. This is probably the clay-rich material used as the basis of the paste for ceramic production on the site. Fig. 8.17b shows another sample from the same series of sherds, with a fine-grained matrix of clay and temper grains but a second group of temper grains of rather constant size is present, suggesting the tempering of the initial day-rich paste material with sand grains. Fig. 8.17c shows a sample where three components are present: fine-grained, medium as in Fig. 8.17b and very large grains. Here, it is not clear what the processes were in preparing the paste. The contrast between temper grain sizes and their distribution is very great and can be identified very easily in a qualitative manner. In these three examples, the materials used are similar in mineralogical composition, coming from the same archaeological site. They were most likely produced from the same resources. These size differences are evident in the black and white (binary) representation of the photographs. Here, it would appear that the large grains were added for tempering in sample (b), possibly the clay resource held medium and small grains as in (a) while the lack of medium grains in (c) suggests a segregation process by sorting, probably by water washing, decantation or levigation.

a

Fig. 8.17 (a-c). Black and white reduction of photographs of thin sections of three samples of St. Marcel Roman common ware samples coming from the same site having very similar production techniques with almost identical clay-temper mineral sources. The range of grain sizes is very indicative of use of large grains to temper a given clay source to different degrees. (**a**) A sample with the smallest temper grains found on the site. This is the basic clay source material. (**b**) A sample with a distinct bimodal distribution of grain sizes.

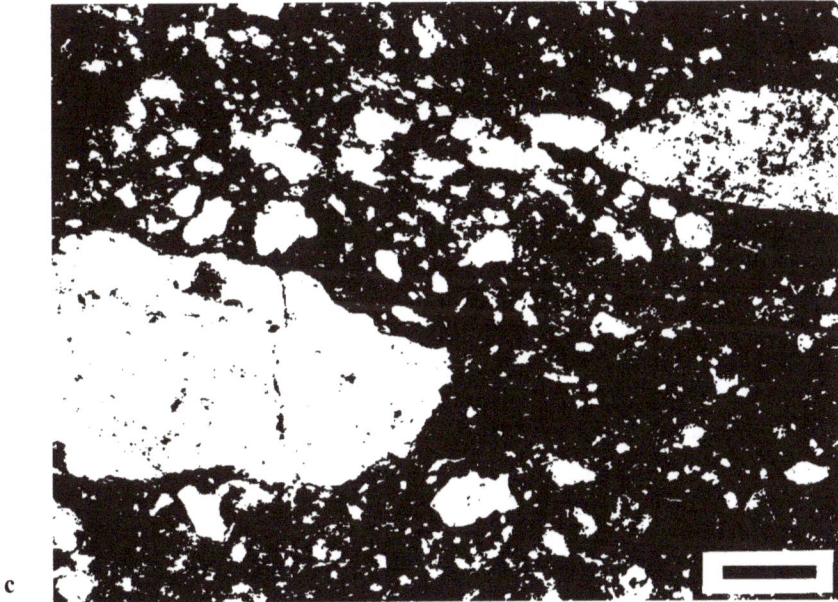

Fig. 8.17 (c) Sample with fine, medium and several very large grains. All the grains have the same mineralogy, quartz and potassic feldspar. *Bars* = 0.3 mm.

8.3
Identification of Different Techniques of Paste Preparation

Some Measures of Temper and Non-Plastics. In the case of the use of crushed rocks, lithic fragments, as clay and temper resources, a model for rock tempering identification can be proposed. Three main conditions must be met to conclude in favour of rock tempering. First, there must be a strong presence of rock fragments of the same kind (same type and same texture). These should be numerically more abundant than the other lithics or be the only lithic present. Second, the granulometry of the rock fragments should be distinct from the granulometry of the other inclusions, particularly in the coarse fractions. Exceptions may occur when the clay is coarse itself and raw materials are crushed and sieved, leveling off granulometry differences. Usually, however, added rock inclusions are larger than natural inclusions in the clay. Third, the angularity of the rock fragments, when they do not come from "soft" stone like tuffs, should be more pronounced than for the other inclusions.

For other mineral temper, such as sand, or clay additions, one should combine careful study of the granulometry, angularity, alteration, and abundance of the inclusions. This can reveal the existence of different sources of raw materials or the geological environment that may have produced the inclusions seen in the paste.

Bimodal grain distribution can also reveal the presence of different raw materials. This has been discussed in Chapter 7, section 7.2.3.

8.4
Paste Texture

The objective of a potter is normally to produce or obtain a homogeneous clay-temper grain mixture material to use as a paste. This necessitates a certain manipulation of the clays and tempers if they are present, as seen in Chapter 6. Usually, the potter or a supplier takes a raw material and breaks it into fine particles in the dry state. This material is then mixed with water to produce a paste. The extent that the clay is crushed, sieved and impurities extracted from it gives a quality of homogeneity necessary for the production of a good ceramic. If the paste product has lumps and grains in it, it cannot be used for fine tableware. If the clays are inhomogeneous, large cavities and pores will be present which give uneven qualities of firing and transformation during the firing process.

The clay and temper grain distribution or arrangement in the paste is what can be called texture. Texture gives an idea of the amount and type of effort which has gone into producing the paste for the ceramic. Figure 8.18a shows a fine-grained ceramic, sigillate ware, which has a homogeneous mix of clays and small temper grains. It is obvious that the preparation of the paste was done with care to make it as homogeneous as possible. By contrast, the Neolithic wares of central France show a high irregularity of grain-size distribution from one portion of the sample to the other. The paste is made from an irregular clay source which was not homogenized after extraction from its source. Lumps of clay are joined to form the paste giving a very irregular texture. Figure 8.18b shows the temper grains in a sample where the clays were certainly not ground, sieved and mixed. The outstanding aspect of the poorly prepared paste is the irregular distribution of the temper grains. Patches of fine grains are found and clusters of larger grains can be seen also.

a

Fig. 8.18 (a) Texture or distribution of grains in a paste. **a** Black and white enhanced photograph of a thin section showing a typical homogeneous sigillate ware temper grain distribution in the clay matrix. The even distribution and grain size gives a regular texture which indicates careful mixing and working of the clay resource. *Bar=0.3mm.*

Fig. 8.18 (b, c next page) (b) Sherd from a Neolithic production area showing a very irregular distribution of the temper grains (shown in *white)* from one portion of the sample to the other. The zone with larger grains has fewer present. This inhomogeneity shows poor mixing of the components used to make the paste. *Bar* = 0.3mm. **(c)** Photograph of a thin section of a Neolithic ware sherd from north-central France showing a nebulous clay texture. This shows a very poorly sized and rather unworked clay resource where there has been almost no attempt to homogenize the clay and temper grains. Most likely all of the material comes from the initial clay source, which was a soil, and no tempering has been done when preparing the clay paste. *Bar* = 0.3 mm.

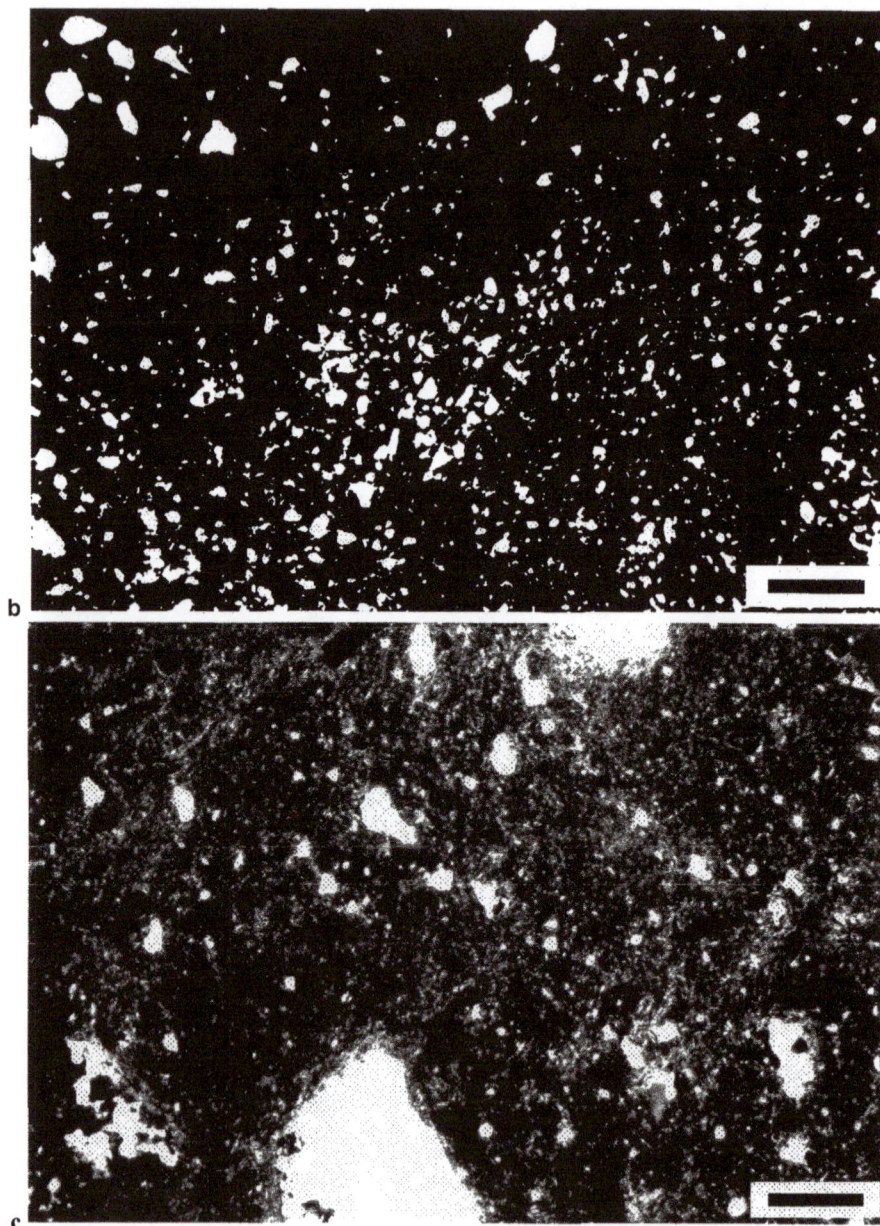

Figure 8.18 (**c**) shows another Neolithic ware sample from north central France where the temper grains are very irregularly distributed in their disposition and grain size. It is evident that a natural clay source with various natural temper grain inclusions was used. No attempt to homogenize the material was made, thus giving a very irregular aspect to the paste structure. The very nebulous and irregular aspect of the orientation of the clays (fine-grained material around the temper grains) indicates that the clays were not sorted nor worked. They maintain much the same

structure as they had in the soil clay source from which they were extracted. They show a nebulosity of grain orientation typical of poorly prepared clay sources from soils. The temper grains are also of highly variable dimension.

8.5
Summary

All these considerations argue for a careful examination of the ceramic constituents, integrating different analytical approaches. In the process of raw material identification, reference material and raw samples are very important. They help establish the regional variability and composition of the clays, soils and geology around the supposed center of production, as well as the clay's properties. It is then possible to see if the local clays are coarse or fine, and what their natural inclusions are, as compared to the inclusions observed in the ceramic paste. Familiarity with the actual pottery tradition also yields valuable information. The assumption of cultural continuity is, of course, not warranted, but traditional ceramic production offers comparative views and an interpretative framework for archaeological data (see Chap. 8). Ethnographic studies and laboratory analyses are, indeed, very important in the construction of our knowledge about making a pot.

Optical observation of ceramic sherds coupled with quantitative methods is a very powerful tool in archaeological research. The best way to communicate aspect and differences is by a quantitative approach. Coarse-grained and fine-grained is not enough. What is coarse for one per- son will be fine for another. Also, it is frequently a surprise to find out exactly the numerical differences for objects estimated by eye. One often fools oneself into seeing what one wishes to see.

A very important point to keep in mind is that the number of grains in a sample is often less important than their mass. The surface area is more of a factor than the grain abundance (number) in the tempering of a plastic clay. This relates to the function of tempering, natural or added by the potter. It is the amount (mass) which determines the tempering action. If the paste is to be fine, small grains are needed. If the paste can support a high charge of tempering material, it will be easier to use large grains. It would seem that the tempering function is one of the relative mass of clays versus non-clay materials.

In any event, using simple tools, scanners, binocular microscopes or more sophisticated ones such as the petrographic microscope with thin sections, one can arrive at some surprising results; but it is the eye, aided or unaided, which does the work.

Bibliography: Analysis Methods for Archaeological Materials and Case Studies

Adan-Bayewitz, D, and Wieder, M (1992) Ceramics from Galilee: a comparison of several techniques for fabric characterization. *J Field Archaeology,* 19:189-205

Betts, IM (1991) Thin section and neutron activation analysis of brick and tile from York and surrounding sites. In: Middleton, A and Freestone, I (eds). *Recent developments in ceramic petrology.* British Museum Occasional Paper 81, London, pp 39-62

Chayes, F (1954) The Theory of Thin-Section Analysis. *J of Geology* 62:92-101.

Centeno, SA, Williams, VI, Little, NC, and Speakman, RJ (2012) Characterization of surface decorations in Prehispanic archaeological ceramics by Raman spectroscopy, FTIR, XRD and XRF, *Vibrational Spectroscopy* 58, 119– 124. doi: 10.1016/j.vibspec.2011.11.004

DeAtley, SP (1985) Mix and match: traditions of glaze paint preparation at Four Mile Run, Arizona. 296-302. In: Kingery, WD (ed). *Ceramics and civilization, vol II. Technology and style,* American Ceramics Soc, Colombus, Ohio, pp 396-302

De La Fuente G., Josa, VG, Castellano, G, Limandri, S, Vera, SD, Días, JF, Suárez, S, Bernardi, G, and Bertolino, S (2020) Chemical and mineralogical characterization of Aguada Portezuelo pottery from Catamarca, north-western Argentina: PIXE, XRD and SEM-EDS studies applied to surface pre- and post-firing paints, slips and pastes. *Archaeometry* 62(2): 247-266

Dickinson, WR (1985) Interpreting provenance relation from detrital modes of sandstones. In: Zuffa, G.G. (ed.), *Provenance of Arenites.* NATO ASI Series, C 148, D. Reidel Publishing Company, Dordrecht, pp 333–363

Dickinson, WR (2001) Petrography and Geological Provenance of Sand Tempers in Prehistoric Potsherds from Fiji and Vanuatu, South Pacific. *Geoarchaeology* 16: 275-322.

Dickinson, WR, Takayama, J, Snow, EA, and Schutler, R (1990) Sand temper of probable Fijian origin in prehistoric potsherds from Tuvalu. *Antiquity* 64:307-312

Druc, I (2015) *Portable digital microscope. Atlas of ceramic paste. Components, texture and technology.* Deep University Press, WI.

Druc, I, Gonzales P, and Inokuchi K (2020) What is that pigment? Surface analysis of early ceramics from Kuntur Wasi, Andes of Peru, by way of Raman microscopy and portable X-ray fluorescence spectroscopy. *Annals of Archaeology* 3(2): 49-60.

Edwards, H, and Vandenabeele P. (2016) Raman spectroscopy in art and archaeology. Philosophical transactions of the Royal Society A-mathematical Physical and Engineering Sciences. http://hdl.handle.net/1854/LU-8510665

Fieller, NRJ, and Nicholson, PT (1991) Grain-size analysis of archaeological pottery; the use of statistical models. In: Middleton, A and Freestone, I (eds). *Recent developments in ceramic petrology.* British Museum Occasional Paper 81, London, pp 71-112

Folk, RL (1965) *Petrology of sedimentary rocks.* The University of Texas. Hemphill's, Austin, Texas

Freire, E., Acevedo, V, Halac, EB, Polla, G, López, M, and Reinoso, M. (2016) X-ray diffraction and Raman spectroscopy study of white decorations on tricolored ceramics from Northwestern Argentina. Spectrochimica Acta Part A, *Molecular and Biomolecular Spectroscopy,* 157: 182–185

Gehres, B., and Querré, G (*2018*) New applications of LA–ICP–MS for sourcing archaeological ceramics: microanalysis of inclusions as fingerprints of their origin. *Archaeometry, doi:* 10.1111/arcm.12338.

Goldstein, DJ, and Shimada, I (2007) Middle Sicán multi-craft production: Resource management and labor organization. In: Shimada, I (ed). *Craft Production in Complex*

Societies: Multicraft and Producer Perspectives. The University of Utah Press, Salt Lake City, pp 44-67

Goren, Y (2014) The operation of a portable petrographic thin section laboratory for field studies. *NYMS* (*New York Microscopical Soc*) *Newsletter*, September 2-17

Heidke, JM, Miksa, EJ, and Wallace, HD (2001) A Petrographic Approach to Sand-Tempered Pottery Provenance Studies. In: Glowacki, DM and Neff H (eds). *Ceramic Production and Circulation in the Greater Southwest: Source Determination by INAA and Complementary Mineralogical Investigations*. Donna M. Costen Institute of Archaeology, Monograph 44. University of California. Los Angeles.

Hunt, A (ed) (2017) *The Oxford handbook of archaeological ceramic analysis.* Oxford University Press, Oxford, pp 724

Ingersoll, R.V., Bulard, T.F., Ford, R.L., Grimn, J.P., Pickle, J.P., Sares, S.W., 1984, The effect of grain size on detrital modes: a test of the Gazzi-Dickinson Point Counting method, *J Sedimentary Petrology*, 54:103-116

Ixer, RA, and Bevins, R (2014) The vexted question of the Stonehenge stones. *British Archaeology*, Sept-Oct issue: 49-55.

Ixer, RA, and Lunt, S (1991) The petrography of certain pre-Spanish pottery from Peru. In: Middleton, A and Freestone, I (eds). *Recent developments in ceramic petrology*. British Museum Occasional Paper 81, London, pp 137-164

Kałaska, M., Druc, I., Chyla, J., Pimentel, R., Syczewski, M., Siuda, R., Makowski, K., Giersz, M. (2020) Application of electron microprobe analysis to identify the origin of ancient pottery production from the Castillo de Huarmey, Peru. *Archaeometry, 62(6)* May https://onlinelibrary.wiley.com/doi/10.1111/arcm.12581

Kamilli, DC, and Steinberg, E (1985) New approaches to mineral analysis of ancient ceramics. In: Rapp, G, and Gifford, JA (ed). *Archaeological geology*. Yale University Press, New Haven, CT, pp 313-330

Kennoyer, JM (2005) Steatite and faience manufacturing at Harrapa: New evidence from Mounde E excavations 2000-2001. *Museum Journal* (National Museum of Pakistan) 3-4 (Jan-Dec 2002): 43-56.

Kennoyer, JM, and Miller, HML (2007) Multiple crafts and socioeconomic associations in the Indus civilization. In: Shimada, I (ed). *Craft Production in Complex Societies: Multicraft and Producer Perspectives*. The University of Utah Press, Salt Lake City, pp 152-183

Killick, D, and Miller, D (2014) Smelting of magnetite and magnetite-ilmenite iron ores in the northern Lowveld, South Africa, ca. 1000 CE to ca. 1880 CE. *J of Archaeol Sc* 43: 239-255

Klemptner, LJ, and Johnson, PF (1986) Technology and the primitive potter: Mississippian pottery development seen through the eyes of a ceramic engineer. In: Kingery, WD (ed). *Ceramics and civilization, vol II. Technology and style.* American Ceramics Soc, Colombus, OH, pp 251- 271

Li, YT (2007) Co-craft and multicraft: Section-mold casting and the organization of craft production at the Shang capital of Anyang. In: Shimada, I (ed). *Craft Production in Complex Societies: Multicraft and Producer Perspectives*. The University of Utah Press, Salt Lake City, pp 184-223

Livingood, PC, and Cornell, AS (2017) The accuracy and feasibility of digital image techniques in the analysis of pottery tempers using sherd edges. In: Ownby, M, Druc, I and Masucci, M (eds). *Integrative approaches in ceramic petrography*, The University of Utah Press, Salt Lake City, UT, pp 196-214

Loughlin, N (1977) Dales ware: a contribution to the study of Roman coarse pottery. In: Peacock, DPS (ed). *Pottery and early commerce*. Academic Press, London, pp 85-128

Magetti, M, Galetti, G, Picon, M, and Wessicken, R (1981) Campanian pottery: the nature of the black coating. *Archaeometry* 23:199-207

Maniatis, Y, Aloupi, E, and Stalios, AD (1993) New evidence for the nature of the Attic black gloss. *Archaeometry* 35: 23-34

Marshall, D, Anglin, CD, and Mumin, H (2004) Ore mineral atlas. Geological Association of Canada, Memorial University of Newfoundland, St John's, Newfoundland, Ca

Mason, RB, and Tite, MS (1997) The beginning of tin-opacification of pottery glazes. *Archaeometry* 39:41-58

Mason, RB (1991) Petrography of lslamic ceramics. In: Middleton, A and Freestone, I, (eds). *Recent developments in ceramic petrology*. British Museum Occasional Paper 81, London, pp 185-210

Mason, RB (1995) Criteria for the petrographic characterization of stone paste ceramics. *Archaeometry* 37:307-321

Matthew, AJ, Woods, AJ, and Oliver, C (1991) Spots before the eyes: new comparison charts for visual percentage estimation in archaeological material, 211-276. In: Middleton, A and Freestone, I (eds). *Recent developments in ceramic petrology*. British Museum Occasional Paper 81, London, pp 211-276

McGovern, PE (1987) Cross cultural craft interaction: the late Bronze Egyptian garri- son at Beth Shan. In: McGovern, PE (ed). Ceramics and civilization, vol IV. Cross-craft and Cross-cultural interaction in ceramics, Am Ceramics Soc Westerville, Ohio, pp 147-196

Middleton, AP, IC Freestone and Leese, MN (1985) Textural Analysis of Ceramic Thin Sections: Evaluation of Grain Sampling Procedures. *Archaeometry* 27: 64-74.

Molera, J, Vendrell-Saz, M and Garcia-Valles, M (1997) Technology and color development of Hispano-Mauresque lead-glaze pottery. *Archaeometry* 39:23-39

Papakosta, V, Lopez-Costas, O, Isaksson, S (2020 online) Multi-method (FTIR, XRD, PXRF) analysis of Ertebølle pottery ceramics from Scania, southern Sweden. *Archaeometry* 62(4): 677-693

Peacock, DPS (1977) Ceramics in Roman and Medieval archaeology. In: Peacock, DPS, (ed). *Pottery and early commerce*. Academic Press, London, pp 21-33

Prieto, G, Druc, I, Monzón, E, Baldeos, J, Watanabe, A, Risco, L, Lezama, R, and Cáceres, P (2019) La cerámica temprana de Gramalote en el valle de Moche: Aproximaciones a sus modos de producción, formas y usos. In Prieto, G, and Boswell, A (eds). *Actas de la Primera Mesa Redonda de Trujillo. Nuevas Perspectivas en la arqueología de los valles de Viru, Moche y Chicama*. Fondo Editorial de la Universidad Nacional de Trujillo, Trujillo, pp.31-69

Prieto, G, Wright, V, Burger, RL, Cooke, CA, Zeballos-Velasquez, EL, Watanave, A, Suchomel MR, and Suescun L (2016) The source, processing and use of red pigment based on hematite and cinnabar at Gramalote, an early Initial Period (1500-1200 cal. B.C.) maritime community, north coast of Peru. *J of Archaeol Sci: Reports* 5: 45-60.

Puente, V, Porto López, JM, Desimone, PM, and Botta, PM (2019) The persistence of the black color in magnetite-based pigments in prehispanic ceramics of the Argentine northwest. *Archaeometry*, 61(5): 1066-1080 DOI:10.1111/arcm.12476.

Ratto, N, Reinoso, M, Basile, M, Freire, E, and Halac, EB, (2020 online) Archaeometrical characterization of pigments and paintings on prehispanic pottery from the regions of Fiambala and Chaschuil (Catamarca, Argentina) *Archaeometry*, DOI: 10.1111/arcm.12591 (Raman and XRD studies)

Reedy, CL (2008) *Thin section petrography of stone and ceramic cultural materials*. Archetype, London, pp 256

Reedy, CL, Anderson, J, Reedy, TJ, and Liu Y (2014) Image analysis in quantitative particle studies of archaeological ceramic thin sections. *Advances in Archaeological Practice* 2(4): 252-268.

Reedy, CL, Vandiver, PB, He, T, Xu, Y, and Wang Y (2017). Research into coal-clay composite ceramics of Sichuan province, China. *MRS Advances* 2(37-38): 2043-2079

Riera-Soto, C, Uribe Rodríguez, M, Menzies, A, and Barraza Bustos, M (2019) Avances en petrografía automatizada: cerámicas tempranas de Guatacondo, norte de Chile (900 a.C-200 d.C.) *Boletín de Arqueología PUCP* 26: 141-157

Rice, P. 2015. *Pottery analysis. A source book second edition.* University of Chicago Press, Chicago, pp 592

Schubert, P (1986) Petrographic modal analysis - a necessary compliment to chemical analysis of ceramic coarse ware. *Archaeometry* 28: 163-178.

Siddall, R (2018) Mineral Pigments in Archaeology: Their Analysis and the Range of Available Materials. *Minerals*, 8 (5), 201.

Stoltman, JB (1989) A quantitative approach to the petrographic analysis of ceramic thin sections. *American Antiquity* 54(1): 147-160

Stoltman, JB (1991) Ceramic Petrography as a Technique for Documenting Cultural Interaction: An Example from the Upper Mississippi Valley. *American Antiquity* 56:103-120.

Tite, MS, Bimson, M, and Cowell, MR (1984) Technological examination of Egyptian blue. In: Lambert, JL, (ed). *Archaeological chemistry III.* Advances in chemistry, vol 205. Am Chemical Soc, pp 215-242

Vandiver, PB, and Kingery, WD (1986) Egyptian faience: the first high-tech ceramic. In: Kingery, WD (ed), *Ceramics and civilization*, vol III. Am Ceramics Soc, Wester-ville, Ohio, pp 19-34

Vandiver, PB, Cort, LA, and Handwerker, CA (1987) Variations in the practice of ceramic technology in different cultures: comparison of Korean and Chinese celadon glazes. In: McGovern, PE, (ed). *Ceramics and civilization*, vol IV. Cross-craft and crossculture interaction in ceramics. Am Ceramics Soc, Westerville, Ohio, pp 347-378

Vandenabeele, P (2013) *Practical Raman Spectroscopy. An introduction.* Wiley & Sons, Chichester, UK pp 168

Whitbread, IK (1991) Image and data processing in ceramic petrology. In: Middleton, A and Freestone, I, (eds). *Recent developments in ceramic petrology.* British Museum Occasional Paper 81, London, pp 369-388

Williams, DF (1977) The Romano-British black burnished industry, an essay of the characterization by heavy mineral analysis.163 -174. In: Peacock, DPS, (ed) *Pottery and early commerce.* Academic Press, London, pp 163-174

Williams, DF (1983) Petrology of ceramics. In: Kemp, DRC, and Harvey, AP (eds). *The petrology of archaeological artifacts*, Clarendon Press, Oxford, pp 301-329

Woods, AJ (1985) Form and fabric and function: some observations on the cooking pot in antiquity. In: Kingery, WD (ed). *Ceramics and civilization*, vol II. Technology and style. Am Ceramics Soc, Colombus, Ohio, pp 157-172

9 Ceramics and Archaeology: Case Studies

In the preceding chapters, we have attempted to give a background dealing with the different aspects of origins of ceramic materials and the different possibilities of combining them to make pots. The information is there but, as any student will ask, what do we do with it? In this chapter we would like to give examples of ceramic investigations in archaeological contexts. After all, the problems to be solved are archaeological, not those of geology or chemistry or physics.

The archaeological questions are usually related to technology and provenance. Is it locally made, how was it made, where can it come from? And derivatives of these questions. Here are instances where the study of the ceramic paste has been made in order to answer several questions posed by the object found in the ground. Examples are from archaeological finds and present-day production.

For provenance studies, knowing the local geology and getting comparative samples of clays and sands is crucial. These give you the baseline against which you will compare the archaeological ceramic data. The previous chapters offered tools and knowledge to better understand the relationships between raw materials and the finish product, taking into account the work of the potter. When evaluating a possible provenance, that is, from which geological formation or environment the raw materials came from, the entire mineral assemblage AND respective (estimated) frequencies in the ceramic thin section, paste or chemical data are to be considered, not only a few particular rock fragments present that could hint toward one or another geological formation. These may greatly help of course, as Anna Shepard pointed, but you should consider if these are accidental, occasional, part of a mix sediment, or indeed the main constituent of a primary deposit. You should also remember that rock fragments and minerals are subject to weathering, alteration and breakage during wind-river-down-hill sliding and other natural transport-related processes or deposits. Different examples can illustrate this point.

A granitic body subjected to frictions during geological 'encounters' (contact metamorphism for example) could become distorted on its fringe, presenting mylonitic texture, with areas of recrystallization, banding, and smaller, elongated quartz crystals. The small granitic fragments seen in a ceramic paste, 0.6 to 2 mm in size, are fractions of the whole and may only show the recrystallized, smaller quartz grains following the same orientation as the biotites and muscovites present in the rock fragment could induce you to think you are in a presence of a quartz-mica schist, while it originates from a mylonitic granite. One is from a metamorphic source, the other is intrusive. This may lead you to quite different regions of provenance for the raw materials, especially if one is not local and the other is.

Another example is that of ceramic pastes of the high Andes of northern Chile, very rich in medium to coarse, 'fresh' biotite fragments, along with plagioclase and quartz of volcanic, pyroclastic origin. How do you explain the abundance of the biotites when none of the comparative samples of soils, clays and rock fragments show this? The answer to this clue was given by a German geologist who had studied extensively the area. This type of assemblage was specific to 'pantanos' -shallow altiplano lagunas- where the biotites concentrate: being lighter (in weight) than other minerals they float and are transported further, while the heavier material stays closer to the source, something akin to levigation, and an outcome of erosional processes. The potters, then, must have gathered their clay or tempering material from these areas, which abound in the altiplano (see Druc and Uribe 2018).

Another aspect of ceramic analysis and data interpretation for provenance or technology studies, is the methodology, and how and where to draw the line between one composition group and another. In other words, the question of group attribution, internal variability, and of what is considered local (see for example Druc 2013). What is accepted true for one context (one archaeological project, research area), might not work for another. Good interpretation of the data is based on well-informed, multi-angle, detective-like work.

The following case studies are examined to illustrate certain analytical aspects, data interpretation, or material identification, leading to an understanding of technology, organization of production or provenance of the objects under study. The cases deal with Roman amphora, iron age pottery, and sigillate ware (Velde) or past and present Andean productions (Druc). They were presented in the initial version of this book and are reproduced again here, as they illustrate well what has been discussed in this book. A few other studies have been added in this chapter, while many other examples are given throughout the book where most appropriate.

9.1
Yellow Garnets and Trafficking Wine [1]

The use of amphorae to transport and conserve wine is a trademark of Roman civilization. Amphora sherds are found on a great number of sites in Europe, associated with signs of Roman occupation as well as those of their predecessors, the Celts. Wine was traded and consumed by the pre-conquest peoples as well as those safely under Roman rule and protection. In the amphorae fragments several types are known; some are of local or regional importance and others are found in many districts and areas. The Dressel types I and 2/4 are those found in many areas under Roman influence and are almost ubiquitous in the period of the 1st century before to the 2nd century after our era. Their form is distinctive and they have been well identified from Italic Rome to the

---

[1] Velde, B and Courtois, L (1983) Yellow garnets in Roman amphorae – a possible tracer of ancient commerce. *J of Arch Sci* 10: 531-639.

region of Naples in many instances, through finds of production as well as frequency in certain areas of the Italian peninsula. Another type of amphora found in the same period and used to transport and conserve wine is called the Graeco-Italic form. This form has a double name in that it is not a true Greek form but one that is close to the classical forms from the near wine producing neighbour. These shapes are indicated in Figure 9.1. In some of the Dressel forms one finds what are called black sands, grains visible to the naked eye which are, in fact, sand temper of pyroxene and amphiboles as well as olivines, minerals commonly found in basalts. All of these minerals contain iron, which gives the dark color. Some of the Graeco-Italic forms contain these grains also. This has been a sort of facies type for some time for archae- ologists. The black sand amphora are known in present-day France, Germany and England. They have been identified in other Mediterranean areas also.

Here, one must interject the fact that in the 1st century B.C. and to a certain extent in the 1st century A.D., Greek wines had a very high reputation. Others were as good, perhaps, but the real Greek goods fetched a high price, or so one could surmise. At this time, the Italic wines had a slightly lower reputation. Then they would fetch lower prices. Thus, one would expect that Greek amphora would be found in sites where people could have paid the price. However, the Graeco-Italic forms are often found in sites which were Celtic, barbarians traded with for the profit of the nascent Empire. One begins to sense something.

In looking at some of the Dressel type I and 2/4 forms containing black sands one finds a special mineral, a yellow garnet. The other minerals are typical of the volcanic cortege, pyroxenes, olivines and some amphiboles, and occasionally one finds yellow garnets. These garnets are not only unusual in their occurrence (garnets are normally found only in metamorphic rocks and not volcanic ones) but they have specific composition which identifies them as being of volcanic origin. They are manganese (Mn)-and titanium (Ti)-rich. This helps distinguish volcanic garnets from the normal metamorphic ones. Thus, when one determines the composition of the yellow garnets by the electron microprobe (see Chap. 10.3.5), the identity and origin of the phase is immediate. In fact, these garnets are found in Dressel type I and 2/4 amphorae as well as in Graeco-Italic ones and some Etruscan forms (earlier productions). If one compares these rare garnets in the black sand amphorae with potential sources from volcanic rocks in the area between Rome and Vesuvius, one finds that they most likely came from either the Roman area and Roccamonfina to the south or that of the Vesuvius volcanic region. Yellow garnets of slightly different compositions are found over these two broad geographic areas.

The compositions are shown in Fig. 9.2 as two broad zones separated by manganese content (weight percent MnO). Analyses of the yellow garnets from different amphorae are given in the diagram also. One can see that there are two loose groups which can be attributed to Roman and Vesuvian area (Naples or Pompeii) production sites. Garnets from both the Dressel I, 2/4 and Graeco-Italic forms are found to lie in these compositional groups. One can conclude that the black sand amphorae were produced on the Italian peninsula in both areas where the garnet-bearing volcanic material formed black sands.

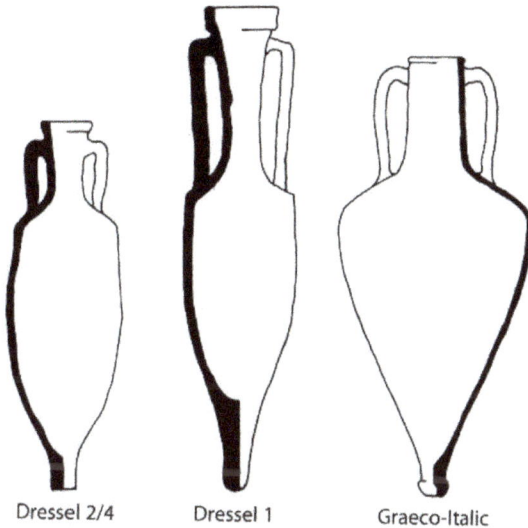

Fig. 9.1. Shapes of the Italic and Graeco-Italic amphorae discussed in text.

Dressel 2/4   Dressel 1   Graeco-Italic

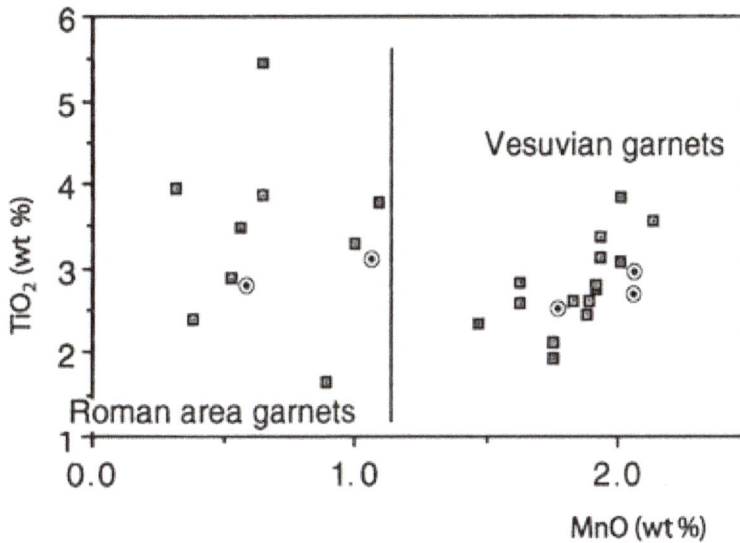

Fig. 9.2. Compositions of yellow garnets found in amphorae. Squares indicate Italic forms and *circled diamonds* indicate Graeco-Italic forms. The dividing line (just slightly greater than 1% MnO) is established on the basis of garnet analyses from rocks found in the Roman and Vesuvius regions. Titanium (TiO_2) content varies in both series of garnets but the determining factor is the manganese content.

The analyzed archaeological finds are from sunken trade vessels found along the French Mediterranean coast and several land sites in Italy and the middle East. In fact, some Etruscan amphorae, preceding the Roman material in age, give black sand yellow garnets also. Not all amphorae with black sands revealed the yellow garnets. Hence, one has an idea that the Roman Dressel type I and 2/4 amphorae coming from the Roman and Neapolitan region were made in some instances from materials from the Vesuvian eruptions near Naples and the Roman comagmatic province south of Rome. However, the Graeco-Italic amphorae also contain garnets from the two provinces even though there are no garnet-bearing volcanics in Greece.

Therefore, black sand-bearing amphorae are likely to have come from Italy (present-day); they are not likely to have come from Greece. Thus, the shapes similar to the Greek forms are more Italo- than Greek. It would seem that there was a certain amount of counterfeiting of amphorae, making Greek shapes in Italy. Given the higher appreciation and most likely price of Greek wine compared to Roman at the time, it seems that there was a certain temptation to procure "bottles" of a certain shape to be used for wine of another origin. The fact that these Graeco-Italic shapes were found in the Celtic areas inhabited by the barbarians to the north seems to indicate a dupery of the brave but somewhat crude northern neighbours of the Roman world. It is not, of course, an unknown practice now, but it is interesting to note that such illegal trade, should there have been laws on the content of packages, was practised in such an early period, and even more interesting to be able to identify such illegal trade at such a posterior date.

This study relies on the composition of specific mineral grains used as a tempering agent, in order to identify the origin of a specific production site or sites of a ceramic material. The method used in the study is the electron microprobe, which is capable of giving a chemical composition of a single sand grain. The study necessitated the preparation of a thin section for microscopic study to find the suspect grains and then the use of the electron microprobe to give a definitive chemical identification of the grains. The process is long and costly. One must have a good idea or motivation before undertaking it. The presence of black sand-grained amphorae and the identification of yellow garnets in black sands led to such a study. This means that the archaeological context and the geological content were known before the study was initiated; these are special conditions. It is not always possible to use such a set of data before undertaking an archaeological investigation. However, when the conditions are present, the likelihood of success is greatly enhanced.

9.2
Iron Age Pottery in Southwestern England and its Geological Sources [2]

A very interesting (and classic) study of the inclusions of rock fragments in Iron Age pottery and the distribution or trade of such objects has been made on material found in southwestern England. The question posed initially was: had there been trade of ceramic objects between the continental Gaulic areas in France, now Brittany, with the adjacent portion of the British Isles? Further: was there commerce between the different sites of pottery production and the sites where the pottery was found in

---

[2] Peacock, EPS (1969) A contribution to the study of Glastonbury ware from southwestern Britain. *Antiquaries Journal,* 49: 49-69.

England? In order to solve this problem, a study of the lithic fragments in the pottery was undertaken. Lithic fragment means simple rock fragment. The tempering agents in this pottery were of such a size as to be composed of rock fragments instead of individual mineral grains, or at times individual mineral grains were used which indicated a special rock origin.

The pottery type is called Glastonbury ware, and is well recognized from its decorative motifs produced in the 1st, 2nd and 3rd centuries of our era. The distribution recovered in archaeological investigation is that of the peninsula of present-day southwestern England. This is indicated in Fig. 9.3. Several types of rock fragments and minerals were identified under the petrographic microscope (see Chap. 10.3.3) using thin sections. We indicate three, which were: (1) gabbro, a rock containing olivine and pyroxene from intrusive non-volcanic sources, (2) Old Red Sandstone, a sedimentary rock composed almost exclusively in this case of quartz grains containing at times fragments of metamorphic rocks, (3) shell fragments from a sedimentary rock of Jurassic age (some 250 million years in age). Here, the petrographic microscope gave a wealth of information on rock and mineral types helping identify rather specific materials. The archaeologist rapidly identified the source areas of the possible rock types in the southwest English countryside.

Figure 9.3 shows the possible source areas for three of the lithic fragments found in the Glastonbury ware samples. In this figure one can find the distribution of all the samples found (Fig. 9.3.1) and three examples of specific sherd temper grain samples in their last resting place before archaeological investigation.

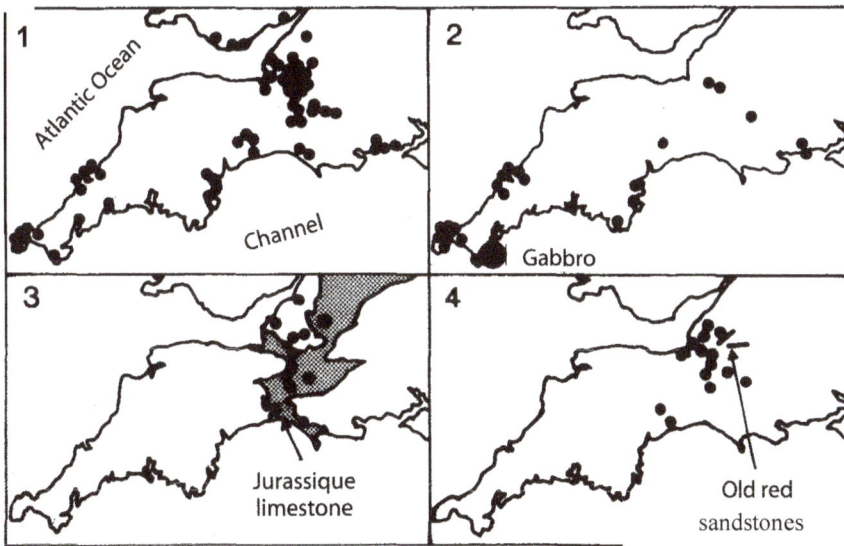

Fig. 9.3 1-4. Distribution of Gastonbury ware ceramics based upon the temper grains of characteristic rock fragments found in them. **1** shows the find sites for this ware in southwestern England. The point to the *lower left* hand of the map, Land's End, is the extremity of the British Isles. **2** Find sites of sherds containing fragments of the basic

magmatic rock gabbro and the only site in which it is found in the area, on the tip of the Island *(large dot)*. **3** Find sites for sherds containing fragments of Jurassic limestone, and the area where this rock is found at the surface which is figured in *grey*. **4** Find sites for sherds containing fragments of the Old Red Sand- stone geologic unit and the outcrop of the sandstone shown by *two thin bars* on the map. The dispersion of samples containing gabbroic lithic fragments is greater than that of the Old Red Sandstone-containing samples. The Jurassic limestone-bearing sherds are found only in areas where the rock is found at the surface. This clearly indicates both production sites and trade or other distribution.

From this, it is evident that there has been a dissemination through trade in some cases and a concentration through local consumption in others on the English side of the Channel. However, no evidence from the lithic and mineral fragments could link these sherds with a possible geological source on the French section of the Continent opposite to that of southwest England. Thus, the initial question of possible trans-Channel trade is answered for samples containing gabbro, Old Red Sandstone and Jurassic limestone lithic fragments. These materials do not exist on the other side of the Channel.

Several of the rock fragment tracers are quite interesting. The samples with a gabbro rock type, found in the southwestern most corner of England, are the most widely distributed (Fig. 9.3.2). They are found in coastal areas and along the Bristol Channel-Thames axis. The Jurassic limestone-carrying sherds are found uniquely within the potential area where these fragments could be incorporated locally into the potter's paste (Fig. 9.3.3). This appears to be a use or commercial zone in which one finds goods of local production and those coming from afar. An intermediate case is that of fragments of the Old Red Sandstone geologic unit. Here, the sherds containing the rock fragments are found in the vicinity of the possible source, but there is some dispersion and hence trade or barter of these pots.

The study is a very good example of the usefulness of petrography. A rapid examination of the mineral and rock fragment species in the Glastonbury ware indicated different groups of material, and a search of the possible geologic sources gave a good idea of possible provenance.

Interestingly, a similar line of inquiry is seen in a recent study conducted by Gehres and Querré (2018) using LA-ICP-MS on Neolithic to Iron Age ceramics from Western France. Based on the fact that each granite has a particular signature, the authors identified (and paired) the composition of biotites in different granites and ceramics. They were able to show the provenance of the ceramics found on the islands and what was the circulation of the ware.

9.3
Whole-Sample Compositions of Some Sigillate Ware Produced in France[3]

One of the most widespread ceramic materials in digs of sites of Roman occupation in Europe is sigillate ware. This is very important because it is easy to identify by its red lustrous surface, and the designs moulded into its surface give a very good idea of its production site. However, there is more information to be gleaned from such material. Several studies have been undertaken in order to characterize this material and to elucidate some specific problems. The initial study was designed to establish the compositions of sigillate ware found on a production site of several of the most abundant and well-known sigillate types found in Roman Gaul. Samples from several production sites in France (Lezoux, 125 samples; Lyon, 79 samples; La Graufesenque, 124 samples; Montans, 57 samples; Banassac, 55 samples) were analyzed for their chemical constituents and compared to a classical "source" site for sigillate production in Italy (Arezzo, 82 samples). The chemical data used below are taken from the tables given by the authors (bottom page note) to illustrate the information chemical analysis can provide.

Initially, one must state that the sigillate ware is a specific production formed by a moulding process. Designs are carved into the moulds which represent vegetal (fruit, flowers, leaves, trees) and animal motifs (lions, ducks, wild boar and the like). These figures often show details of much less than a millimeter in dimension. The paste must be very fine and homogeneous to be formed into a mould. The 2nd striking aspect of sigillate production is the sheer quantity of it. One site, La Graufesenque, produced millions of pieces. This material can be found in the far corners of the Roman trading zone.

The major characteristic of good-quality sigillate ware is that the temper material is almost invisible. The non-clay particles are very fine. The material is extremely well mixed and homogeneous. The surface is covered by a thin (as thin as 30 μm) vitrous coating of clay composition material and the whole object is in a thoroughly oxidized state. The pottery is "well fired" in that it is solid, hard and impermeable (e.g. see Eramo and Maggetti 2013 regarding Roman ware firing). The question then is: can one find differences in production techniques in the different sites?

In order to investigate this type of material, there is only one solution, which is whole or bulk sample analysis. The grain size is so small that optical microscope and electron microprobe are not at all useful to investigate the identity of the non-clay grains. This method was adopted in the studies mentioned here. The importance of

---

[3] Picon, M, Vichy, M, and Meille, E (1971) Composition of the Lezoux, Lyon and Arezzo samian ware. *Archaeometry* 13: 191- 208. Picon, M, and Lasfargues, J (1974) Revue Arch. Est et du Centre Est, 25: 61-69 Picon, M and Garnier, J (1974) Un atelier d'Atevis a Lyon, *Revue Arch. Est et du Centre Est,* 25, pp 71-76. Picon, M, Carre, C, Cordoliani, ML, Vichy, M, Hernandez, JA, and Mignard, JL (1975) La Graufesenque, Banassac and Montans terra sigillata. *Archaeometry,* 17, pp 191-199.

composition in the art of ceramic manufacture is put to work. In looking at the chemical compositions of the ceramic sherds, one should keep in mind that the different elements analyzed represent different elements in the components that make up a paste. In the case of sigillate ware the situation is rather simple. All non-clay grains are of small dimension and probably came into the paste with the clay source. There is no evidence of added temper and one can assume that the clay resource was purified to give such a fine-grained material. In fact the manufacturing process dictates fine temper grains as the clay paste must express the designs sculptured into the moulds and hence large grains are detrimental. Then one can expect a purified clay source and perhaps a fusing agent. This last element is quite evident. However, one can associate or compare the relative abundance of several chemical elements.

Considering the major elements in ceramics, the silica (SiO_2) will probably represent the non-clay temper grains associated with the clay source. The potassium (K_2O) will indicate the micaceous illite clay mineral content of the clay. It will be associated with alumina (Al_2O_3) in the clay mineral. Iron oxide (Fe_2O_3) can be associated with the clay source or possible addition as a coloring agent. The samples fire very red. Titanium (TiO_2) should be an accessory part of the clay source, being present in the form of oxides. Magnesium (MgO) could be associated with the fusing agent or associated with the clays.

In the samples reported here there are several geographic areas of the 2nd- to 4th-century production: Arezzo, northern Italy; the Lyon and Lezoux sites in the northeastern part of the French Massif Central and the La Graufesenque, Montans and Banassac sites, which are on the southern edge of the mountainous and hilly areas of central France. One can try to distinguish between these different producing regions using their chemical characteristics of the ceramic produced there.

9.3.1
Lezoux Samples

The Lezoux site material available covered production over four centuries, and not only fine sigillate ware was available to analyze but also common ware was found on the production site of the Lezoux sigillate ceramics works. The first striking thing is the difference in lime content (CaO). In the early sigillate ware the early samples were of low lime content, near 2 %, whereas the more recent materials were in the 8-10 % range. This indicates the use of lime as a fusing agent. Increasing the CaO content of a paste allows the destruction of clays, and the creation of a more sturdy, impermeable, and resistant ceramic at lower temperature. It seems that the use of lime was practised in the period of great production and widespread commerce of the Lezoux ware, 2-4th centuries A.D. On the other hand, the calcium content of the common ware from the 2nd and 4th centuries is systematically lower than in the sigillate ware. The addition of lime was obviously reserved for the finer, more widely traded, ware.

The common ware from the same site shows similar lime content for sigillate production in the 1st century A.D., but there is a decided difference in later production. These compositional relations of CaO (lime) are shown in Fig. 9.4.

In looking at the analyses for all the sites, the tendency for fancy sigillate ware from most sites is to have a high lime content. The Lyon, Arezzo, La Graufesenque, Montans and Banassac sigillate wares have CaO contents of greater than 7 % for the most part. The lime content is then a sort of hall-mark for this production. It can be either an added component, from the firing of calcium carbonate in the form of carbonate rocks or from firing seashells or an initial component of the clay-rich source, as is the case at the La Graufesenque site.

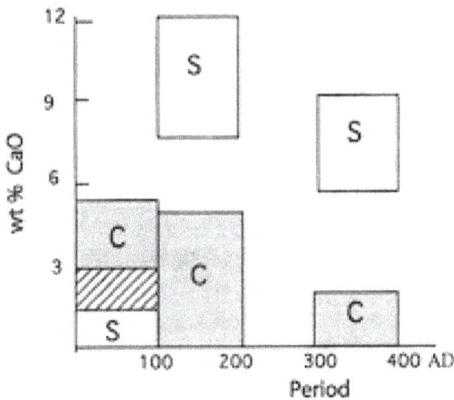

Fig. 9.4. Lime content (CaO) of Lezoux sigillate ware by century of production.

This gives us a background of the type of ceramic production that sigillate ware implied. The use of lime as a fusion agent lowers the temperature of firing to produce a hard, fused paste resisting knife cuts and shocks common in the handling of table ware.

9.3.2
Identifying Production Areas

Given the great amount of information concerning the ceramics of sigillate ware found on production sites, it is tempting to define the differences between the basic major elemental variables from the different productions. The decoration of the sigillate ware is usually sufficient to identify the sample and give a production age, however, it is instructive to see how one can attribute a production given the differences in basic chemical components of the different materials.

Individual Element Variations. The elements analyzed for the material considered are magnesium Mg, aluminium Al, silicon Si, potassium K, calcium Ca, titanium Ti and iron Fe. These elements represent some 90 % or more of the elements present. Water is not included in the analyses because its presence can be due to postproduction effects, such as burial and alteration. CaO: the content of CaO of the different sigillate sample sets, which is most likely a fusing agent added independently of the clay and tempering source (if used) is given in Fig. 9.5.

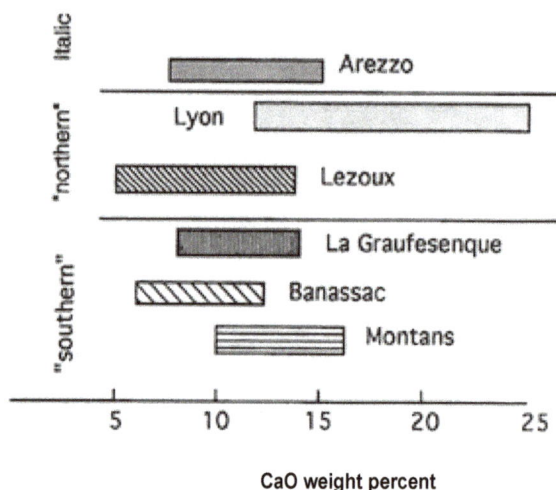

Fig. 9.5. Lime content (CaO) of different sigillate productions from five sites in or around the French Massif Central and from Arezzo, Italy. The sites from present-day France are labelled *southern* and *northern* to distinguish the general areas from which they come.

All the data are given in ranges of composition found representing 80 % of the data points. This is done because the numbers of samples analyzed, although at times important, probably does not represent a full range of possible compositional range. It is clear from inspection of the figure that the range found for CaO content for each of the production sites cannot be very useful in deter- mining a given production. The spread, or possibilities, of compositional variation is rather large and overlapping in all instances. The major part of the samples from the different sites in the range of 8-15% CaO. The potters in each area used rather similar compositions for the fusing agent.

A first step is to look at other elemental concentrations in order to describe the productions which would be based upon the clay sources. One can compare the ranges of compositional variation for the different elements, initially element by element, then in pairs in order to find differences which would be useful to define the different groups of sigillate sherds. K_2O: this element seems more useful than CaO for these materials in that some sample groups show all of their values in a range not found in some others, Arezzo and Banassac, for example (Fig. 9.6). $Al2O3$: this element gives another grouping, some productions are distinct and others are strongly overlapped. Lyon ware is strikingly alumina-poor (Fig. 9.7). TiO_2 titanium oxide shows an even wider difference between the productions. TiO_2 is an accessory element not present in most common silicate minerals. It is concentrated in high-temperature oxides which occur irregularly in clay and other sedimentary deposits. It is perhaps the one element that shows best the minor, chance differences in provenance (Fig. 9.8). Fe_2O_3 iron content does not seem to be of much use in that, as is the case for CaO, most of the sample sets overlap in the range of values they present. One exception is the Arezzo production which is iron-rich.

This survey of basic individual elemental differences gives a good idea of potential tracers for distinguishing the production sites. This should be enhanced by using two elements at a time to distinguish different productions.

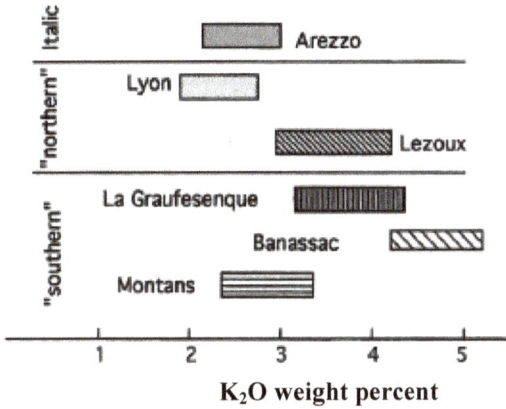

Fig. 9.6. Potassium oxide content (K_2O) of sigillate ware samples from different production sites.

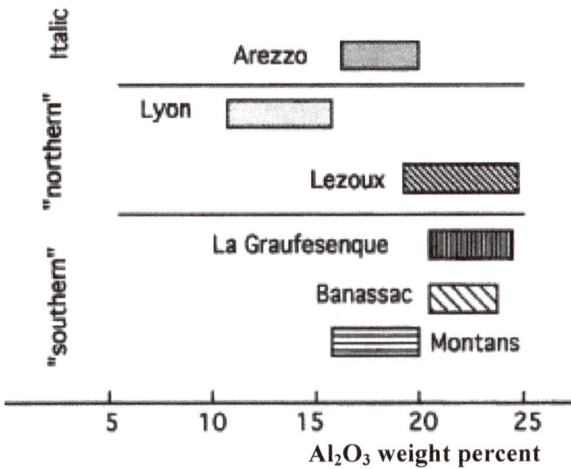

Fig. 9.7. Alumina (Al2O3) content of sigillate ware from different production sites.

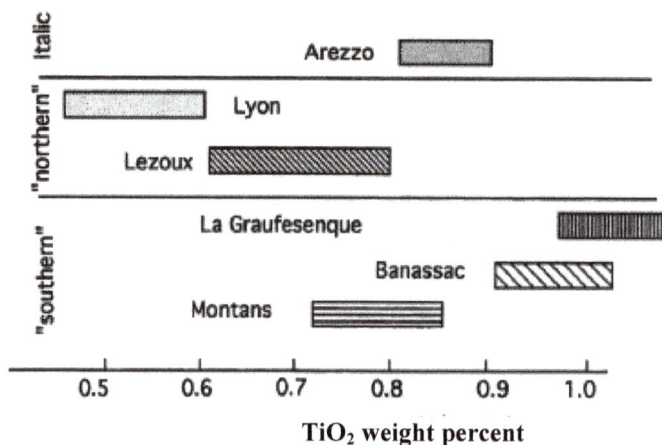

Fig. 9.8. Titania (TiO_2) content of sigillate ware from different production sites.

Variations of Element Pairs. Instead of looking at the distribution of elements one by one for the different production sites, one can observe the variations of elements in pairs. Let us look at the mica relation, K_2O against Al_2O_3 (Fig. 9.9). These two elements are coupled in micaceous clay sources. If kaolinite, for instance, is present, there will be no relation, and K_2O should be of low abundance. In the sigillate ware samples, the K_2O content is high (relative to other pastes), and one can assume that K_2O is present in illite clay minerals.

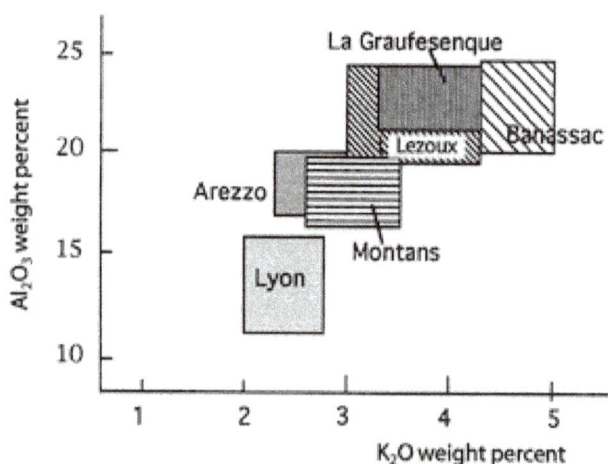

Fig. 9.9. Relations of alumina (Al2O3) and potassium oxide (K2O} for the different production sites. The area of compositional variation is designated as a *rectangle* covering the scatter of the observed analysis points. There is obvious overlap for many sites in this diagram.

The plot of these two elements shows a clear trend for the different sigillate productions, the higher the K_2O content the higher the Al_2O_3 content. In this plot of the samples investigated, the Lyon samples seem to contain the lowest illite or mica component, and as such distinguish themselves from the others. There is no overlap in compositional range for these samples and the other groups. The same is true for the Banassac group, but in the reverse sense; they are the most potassic and aluminous, containing the highest illite content.

We can use the potassium ($K2O$) content to represent the illite content and attempt to find a distinction with the titanium (TiO_2) content which, as a single element, seemed useful to separate the groups. Three production sites can be distinguished by a K_2O-TiO_2 plot, Banassac, La Graufesenque and Lyon, but the other groups overlap with one another so as to lose any discrimination in such a plot of chemical abundance (Fig. 9.10).

It is apparent from simple analysis of individual elemental abundance that $K2O$ is a reasonably good discriminant as well as $TiO2$. Using these two values, the production sites in what is now France can be distinguished and only the Arezzo samples overlap with Montans sample compositions. Thus, the best discriminants are potassium ($K2O$) combined with titanium ($TiO2$). These elements are not in any way particularly associated in a given mineral component but in the context of clay sources, i.e. an assemblage of materials; there is a good discrimination when these elements are compared.

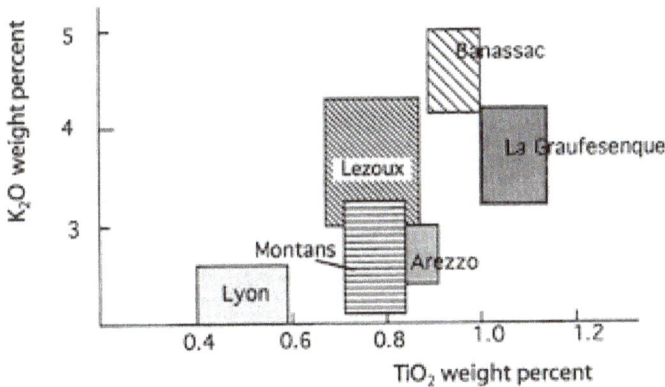

Fig. 9.10. Potassium oxide (K_2O) versus titania (TIO_2) contents for sigillate ware productions. This representation indicates that a better separation of the different productions can be effected using these two elements as a comparative basis

Regional Variation. One can ask the question of whether or not regional variations in clays are discernible, i.e. the northern samples (Lezoux, Lyon) and southern Massif Central (La Graufesenque, Montans, Banassac) compared to the Arezzo sample from present-day Italy. In fact, it seems not. The individual elemental variation and binary (two-element variation) diagrams show no distinctions. There is a continuum of compositions in most cases, taken either as individual elemental distributions or taken in pairs. It seems that each site has specific characteristics which cannot be linked to a general geographic trend.

This is, of course, to be expected in that the geology of the sources is not linked to present-day topographic or geographic situations. The source rocks are independent, in general, of political boundaries.

This exercise is useful to show that one can make use of chemical data using the raw data and a piece of paper. It is not necessary to use a discriminant program to tell you that there is a way of separating the production groups. It is much better to see what the mechanism is and how it works than to let a computer program give you the results in a numerical and non-chemical display. The fun of the game is understanding the problem, and not just getting the result.

9.3.3
Lezoux Coarse or Common Ware

Let us look specifically at one production site, Lezoux, where common ware was analyzed as a function of its date of production. Initially, one can consider that the CaO is likely to be a variable; the fusing agent is added as needed. However, it was seen that the CaO content of the common ware is lower than the sigillate ware, and thus the use of a fusing agent, if practised, was parsimonious. Silica (SiO_2) will probably represent the tempering grains of quartz, which will normally be larger than those of the sigillate ware. Alumina (Al_2O_3) and potassium (K_2O) will probably be related as they are in the sigillate samples, where they indicate the presence of micaceous illite which is the clay component of the paste. If mica or illite is not present, alumina will be independent of potassium, for example when kaolinite is the major clay. Iron (Fe) may or may not be associated with elements in the clay matrix. Thus, one can assign, at least as a working hypothesis, several elements to different components which make up the clay paste, as was done in the section of the sigillate ware. In the 52 Lezoux samples reported for common ware samples from the first, second and fourth centuries, there is no good correlation between CaO and another element. This suggests that the fusing agent was added independently of the other components. However, silica and alumina are correlated (Fig. 9.11). This indicates that the quartz tempering agent was associated with the clays or was in low abundance compared to them, as is the case for the sigillate ware. Silicon appears as a pure element in quartz but also as a major element in most clays. If it were added or included in a variable fashion in the paste, the Si would not correlate with Al. Hence, the clay and temper (quartz) seem to be related in abundance but not the fusing agent CaO (lime).

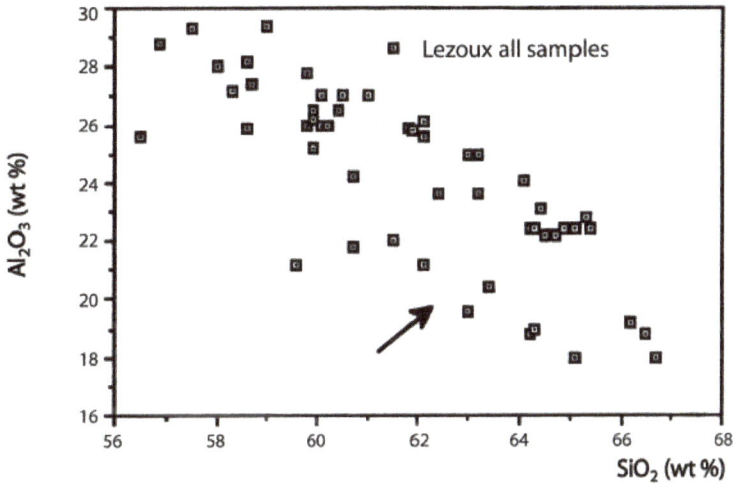

Fig. 9.11. Alumina (Al2O3) and silica (SiO2) relations for sigillate ware from the Lezoux site. One major trend is evident and a minor one is also apparent slightly below the major trend *(arrow).*

In Figure 9.11 most samples follow the same trend or alignment. However, some are below "the line", being less aluminous for the same silica content. The arrow in the figure indicates the anomalous samples. Inspection of the data tables indicates that these outlying samples are from 2nd-century production. When only 2nd-century production is plotted, there are two groups, one which follows the major trend of Lezoux common ware production over the entire period, and the outlier group. In Figure 9.12 it is evident that there are two series of samples, one parallel to the other. This suggests that there were two sources of clays or clays plus temper in the common ware productions in this period. A comparison of alumina (Al_2O_3) and potassium content shows a good correlation also but the distinction of two series is not evident. This indicates that the major clay minerals were illitic or micaceous in nature and quartz content separated the two sources giving two series of ceramic productions.

Three Element Comparisons. In geological research, it is common to compare three elements simultaneously. This gives a triangular plot of data. However, for some reason, it is rather difficult to find a common spread sheet program which does the job for three elements. One must search for a geological computer program. However, it is relatively easy to compare three elements without such sophisticated software. One can use ratios. If one divides an elemental concentration by another and a third by the same values, a sort of three-component diagram can be produced. This operation is very easy to do with spread sheets, and hence should be used commonly.

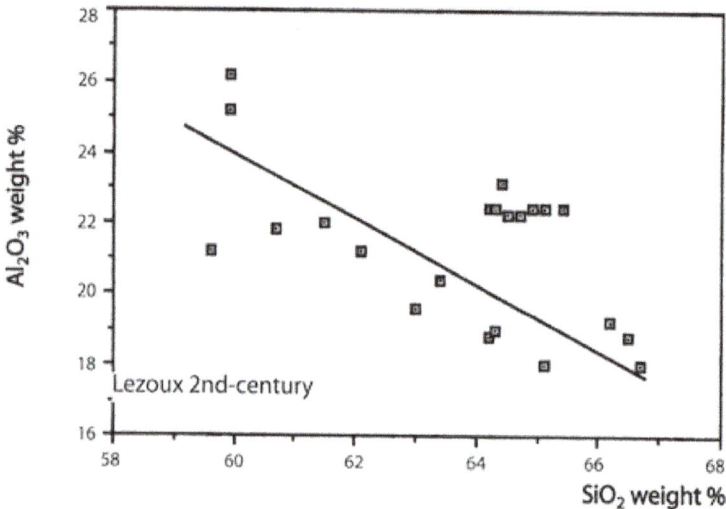

Fig. 9.12. Alumina and silica values for 2nd century Lezoux sigillate ware sherds. Two trends are distinct, identifying the source of the lower trend observed in Fig. 9.11.

We will use the Lezoux common ware again as an example. As seen be- fore, potassium (K_2O) and titanium (TiO_2 were good discriminants; and, to a lesser extent, one could use aluminum oxide (Al_2O_3). If we divide K_2O and TiO_2 by Al_2O_3 content, in the left hand lower corner of the graph one has relative alumina content and the ordinates express Ti and K content (Fig. 9.13). In this plot, the 4th-century samples are easily distinguished in a rather tight group, but the 2nd-century examples are more scattered. As seen before in Si-Al comparisons, there are two series of compositions for 2nd-century common ware. In the Ti/Al-K/Al comparison one series of 2nd-century samples is shown on the extreme right of the diagram and the other in the bottom center. Two samples of 2nd-century manufacture fall in the 4th-century compositional field. In this type of representation there is even more information concerning the different compositions and relations in each sample. In general, the more data one uses, the more definition one has. However, this axiom has limits. When an analysis considers six or seven variables simultaneously, one can distinguish groups but the idea of an individual sample is lost. For example, it would be very difficult to distinguish the two Lezoux 2nd-century common ware samples "buried" in the 4th-century compositional field.

As a general rule, it is very useful to look at any set of data by oneself. In using elements one by one, it is possible to gain an idea of the types of variation one will see and which elements will give the best discriminating power. Then one can combine them one by one. This is very easy with computer power. It takes only a second or two to compare examples in a large set of data. One should not hesitate to work out these differences by oneself instead of searching for an expert with a more complex analysis method. The fun of investigation is in doing it yourself.

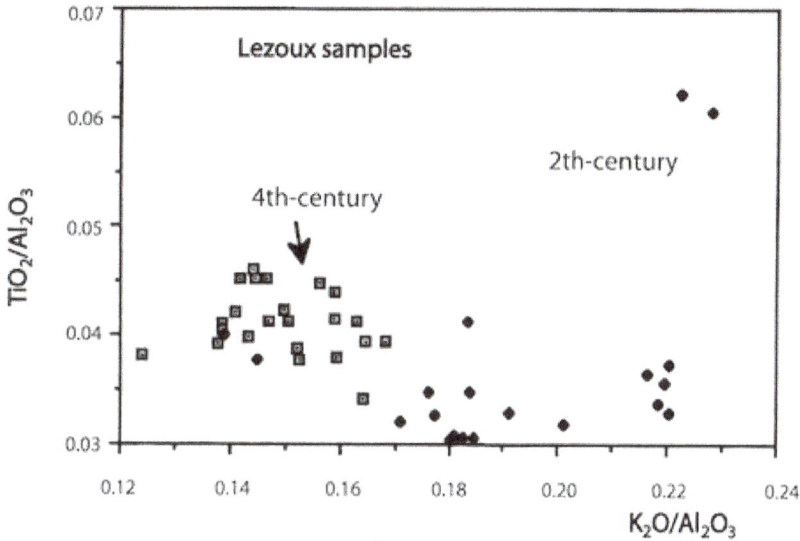

Fig. 9.13. Relations of three elements compared as ratios with alumina as the base element for the Lezoux samples. K/Al oxide ratios separate the samples into a 4th- century group *(squares)* and two 2nd-century A. D. groups (diamonds)

9.3.4
Specific Problems of Archaeological Interest Using Sigillate Ware Data

All this is fine, but what use can be made of it for archaeological purposes? First, one must remember that one cannot generalize, give a general description, without a certain amount of analytical background. If one wishes to describe the Arezzo, Italian-source sigillate ceramics, it is necessary to have more than two samples analyzed. In fact, representativity should be based upon about 100 samples or more. However, this is not usually possible, and some compromise must be made. In the studies of the sigillate ware the sampling is reasonable, from 50 to 130 samples per site. However, there can be some exceptions to the ranges of compositions. Nevertheless, the data given by the authors is correct. Thus, one has a good idea of the compositional variation that one can expect from production in each site.

Given this, several studies were conducted using this data base in order to solve specific questions of provenance for a given object.

9.3.4.1
Arezzo Moulds

A number of moulds for the production of sigillate ware lamps were found in the archaeological finds in the Lyon sites. These have strong parentage, in the decorative motif, with some of the sigillate production known to have come from Arezzo. The question, is then, do the ceramic productions made with these moulds found in the Lyon sites have the Lyon paste composition or that of Arezzo? Figure 9.14 shows the discriminants of potassium (K_2O) and titanium (TiO_2) which seemed successful in distinguishing the sigillate ceramics from these two production areas (Fig. 9.14). It is evident that the large portion of the sigillate ware with the Arezzo type decoration from the moulds found in the Lyon work sites are of the Lyon composition. However, two samples belong to the Arezzo compositional zone. It would seem, then, that some of the Arezzo pottery was made from these moulds, and trans- ported to Lyon.

Thus, one can propose close ties between the Arezzo production, occur- ring in the same period, with the Lyon factories. This is deduced not only on the basis of stylistic similarity but on the analysis of the materials them- selves which were produced in Lyon and Arezzo.

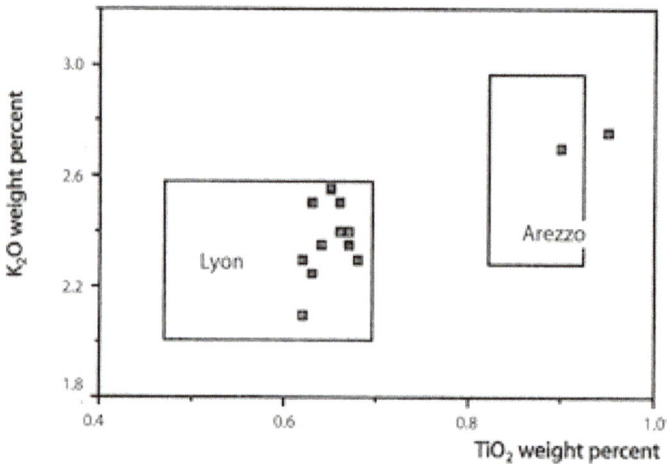

Fig. 9.14. Compositions of samples made from Arezzo moulds found in Lyon. It is obvious that most were made with Lyon clays, at least according to the potassium oxide (K2O) and titania (TiO2) contents. *Rectangles* show the ranges found for samples from the Arezzo and Lyon workshop sherds

9.3.4.2
Atevis Workshop

A well-known commercial stamp on sigillate pottery is that of Atevis. It is known in Arezzo and Pisa, in present-day Italy. Archaeological finds in the present French city of Strasbourg revealed several samples bearing the Atevis stamp. The question is obvious: local production or importation? The chemical analysis of the sherds revealed quickly that the material did not come from Arezzo (Fig. 9.15) but when compared with the Lyon work- shop compositions, it seems likely to have been made in this site. Thus, lacking evidence of a workshop in Strasbourg at this period, but having evidence of a workshop in Lyon, it appears that the production of the Ate- vis materials did not occur in Arezzo but probably in Lyon. Of course, it is possible that other workshops have the same characteristics in potassium content as do the Lyon sites, but it is still likely that the Atevis production occurred in Lyon as well as in Arezzo. Given the use of the same moulds in Arezzo and Lyon (preceding paragraph) it is likely that the Atevis link existed also.

These examples of the use of bulk (analysis of the whole sample) demonstrate the elements of the use of chemistry to solve some typical archaeological problems. The initial step in any such study is to form a good basis of analysis on well-known material representing the type of samples (production from a given site, those samples with a specific shape or decoration, etc.). Given a good basis for comparison (several dozens of different analyses), it is possible to compare a few or one sample with different source types. If several elemental abundances put the sample in question in a given group, where there is no overlap with another group, then one can assign a provenance to the sample. This method is based upon elementary reasoning, but such simple methods are often more useful than more complex ones which are based on comparisons in an absolute sense that might rely on a more limited number of data points.

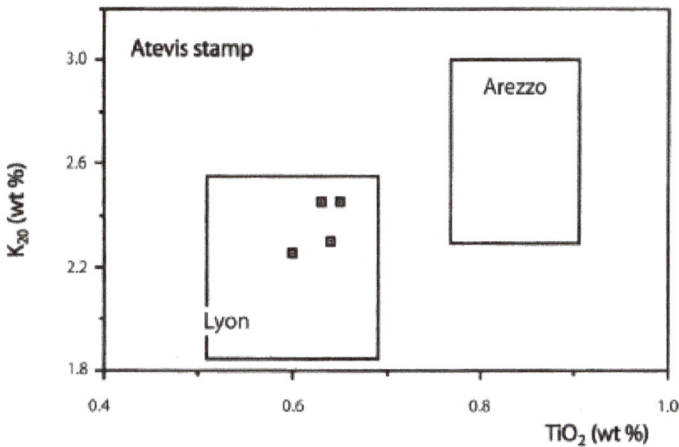

Fig. 9.15. Compositions of samples with Arezzo stamps found in Strasbourg (France) compared to the Arezzo and Lyon compositions *(rectangles)* in potassium (K_2O) and titania (TiO2) components.

9.4
Prehistoric Peru [4]

Now we will turn our attention to the South American continent to investigate pottery making in the past and present using ceramic petrology and chemistry. Here, we have an integrated study of materials and results in the same context over a long period of time. Different analysis methods are illustrated here, that have been used to answer questions on ceramic production and distribution.

The archaeological context of this example is the site of Chavin de Huantar, in the Peruvian Andes, during the 1st millennium B.C. At that time, Chavin de Huantar was an important ceremonial center. It attracted many visitors, judging from the religious offerings found in the temple galleries, in particular beautiful ceramic bottles and vases. It was also an active economic center, importing many long-distance goods. By the year 300 B.C., a large part of Peru had been influenced by the Chavin culture. Many sites in the surrounding valleys and to the Pacific coast showed ceramics of Chavin style, with seal-impressed decoration and strings of circle-and-dots, or with a more complex iconography related to religious ideology. The presence of these ceramics raises many questions. How were these Chavin-style ceramics produced? Were they imported or imitated? Was production activity centered in specialized workshops? Did Chavin de Huantar play a role in the production and distribution of Chavin-style ceramics?

Ceramic studies habitually tackle such questions. Mineral studies are often used to explore production processes (paste composition, tempering, manufacture) while chemical studies traditionally investigate provenance. Provenance studies can settle the imitation/importation question, as seen in the Atevis example above, distinguishing non-local pastes from local ones. Common methods are neutron activation and X-ray fluorescence, but simple petrographic studies can also be successful to discriminate between local and non-local composition based on the analysis of the non-plastic inclusions. In both cases, the local geology, the sampling of raw materials around the site, and ceramics of ascertained local production determine what is local. The criterion of abundance can also be used, for that matter. This criterion simply states that the most abundant material found in one site must be local, on the basis of least-cost principles and economic logic. However, this criterion should be carefully employed, as many counter examples are known, even for utilitarian ware (see a review of 'what is local' in Druc 2013, and Arnold 1985 for common distances to clays and temper materials based on worldwide ethnographic studies). Once the non-local ceramics have been identified, one can try to find their provenance, or see if they come from one of the sites sampled, thus proving or disproving intersite and interregional ceramic exchange.

---

[4] Druc, IC (1998) Ceramic production and distribution in the Chavín sphere of influence. British Archaeological Reports, International Series 731, Oxford.

Druc, IC (2004) Ceramic Diversity in Chavín de Huantar. *Latin American Antiquity* 15(3): 344-363.

Aside from the provenance question, it is interesting to know if different productions can be identified by the presence of different paste groups; if there was ceramic specialization in the paste preparation or type of ware produced; if a certain paste was reserved for a certain type of ware. This can be studied under a petrographic microscope. The texture, granulometry, and inclusion types, for example, will reveal paste preparation differences and the use of a special type of temper in relation to the function of a ware. Ceramics with the same mineral characteristics are then grouped together, and this classification can further be correlated with stylistic (decoration and ware forms), archaeological or chemical data.

Therefore, a research program was designed in order to better understand the production and distribution of the ceramics in the Chavin sphere of influence during this time period. Ceramics from different sites located in different Andean and coastal regions of Peru were sampled and analyzed. Only part of this study will be presented here to illustrate the uses and limits of ceramic characterization. We will take the ceramic sample of the ceremonial center of Chavin de Huantar, consisting of 126 sherds. These sherds come from digs in the ceremonial and residential areas of the site. They represent different ceramic types, utilitarian and fine ware, undecorated and decorated, bowls, bottles, cooking pots (ollas) and jars. The sherds were thin sectioned for petrographic analysis and another 100 to 300 mg of ceramic powder were pressed for X-ray fluorescence analysis. Local clays and soils were also sampled for comparison, and ethnographic records were made of modern ceramic activity in the region. Eventually, regional geological maps helped locate the possible sources of raw materials used by the potters, based on the mineral composition observed in the ceramics. Neutron activation was later conducted to see intra-site differences and temporal changes in composition at the site level of Chavin de Huantar.

9.4.1
Petrographic Analysis

Petrographic analysis was conducted first. It showed the wide diversity of mineral composition in the ceramic material found on the site - a diversity expected to be seen in the chemical analysis results. Many paste groups were identified on the basis of non-plastic grain types and granulometry. The majority of samples were distributed into three groups, while the rest (about 30%) showed rare compositional characteristics.

One of the major paste groups is characterized by glassy welded tuff (material largely in a glassy state produced by a volcanic eruption) inclusions of fine-to-coarse sand size. Some thin sections contain up to 40% of these tuff fragments, as well as large quartz grains with corrosion gulfs, twinned plagioclase, volcanic green hornblende, and large biotite flakes. These minerals are also found as phenocrysts in the tuff fragments, showing parental relationship between the volcanic fragments and the individual crystals in the matrix (i.e. they have the same geologic origin). No other rock type than the welded tuff is present. The mafic (iron, Fe, and magnesium, Mg-bearing) inclusions are unaltered, the quartz and plagioclase are clear and angular. These observations argue for the use of a volcanic raw material source close to its geological origin. Indeed, the materials show no sign of weathering alteration or water or air transport before having been picked out by the potters (Fig. 9.16).

Other thin sections of ceramics have only a few or no welded tuff inclusions but crypto- to microcrystalline rhyolithic fragments are present. Aside from this difference, the paste color, the texture, angularity and granulometry of the inclusions are the same as for the welded tuff group. Considering this, a pyroclastic, tuff layer occurrence was proposed as the possible source of provenance of the raw materials (at least as tempering agent). Indeed, pyroclastic material can include consolidated volcanic ash as well as more crystallized volcanic fragments and individual crystals all of which come from volcanic eruptions. This information helped locate on the geological map the closest volcanic sources to the site of study. These are small rhyolithic outcrops near copper mines, about 12 km NE of the site, which may well have been used for ceramic tempering.

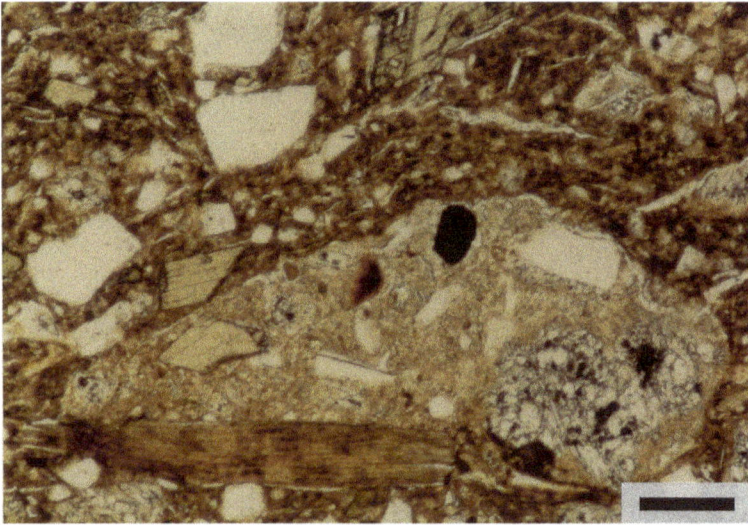

Fig. 9.16. Pyroclastic paste. Plain polarized light. *Bar* = 0.3 mm. The figure shows a very coarse lava fragment which contains small to medium-sized welded tuff fragments. Also one can see a large unoxidized biotite flake on the edge of the fragment (parallel cleavage lines apparent) as well as plagioclase and quartz crystals, green hornblende, and one opaque mineral. The clay matrix also contains tuff fragments, plagioclase, quartz, hornblende and biotite as individual crystals of fine-to-medium sand size.

Interestingly enough, the ceramic fragments with this volcanic paste pertain to coarse cooking pots and jars, a few bowls and two bottles (of a total of 32 ceramics). These ceramics are usually undecorated, and some are slipped. If one remembers the importance of thermal shock resistance and permeability properties for cooking and liquid container wares (see Chap. 6.3.3), the use of coarse, unseriated volcanic material makes sense.

Another particularity observed in the ceramic material of Chavin de Huantar is the granulometry and paste composition of the bottles and of some fine bowls. The majority of the bottles are made with a paste con- taining few inclusions (15-18 %), with very few or no rock fragments. The principal mineral inclusions are quartz, alkali feldspar, plagioclase, pyroxene, amphibole and biotite (iron- and magnesium-bearing mica) crystals, set in the clay matrix. Grain size is fine and homogeneous. Inclusions are well distributed, no voids are visible, and the paste is well compacted (Fig. 9.17).

Fig. 9.17. Fine paste. Plane polarized light. Decorated bottle, Chavin de Huantar. The paste has been well compacted and homogenized for these productions. It contains few non-plastic inclusions of fine sand size and smaller: a majority of quartz and feldspar (few plagioclase), some volcanic and sedimentary fragments, hornblende and minute biotite flakes. *Bar*= 0.3 mm.

The reason for using such a fine paste may be found in the type of ware produced. Bottle construction is different from that of cooking wares. Bottles have a globular body and a long tubular neck. Some of the bottles found at Chavin are also stirrup-spouted. The walls are thin (0-4-0.6 cm) but strong. Chavin bottles were usually fired in a reducing atmosphere, at least at the end of the firing cycle, to yield a black or grey-colored ware. They are often highly polished and decorated with an incision, stamping, applique or other texturing process. For the bottle form, wall thickness, bending of the stirrup, and small diameter of the spout require the use of a plastic paste with good workability. This can be achieved by using a well-homogenized clay with few and small inclusions. This is the paste quality observed for our bottle fragments. The care given to the finishing stage is also distinctive. It is both aesthetic and functional, as the polishing pushes the inclusions below the surface, offering a nice surface and enhancing impermeability.

An interesting point here is that the same mineral composition is observed for another group, but the granulometry and percentage of inclusions are different. In this latter group, the inclusions are coarser and more abundant, and there are many rock fragments. These are of orthoquartzite (metamorphic sandstone) composition, but also include a few sedimentary rock fragments (siltstones and quartzites) and fine-grained to trachytic volcanic clasts. Ceramics with this coarse paste type are common and occur in all categories (bowls, jars, cooking pots and a few bottles), decorated and undecorated. The similarity of composition suggests that the fine paste group (or part of it) is a refined version of the coarse paste composition. Thus, the raw material acquisition area for these two paste groups - the fine paste and the coarse paste group - must be the same. However, the paste preparation differs. By area, we mean the same geological region, but not necessarily the same source or quarry. There are, indeed, small compositional differences that indicate a plurality of quarries. As clay and temper sources are usually highly variable, this is normal. We see here the limits of provenance studies and the necessity of defining a production and acquisition area rather than a specific source for the raw material used. In this order, the lithic fragments present in the paste suggest a raw material acquisition area near intrusive outcrops of intermediate to basic composition.

The few sedimentary and volcanic fragments in the ceramic paste may be accidental or have been present in the clay as natural inclusions, as they are often smaller than the intrusive fragments. The fine paste may contain more clay and less or no temper. In this case, the mafic (iron, magnesium-bearing, pyroxene and amphibole) minerals are seen as natural inclusions and remnants of coarser rock fragments. The clay material of this fine paste group must have been finely ground and sieved, and the paste has been well kneaded. This shows a degree of specialization in paste preparation in relation to the production of a certain type of ware. It is difficult to say if the production was done in one (or more) specialized workshop(s), aside from the workshops producing regular utilitarian ware. It may be so, as this practice is attested ethnographically and for other time periods. Also, small compositional differences (i.e. in the amount and type of mafic minerals) are seen in the fine paste group, suggesting that more than one workshop was involved in fine ware production.

Another petrographic group, of the rare compositional type with only three examples, shows a completely different paste composition. It is highly micaceous, with metamorphic quartz-muscovite fragments and long individual muscovite flakes in the clay matrix (Fig. 9.18). This mineral composition is not found around the site, nor is it found in the surrounding region. The overall paste texture is totally different from the rest of the ceramic corpus found in Chavin de Huantar. The paucity of the fragments with this paste composition, the paste texture, as well as the absence of a similar metamorphic geological sequence in the region argue for a non-local provenance, as yet unknown.

Fig. 9.18. Example of a paste containing quartz and muscovite grains. Cross polarized light. *Bar* = 0.18 mm. This is a non local, decorated bowl, found in Chavin de Huantar, Peru. The paste contains some large quartz- muscovite schist fragments, as well as quartz and muscovite (long, cleavage plane lined) grains as individual minerals. There is no particle orientation in the paste and voids are present, indicating coarse kneading. This micaceous composition also shows on the polished surface of the ware, where the mica flakes are visible to the naked eye.

9.4.2
Modal Analysis

The above observations are based on qualitative mineral analysis. We see how much information petrography already yields on ceramic production and characterization for this particular site. For intersample comparison, however, quantitative data are needed. The mineral composition can be studied by conducting quantitative petrographic analysis. One method is modal analysis for estimating the percent of non-plastics per mineral types and sizes, the other is image analysis, for granulometric distributions. Both methods aim at producing composition profiles for comparative purposes and ceramic classification.

Direct observation is necessary when minerals cannot be separated on the basic of grey levels detected in black and white reproductions of the ceramic internal structure. This case is important when ferro-magnesian minerals are present in the paste. If only quartz and feldspar compose the temper grain population, one can separate them from the clay matrix using a simple black and white photographic method. However, when minerals are more colored, their hues can be those near that of the paste, especially in a black and white reproduction. In these cases, it is necessary to use the expertise of a knowledge of petrography to identify, one by one, the species and hence the mineral grains. Hence, modal analysis is the only way to separate the objects.

Modal analysis was done by ribbon count. Other counting methods exist, such as point counting (a good description is given in Stoltman 1989) and area count, but ruban counting was the most appropriate for the instrument available and the corpus studied (i.e. ceramics with highly varied composition and granulometry). In the ruban count method, the thin section is moved along an arbitrary line or band, and all sand-size grains situated on this line are identified and reported on a classification grid by type and size. The principle is the same with the other methods, except that counting is done at regular intervals (point counting) or includes all grains in a given area (area count). For point-counting analysis, the microscope can also be coupled with an electric counter, with keys corresponding to different inclusion types. The counter registers and adds up the counts of the inclusions when the appropriate key is pressed (e.g. fine quartz corresponds to key 1, medium tuff to key 2, and so on).

The built-in ruler in the reticula of the microscope allows for the measurement of the diameter and length of the non-plastic grains. According to these measures, the grains are classified on a geological scale. The scale used here is Udden-Wentworth, a very common scale which groups the silt- and sand-size fractions in seven granulometric classes with limits expressed in millimeters. Another scale widely used to present granulometric data is Krumbein's phi scale (Folk, 1965), a logarithmic transformation of Udden-Wentworth (Table 9.1).

Once the counting is done, the totals of the modal categories (fine quartz, medium quartz, coarse metamorphic fragments, etc.) are entered into a spread sheet to compute statistical analysis. Simple to more sophisticated analyses can be done to measure inter- and intracategory variability, compare groups, or see which mode (i.e.type and size of grains) best distinguishes one ceramic production from another.

Table 9.1. Granulometric scales

	Type of material	Maximum size (mm)	phi scale Krumbein	Category U-W
Clay	Clay	0.002	8	1
Resourc	Silt	0.03	5	2
	Coarse silt	0.063	4	
Sand	Very fine sand	0.13	3	
	Fine sand	0.25	2	3
	Medium sand	0.5	1	4
	Coarse sand	1	0	5
	Very coarse sand	2	-1	6
Gravel	Gravel	4	-2	7

Figure 9.19 shows the histogram of the frequency of grain categories per paste group. The histogram confirms the intrasite differences and the groups' constitution

observed earlier by simple petrographic inspection. It is here better apprehended by the graphic representation. For example, we clearly see the high mica (muscovite) content of the micaceous non-local group (Ch-E), and the strong presence of clear inclusions in the fine paste group (Ch-F) discussed in the petrographic section above.

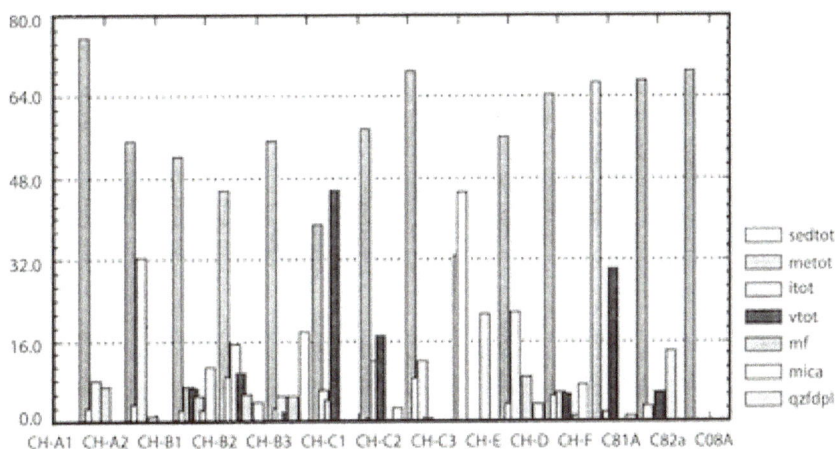

Fig. 9.19. Modal analysis. Frequency of lithic and mineral types per ceramic paste group. This histogram helps visualize intergroup differences. The inclusion types are general to allow overall comparison. Abbreviations on the right of the graph stand for total of sedimentary rock fragments *(sedtot);* metamorphic *(metot);* intrusive (itot); volcanic *(vtot);* mafic individual minerals *(mj);* micas *(mica);* quartz, feldspar and plagioclase *(qzfdpl).*

To show intersite compositional differences, using the mineral variables of the modal analysis, a principal component analysis (PCA) was conducted. Here we look at the overall mineral composition of the ceramics from five different archaeological sites. The analysis sorted out the variables that best distinguish one site from the other. The first two components or factors explain 67% of the total intersite mineral variance, which means that the corresponding plot (Fig. 9.20) is a fairly good representation of the sites' mineral characteristics and intersite differences. The relative position of the sites in the graph indicates how different they are, while the position of the modal variables allows us to understand which variable is responsible for theses differences. The further a variable is from the origin, the more determinant it is in the ceramic composition profile of a site. We see by looking at the plot that each site is characterized by a specific ceramic compositional profile. Qualitative analysis already pointed to these differences, but modal analysis provides a quantitative account of them. The two types of analysis are complementary.

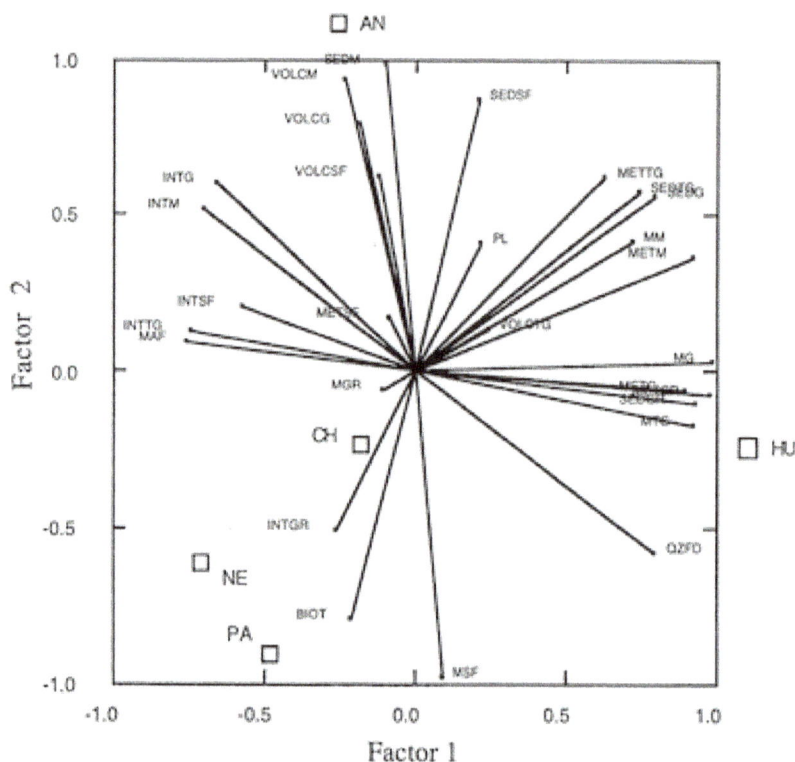

Fig. 9.20. Plot of principal component scores (first two factors) of the ceramic mineral composition per site, based on modal variables of the petrographic analysis. The lithic (rock fragment) and mineral types reported in the modal analysis are used in PCA to highlight mineral differences between the ceramics of various sites. The graph illustrates the mineral tendencies for each site (*CH, AN, HU, PA, NE*). The ceramics of site *AN* (Ancon) are characterized by volcanic and sedimentary fragments of all sand sizes (f fine; *m* medium; *g* coarse). This profile is opposed to the more intrusive composition of the ceramics of sites PA and NE (Pallka and Nepeña). Both sites are in a similar geological environment. They present similar mineral (and chemical) profiles, as shown by their proximity on the graph.

9.4.3
Chemical Analysis - XRF and INAA

The preceding sections illustrated the use of mineral analysis in answering questions of ceramic production. Local and non-local ceramics were identified, but for provenance, chemical analysis is often more efficient. The chemical composition of a ceramic is like its fingerprint; it bears the characteristics of the region of origin, that is, of the region where the raw materials constituting the ceramic have been taken. These characteristics are a particular combination of major, minor and trace elements (or of minerals and rock fragments when speaking of mineral composition). If two ceramics are made with the same raw materials, they should yield the same composition profile, thus indicating the same provenance. They can further be compared to samples of raw materials from a known acquisition area. If the results match, it confirms the origin of the ceramics, assuming the sample variability and chemical differences between raw and fired material are minimum. Local ceramics should cluster together, while non-local ware should stand out.

In the example with the pre-Hispanic corpus, chemical analysis was done first with an energy dispersive X-ray fluorescence instrument (EDXRF). A later analysis used INAA and will be presented further below. For EDXRF, the device used allows the measurement of several elements at once, under varying analytical conditions. Up to 16 samples can be analyzed in one batch. The whole process takes about 6 to 8 h, depending on the conditions of analysis. The intensity spectra are treated and compared to standards to calculate the elemental concentrations. These concentrations are normalized and saved for statistical analysis. It is then left to statistical analysis to discover patterns in the compositional profiles, that is, similarities between ceramics allowing their clustering or attribution to a specific chemical group.

In the Chavin corpus, many ceramics had been found to be non-local (chemically speaking). To see if these ceramics originated from one of the other sites sampled, all samples were put together and subjected to a discriminant analysis. Individual analysis, by site, had shown the characteristics of each site, the different compositional groups, the elements responsible for these differences and so on. At a higher level of analysis, including all sites, intrasite differences are ignored to concentrate on intersite similarities. The sites served as reference groups (whose internal homogeneity had been controlled) to which the compositionally different ceramics were compared. The discriminant analysis yielded the probabilities of site attribution. For example, one ceramic from Chavin with atypical composition had an 8% probability of belonging to the main Chavin group (called local), while it showed a 92% probability of belonging to the group from site B, in the next valley. When probabilities approach 96-98%, we can say with some confidence that the provenance of this ceramic could be site B. In this way, several ceramics have been traced back to their presumed place of production, indicating that intervalley ceramic exchange had taken place. Of course, there is always a possibility that another site, unsampled, exists, with ceramics of a chemical composition more akin to the atypical ceramic. In this case, petrography can help identify the other possible regions of provenance presenting a geology similar to the one observed in thin sections. Also, one should always keep in mind the relativity of statistics and sample representativity.

Ceramic analysis is a tool, a means and not an end. Other data must be integrated to make sense of the results and reach higher explicative power, such as archaeological, geological and stylistic data. For example, stylistic analysis of the ceramics showing non-local origin revealed that nearly all are decorated bowls and bottles. Intervalley and long-distance exchange is thus associated with a special type of ware (i.e. fine ware). These bowls and bottles, however, do not show similar paste characteristics, indicating that they were not produced by the same workshop. There must have been a diversity of pottery-making workshops and no specialized ceramic production and distributional networks can be attested here. Also, ceramic distribution shows that a small quantity of pots from the other sites of study were brought to the ceremonial center of Chavin de Huantar, but none was exported from that site to the other regions. The religious importance of Chavin de Huantar could have attracted visitors, not pottery production. The cold and humid climate, high altitude and sparse clay resources of the Chavin de Huantar region make the site a poor candidate for being a large ceramic-producing center. All in all, the EDXRF analysis suggested that at least 30 % of the ceramics found in Chavín de Huantar were of non-local production. Most of these were bottles and fine wares, probably brought to the site as gifts or offerings.

Although XRF worked well at the inter-regional level, at the intrasite level it would not distinguish groups nor partition the corpus as accurately as petrographic classification. Thus, INAA was conducted with samples from Chavín de Huantar and four other sites for comparative purposes (Ron Bishop and Jim Blackman conducted the chemical analysis at the Smithsonian laboratory). The samples were prepared from 200 to 400 mg of paste drilled from the ceramic fragment cross-sections and from pressed powder pellets. Standards were run at the same time. 31 elements were measured, and 18 elements were retained for the statistical analysis: Na, K, Sc, Cr, Fe, Rb, Cs, Ba, La, Ce, Sm, Eu, Tb, Yb, Lu, Hf, Ta, Th. Elements were chosen according to smallest analytical error, expected values in ceramics, and inspection of the raw data. The alkali Cs and Ba were important discriminators for this specific set of samples and were thus kept for the statistics. Raw data was log-transformed and data exploration was conducted via multivariate analyses using different clustering methods and principal component analysis (PCA). The average linkage method allowed for more homogeneous clusters in terms of the mineral composition of the samples. As cluster analysis imposes different patterning depending on the method used an external validation of the results is required, in this case mineralogy and ware style. In principal component analysis, the variance-covariance matrix was preferred over the use of a correlation matrix. The latter centers the distribution, which is not recommended for very heterogeneous samples, as is the case with the ceramic compositions from Chavín de Huantar.

In this study, one element in particular, cesiumn (Cs), discriminated well the samples in terms of their provenance. High Cs content (above 11 ppm) proved to be an indicator of non-local provenance, which was correlated with information on style, mineralogy, and archaeological provenance of the sherds.

Cesium presence is linked to the geological environment of the resource area for ceramic production. Cesium may be found in late stage-formation crystals in acidic rocks, leucite, feldspar and similar minerals, and as traces in black mica (biotite) and white mica (muscovite) in pegmatite, but not in metamorphic rocks. High cesium contents may also point to the use of micaceous clay to produce the ceramics, due to the preferential absorption of Cs associated with micaceous clays through geological history (Ron Bishop, personal communication 1998). Also, chemical tendency and associated mineralogy suggest that Cs is a component of acid volcanic rocks, while it is of low abundance in ceramics with intermediate to basic intrusive paste.

Thus, ceramics with low Cs content (below 11 ppm) would possibly be local productions. We see that they fall into the mineral group displaying basic to intermediate intrusive rock fragments, into the sedimentary-intrusive group, and into the volcanic-temper group. The region is characterized by intrusive-granodioritic and sedimentary geology, which explain the lower Cs content in local Chavín ceramics.

At the intrasite level, INAA studies allowed the compositional diversity observed with petrography to be linked to production scenarios changing over time, and to a considerable amount of non-local wares of different origins. According to chemical and mineral analyses, various centers of production were supplying the settlement and ceremonial center of Chavín de Huantar. Several workshops were probably active in each area around the site, as suggested by the internal heterogeneity of the compositional groups. The volcanic temper was used during the first occupation phase at Chavín de Huantar, and ollas and bowls were produced. Later in time, during maximum extension of the site, the tendency was to use materials from the slope of the White Cordillera, rich in intrusive rocks, to produce a wider range of ceramic forms, including bottles and jars. A few workshops must have mined sedimentary material from the valley bottom to produce fine paste bottles and decorated bowls throughout the existence of the site. However, this production must have been low.

So, a dramatic shift is observed in the raw materials used, at the end of the first occupation phase, while production became more diversified and intense as the site and its population grew larger. This resource procurement change cannot be explained based on paste analysis alone. Resource exhaustion, landslides burying resources, or technological changes are possibilities, but we know from ethnographic studies that potters tend to be conservative in the technology they use. They adapt and change styles or forms, not raw materials. So it is probable that new potters introduced new manufacturing traditions to respond to a higher demand as the site expanded.

We see in this Peruvian case study how different analysis techniques are integrated to answer specific questions on ceramic production and distribution. Petrography, modal analysis of thin sections, chemical studies, as well as local geology, stylistic comparisons and archaeology all combine to yield a better picture of the past ceramic activity. Data interpretation and the full potential of characterization benefit from different points of view.

9.5
Modern Ceramic Production in the Andes

Ceramic production studies can make use of ethnoarchaeological investigation. One example of the clay-temper-pot relationship is given here to illustrate the use of petrography for production studies. The analysis of the materials used by the potters serve for the investigation of the mineral relationship between the different constituents of the ceramic paste, with an emphasis on the compositional and granulometric changes occurring during the manufacturing process as a result of the preparation and mixing of the constituents. This example is based on a larger study of the modern ceramic production of Andean potters.[5] To show how the inclusions of both clay and temper can be recognized in the paste, the study reports on analyses of samples of unfired clay and temper, unfired clay-temper mix and fragments of fired pots. Interviews with potters and field observation helped relate the petrographic observation of paste texture and granulometry to manufacturing processes. Such analysis of modern ware helps in building a production scenario and a conceptual framework for the interpretation of the analysis of prehistoric ceramics for a specific region of study: what granulometry is to be expected, what composition, and what texture, assuming some continuity in the ceramic tradition (see Cantin and Mayor 2018 for some cautionary tales). It can also help in distinguishing clay from temper in ceramic fragments from archaeological contexts.

9.5.1
Production Setting

The four samples analyzed here come from one potter of one ceramic-producing village in the Callej6n de Huaylas, a high inter-Andean valley in North-Central Peru. These samples are from the raw materials used, the mix and the fired product. The potters of the Callej6n still produce ceramics in a traditional way, that is, without wheel or electric kiln. The wheel, kiln and glazes are, however, used by a new generation of potters to pro- duce tourist ware, teapots, ashtrays and the like. The clay used is different and refined by decantation. These are recently introduced materials and processes and thus of no concern here.

The ware produced by the traditional potters is utilitarian: cooking pots and jars of different sizes, and maize toasters. They are made with the coiling technique, using a few big coils (about 5 cm in diameter) thinned with the hand and finished with the paddle-and-anvil technique. Pots are pebble-polished but not slipped. A white painted line often decorates the shoulder of the pots. The firing is done in an open fire, on the ground, in a field or in the potter's backyard. Pots are piled over wood and covered with straw. The firing is short (about 2 h), and the temperature is not very high. The pots are ready when they have the correct color and sound right when hit. As is evident, the process is not highly controlled; it is largely based on the practical experience of the potter. Thus, often a ceramic may be incompletely oxidized or a pot could present fire marks (dark spots) on the outside.

---

[5] Druc, IC, and Gwyn, H (1998) From clay to pots: a petrographic analysis of ceramic production in the Callejon de Huaylas, north central Andes, Peru. *J of Arch Sci,* 25:707-718; Druc, IC (1996) De la ethnografia hacia la arqueologia: aportes de entrevistas con ceramistas de Ancash (Peru) para la caracterizaci6n de la ceramica prehispanica. *Bulletin de l'Institut français d'études andines* 25(1): 17-41. 2005. Producción cerámica y etnoarqueología en Conchucos, Ancash, Perú. Instituto Runa, Lima. 2007. Encounters with Andean Potters. E-book with 14 min video. National Social Science Press, CA.

Most important for our purpose, however, are the raw materials the traditional potters use and how they are prepared. The potters use two clays, or earths, as they call them, distinguished by their color: a yellow one and a black one. The proportions vary between 1 to 3 parts black clay to 1part yellow. Neither of these clays was designated as temper by the potters, so one goal of the analysis was to determine which of the two clays serves as temper. Given the functions of tempering, that is, modulating clay, modifying its plasticity, and adding non-plastic inclusions to open the paste (cf. Chap. 7), it is to be expected that one of the two raw materials will be coarser or of different composition or have a higher non-plastic content. This is indeed the case, as will be seen in the example detailed below.

Distinguishing the clay from the added material in a ceramic fragment is not an easy task, especially when two clay sources are used. The job may be to recognize the presence of two different constituents, or two different sources for the inclusions in the paste. The petrographic microscope allows for the study of non-plastic inclusions larger than clay and fine silt particles. These inclusions yield information on the parent material and the geological environment of the source of the raw material, information which can lead to provenance studies. The main point is thus to determine which inclusions pertain to the clay, and which are added by the potter as second constituent.

Different clay sources show different compositions. In our case, the black clay source is 7 h away on foot (about 9 km) from the place of production, high in the mountains, near the limit of old glacial moraines. In contrast, the yellow earth is obtained near the potter's village in the valley. The geology of the region is composed of acid to intermediate composition intrusive rocks of granitic to granodioritic and tonalitic composition, (the White Cordillera), which are intrusive into the volcanic and sedimentary sequences found in the valley. The valley bottom is at 3200 m at its upper end, while glacier, glacier lakes and moraines are found above 4800 m. This glacier context is responsible for much of the tills and alluvion found in the valley.

The raw clay used by the potters is a hard and compact sediment when in the dried state. It must be crushed with a hammer or a stone before use, and it only becomes plastic when hydrated –a characteristic of shale, which proved to be the main component of the black "earth". The black component is due to organic material in the sediments compacted into a rock. The clay-rich sedimentary rock yields plastic material when finely ground. One also finds rock fragments which function as temper grains. The yellow earth found in the valley is less compacted and less hard to crush than the black earth. Clay (black resource) and temper (yellow resource) are ground separately and sieved, the large inclusions being removed by hand. They are then mixed together with water and kneaded to reach the desired homogeneity and paste consistency to be formed into coils to build cooking pots. Raw material composition and preparation influence paste granulometry and texture.

The above information is important in interpreting mineral and chemical analysis data. Ceramic production depends upon many factors, such as geology, ecology, tradition, socioeconomics, which all have an impact on the finished product. These factors cannot be forgotten if one aims at reconstructing human activity.

9.5.2
Petrographic Analysis

The analysis shows that both clay resource materials contain non-plastic inclusions (minerals and lithics or rock fragments), but in differing proportions and composition types. The black clay is richer in clay particles or clay minerals in the form of compacted or cemented clay sediments (shale fragments), while the yellow clay contains many quartz and feldspar mineral grains, as well as sandstones but almost no clay minerals nor clay-sized material. For this reason, we call the yellow clay the temper. It contains little clay mineral material and thus has less plasticity when mixed with water to form a paste.

9.5.2.1
The Unprepared Black Clay

This sample comes from unprepared material that has not been crushed or sieved by the potter. Consequently, the loose grains in the thin section are very large lithic fragments with a few individual minerals (Fig. 9.21). These are mainly quartz, K-feldspar, plagioclase, and biotite grains that may have been detached from the lithics (rock fragments) during the clay extraction and transport or the thin section preparation. The grains are set in epoxy on the glass slide, and their distribution is not related to the original material (no paste texture here to analyze). The lithic composition is varied but shales are predominant. The other lithic fragments are from claystone, siltstone, fine quartz-arenite, clay-rich oxidized pellets and some sandstone and volcanic trachytic clasts surrounded by clay or occurring as inclusions in shale fragments. Some siltstones grade to claystones, and there are shales with silt and oxidized beddings. The lithics are of medium to very coarse sand size, with many shales of granule size (up to 4 mm long). The lithic fragments are rounded to subrounded. The composition and compaction of material, granulometry and grain shape identify the source as an unclassed consolidated clay-shale-rich deposit.

9.5.2.2
The Unprepared Yellow Temper

The yellow earth (Fig. 9.22) shows large fragments of quartz sandstone with mica, cement, orthoconglomerate, sandstone, wacke, and some clay pellets and secondary calcite grains. There are no shale, volcanic or intrusive rock fragments present. A majority of monomineral grains are visible: angular quartz with rolling extinction, feldspar, zoned plagioclase, fine mica flakes, pyroxene and opaque minerals. Quartz represents about 70% of the inclusions. Granulometry readings range from silt to fine sand size, with one quartz grain of very coarse sand size. Inclusions are subrounded to subangular, of silt to coarse sand size. Lithic fragments are generally of medium to coarse sand size, except some quartz arenite and quartzite fragments of very coarse to granule size. This material is much sandier than the black clay. It is an unsorted sediment of mix composition, with loose grains in a clay-sand matrix. It is unconsolidated, not yet a rock.

Fig. 9.21. Unprepared clay (roughly crushed, unsieved) with shale fragments, quartz, feldspar, plagioclase, biotite, occasional rounded volcanic clasts, iron nodules, clay pellets, and loose minerals in unconsolidated clayey soil. xpl. *Bar* = 0.6 mm. Vicus-Pashpa clay, White Cordillera Peru. This clay is eluvial, resulting from in-situ weathering and gravitational accumulation of ancient glacier-covered sediments.

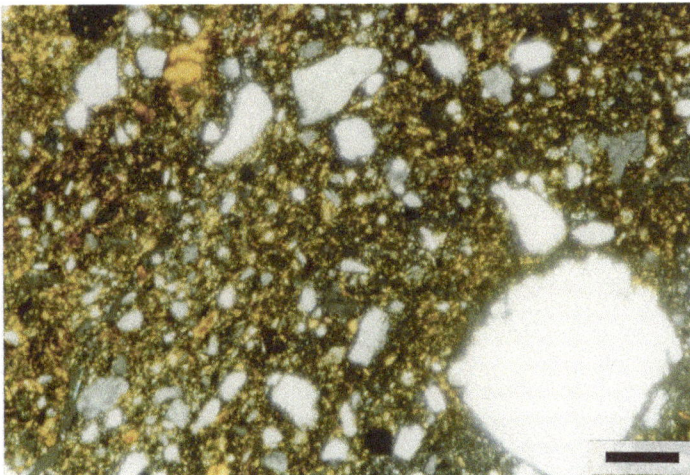

Fig. 9.22. Unprepared yellow temper: unsorted quartz and feldspar grains and very fine mica flakes in the matrix. The coarse grain lower right is an alkali-feldspar. Plane polarized light. *Bar* = 1.8 mm

9.5.2.3
The Clay-Temper Mix

As stated before, the preparation of the raw materials includes crushing and sieving to remove the coarser inclusions. Materials are not finely ground and no decantation or levigation is performed. The sample of clay temper mix analyzed shows a difference in the granulometry of the inelusions when compared to the unprepared raw materials (Fig. 9.23).

Fig. 9.23. Clay-temper mix. Plane polarized light. *Bar=* 0.5 mm. Shale and sandstone fragments are the coarse non-plastics in the clay matrix, surrounded by fine to medium monomineral grains (quartz and feldspar appear as *white grains* in the photograph)

None of the granular grains (above 2 mm) is visible, and very coarse fragments are rare. They have been crushed or picked away. The maximum granulometry of the inclusions in the mix reflects the canvas sieve lattice size, that is, about 2 mm. The measure of the sphericity factor done by image analysis shows that grain angularity is not significantly enhanced by the grinding process. That is, no sharp differences are observable in the shape of the inclusions in the raw material before and after milling. The mix has not yet been homogenized and compacted, either. Large voids and clots of inclusions are visible. The color of the mix is yellow in plane polarized light (PPL) under the petrographic microscope.

9.5.3
The Fired Pot Fragment

The rock fragments in the ceramic paste are shales, quartz sandstone, volcanic rock and clay-mica rock fragments, quartzite and schist, and basaltic volcanic fragments. The quartz schist and tectonized quartzite were not present in the samples of raw materials taken for analysis. Their occurrence in the ceramic paste illustrates the sample variability that may exist. Ideally, a large sampling should be done, but even then accessory minerals and occasional grains may be missed. A fairly good overall geological profile, however, can be obtained with a small sampling. The metamorphic fragments are rare and rounded, indicative of a secondary sedimentary setting and a weathered and transported material.

The lithic fragments are of coarse silt to coarse sand size, with few of very coarse sand size. The individual minerals consist largely of quartz and feldspar, with many plagioclases. The mafic minerals are amphiboles and fine mica flakes. Some opaque minerals are present. The paste matrix is red in crossed polarizers (XPL) under the petrographic microscope compared to the yellow color of the unfired material. The paste is also more compacted and homogeneous as a result of paste kneading and forming of the ware, with compaction due to hand pressure and paddling. Granulometry is the same as in the mix, finer than in the raw materials (Fig. 9.24).

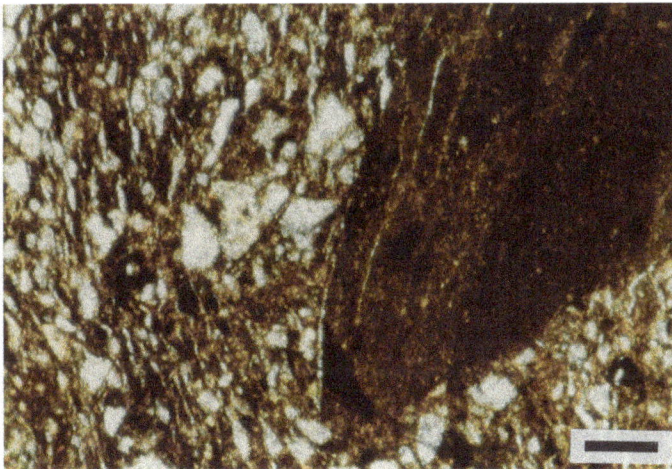

Fig. 9.24. Cooking pot thin section. Crossed polarized light. Bar = 0.18 mm. This figure, at a high magnification, shows a large shale fragment in a clay matrix with many small quartz and feldspar grains. Many drying cracks (elongated pores) are visible, which were not apparent in the unfired clay-temper mix

The contribution of the different materials to the ceramic paste can be determined from the preceding analysis of the clay and temper used by the potter. Thus, the black clay contribution consists primarily of sedimentary fragments (shales), secondarily of metamorphic and volcanic fragments, clay pellets, and some clear and mafic minerals. The yellow temper assemblage accounts for the majority of the individual inclusions, the clay pellets and the quartzite fragments. The coarse quartz grains come from the yellow earth, since there is no quartz and feldspar of similar size in the black clay source. The fine-grained volcanic fragments are found only in the clay and therefore come from this material. Both constituents are responsible for clay particles, sedimentary fragments and individual quartz and feldspar inclusions.

9.5.4
Image Analysis

We now see how these petrographic observations and the respective contributions of the clay and temper non-plastics can be quantified. Similar to the scanner system (Chap. 8.1.1), image analysis allows for the automatic counting and measuring of objects. The clay and temper contributions can be measured in surface areas, rather than in number of grains. This is only an estimate of their proportion in the paste, as a thin section offers two-dimensional views of three-dimensional objects which have been randomly cut. Volume and percentage of inclusions are thus only approximations of the real figures.

The instrument used here is a petrographic microscope with a video camera coupled to a video screen and a computer. The image analysis program captures the image of part of the thin section through the video interface. The image is in pixels and must be treated to enhance its resolution and delineate the objects to be analyzed. The treatment consists of adjusting or modifying the grey levels (the image is in half-tones) to obtain a better grain-to-matrix contrast. Grain contours are corrected (i.e. if two grains connect, they may be counted as one object), layer coloring and other technical twists are performed to prepare the image for the analysis. Object counting and measuring can then proceed. In a few seconds all the objects (i.e. the minerals and rock fragments) on the screen are numbered and measured. The thin section is then moved so another area can be analyzed. Modal analysis can be done this way, but the analyst must first identify the different grain types by marking them before the counting starts. Also, a larger surface of the thin section must be analyzed to get a good representation of the mineral types present. The process is longer and less automatic than simple granulometric analysis. Subprograms have, however, been created to automatize these operations (see Schmitt 1993).

Granulometric analysis by imaging program yields a temper-to-clay surface count in percentage of the area occupied by the non-plastic inclusions in the image field. The inclusions are attributed to the black clay and the temper in view of the mineral types observed in the original raw materials. The shale fragments are attributed to the clay and all individual clear grains (quartz and feldspar) are classed as coming from the temper. This introduces a bias in favour of the yellow temper. It must also be remembered that clay and fine silt particles cannot be distinguished due to the instrument resolution, and are not accounted for in the temper-to-clay surface counts. In any case, these attributions would be based on proportions as observed in ethnographic setting (how much temper the potter adds to the clay). Measurement of the exact participation of the non-plastics of the clay and temper material in the ceramic paste is quite impossible.

Nevertheless, the following figures are very interesting (Table 9.2). The counts were made at four different points in the thin section of the fired pot. Each image captured by the video attached to the microscope represents a very small area of the total ceramic thin section (639 x 480 pixels or 2.65 mm^2). The thin section, in turn, is a small sample of the whole pot. This should be kept in mind when generalizing the results of the analyses. The variability observed in the following counts per area reflects the inhomogeneity and coarseness of the material. The mean area count is a better representation of the percentage of clay and temper inclusions in the ceramic paste.

Table 9.2. Area counts of non-plastics in modern cooking pot

	Black Clay (shale)	Yellow Temper (quartz and feldspar)
Area 1	20.43% (4 objects)	35.18% temper (354 objects)
Area 2	21.66% (3 objects)	16.15% (458 objects)
Area 3	22.2 % (3 objects)	11.32% (344 objects)
Area 4	38.3 % (5 objects)	11.55% (296 objects)
Mean area:	25.66%	18.55%

In general, there is more clay than temper material in surface area covered by the non-plastics (the plastic portion of the paste, that is the clay particles or the clay matrix, makes up the rest of the image field, with a mean percentage of 55-79 %). Grains attributed to the temper are however much more numerous and smaller than the shale fragments coming from the clay. The large temper percentage in area 1 represents an area of the thin section with many quartz minerals and few clay pellets and shales. In contrast, the last area analyzed presents a large clay percentage due to very coarse shale fragments taking up much of the surface area in the image field. These figures imperfectly reflect the clay-temper proportions used by the potter (1:1), even considering the bias in favour of the temper; but one should remember that the proportion of non-plastics should be balanced by the plastic part of each constituent, which we cannot measure. Thus exact clay-to-temper proportions cannot be deduced from these analyses when two clays are used in the ceramic production process. However, the technique can be used to estimate the proportion of non-plastics in the paste, as seen in the pre-Hispanic example.

These results do not exclude looking for the acquisition area of the different raw materials used to produce ancient ceramics, providing at least one constituent could be identified in the ceramic material. We see from the image analysis results and from simple qualitative analysis that a small number of inclusions are much coarser and of different mineral type than the majority of the grains. This reveals the presence of two different constituents in the paste. In this example, the coarser fraction is to be attributed to the temper and the finer fraction to the clay material. This can serve as a production model or a comparative one when investigating ancient ceramics of the same region.

Finally, looking at the mineral composition of the ceramic thin section, the high number of mineral and arenaceous inclusions, mainly quartz and feldspar, could lead one to infer the addition of sand to temper the clay base. As the analysis of the raw materials has shown, this would be wrong. A coarse clay is used here as tempering material and not sand. Instances like this argue strongly for the use of comparative material and sampling of clay beds to attain familiarity with their diversity and variability around the site or region of study.

This brief example presents an aspect of ceramic production studies using petrography and image analysis. Such study can be further enlarged to investigate interpotter, intravillage, or production variability, or the incidence of grain angularity due to grinding, a subject not much studied in ceramic production.

9.6
Clay Characterization by SEM (Scanning Electron Microscope)

A final example of analysis using some of the pre-Hispanic samples from Peru is the characterization of the clay matrix by SEM-EDS (SEM with Energy Dispersive X-Ray spectroscopy capacity). To study the clay or the clay matrix in the ceramic paste requires an instrument whose resolution and magnification are more powerful than a simple petrographic microscope. The SEM (Chap. 10.3.5.1) is such an instrument. High magnification and an X-ray fluorescent radiation microanalysis functions allow one to study the microstructure and elemental composition of the grains on which the electron beam is focused. Particle and grain identification is based on chemical composition and not on optical properties as in petrography. Another function, not used here, is elemental mapping (establishing a two-dimensional intensity map of elemental concentration) to visualize the distribution of the different elements in the sample.

The example presented here illustrates the X-ray analysis function of the SEM to distinguish different ceramics based on the Si/Al ratio of their clay matrix. To understand the analysis done, the basic principle of the SEM will first be recalled. In this type of microscopy (see Chapter 10 section 10.3.5.1), a beam of electrons scans the surface of the sample, which is set in a closed chamber in the microscope under high vacuum. The interactions with the sample produce electron energy intensities which are displayed on a cathode screen. An image is thus produced. Magnification can be up to 300 000 times the object on which the beam is focused. The electron image can be obtained in secondary electron emission or backscattered electron mode, according to the work to be done. Chemical analysis can be made by analyzing the X-radiation excited by the electron beam in the minerals. Consequently, the X-rays yield information on the composition of the area analyzed and the shape of the grains present. As with XRF analysis, the X-ray analysis performed on SEM can use an energy-dispersive or a wavelength spectrometer. The samples used in this analysis are uncovered petrographic thin sections. This option allows one to make a direct comparison with petro- graphic observation while preserving the macrostructure of the ceramic paste, which is not the case if the sample is reduced to powder. Since the sample is a thin section, there is no need for polishing. Also, the study conducted does not require a very high polish, as the image quality is not essential for microanalysis. The thin section is metallized to offer a conducting surface. This is done by applying a thin film of gold-palladium of 100 to 200 Å. A carbon coating could also be used for that matter. The sample should also withstand high vacuum and present no traces of humidity.

The microanalysis is conducted by positioning the electron beam on an area free of non-plastic grains, i.e. the clay mineral-rich portion of the sample. The intensity of the peaks in the elemental spectrum is measured and the concentrations of different chemical elements present in the sample are calculated using corrections accounting for the physics of interaction of the electron beam with the sample and the interactions of the X-radiation leaving the sample. The analysis was performed at three different points in the clay matrix of the sample in order to control compositional homogeneity. As an example, one of the spots analyzed yields the elemental contents presented in Table 9.3. Elements to be measured were chosen to correspond to the usual composition of clays. Percentage of element contents is calculated and normalized using an internal standard. This analysis does not take into account the oxygen content, and hence only the cations are considered.

Table 9.3. Link AN 10000 X-ray analysis – locus 1 ceramic 3780 b

Element	Correction factor	Weight percent	Atom percent
Na	0.886	0.158	0.207
Mg	0.872	1.262	1.566
Al	0.932	25.011	27.968
Si	0.724	53.489	57.450
p	0.654	0.210	0.205
Ca	0.883	2.938	2.212
Ti	0.808	0.810	0.510
Fe	0.843	11.120	6.007
K	0.983	5.022	3.875
Total		100.019	100.000

We are interested here in the silicon and aluminium contents, which are major clay constituents. It is the Si/Al ratio which will characterize the clay type. For example, montmorillonites have Si/Al ratios between 2.58 and 3.47, while illites have a ratio between 1.53 and 2.62, and kaolin clays have a Si/Al ratio close to 1(Newman, 1987).

Table 9.4. Si/Al ratios of clay matrix in pre-Hispanic ceramics

Samples	Si/Al	Locus 1	Locus 2	Locus 3	Mean	SD
No.3708A	Bowl	4.91	4.20	4.57	4.56	0.35
No.3742d	Bottle	2.006	2.005	1.98	1.997	0.01
No.3755a	Bowl	1.90	1.93	1.91	1.91	0.01
No.3757a	Bowl	2.84	2.69	2.84	2.79	0.09
No.3764b	Jar	2.64	2.13	1.98	2.25	0.35
No. 3780a	Cooking	2.62	2.58	2.64	2.61	0.03
No.3780b	Cooking	2.14	2.30	2.27	2.24	0.08
No.3780H	Black bowl	2.62	2.79	2.61	2.67	0.10

Table 9.4 lists the Si/Al ratios for eight samples. The mean and standard deviation are given and show the variability in the clay matrix composition. Six ceramics show results comparable to illite-montmorillonite clays. This suggests the use of these clay types as raw material for the ceramic production from the region of study. Petrographic analysis shows that, al- though the clays may be similar, the mineral composition of these wares is different, with volcanic to granodioritic rock fragments. This may indicate a certain homogeneity in clay resources or the use of similar clay sources, with the addition of a second raw material proper to each ceramic work- shop. These ceramics come from the same archaeological site and time period. Abundance criterion alone (six out of eight samples) could classify these ceramics as local, but it is a little risky to base a conclusion on such a small sample population. Petrographic and X-ray fluorescence analyses, however, confirm their local character. The interesting aspect of this analysis is actually the intersample comparison which allows the researcher to pinpoint two ceramics made with different clays, and probably of different provenance.

The first ceramic (no. 3708 A) presents Si/Al ratios that are totally atypical, much higher than for the other ceramics or one of the local clay ratio (1.91). This ceramic, a decorated bowl, differs stylistically as well as in its paste composition and preparation. The petrographic analysis shows a very fine compacted paste, with rare non-plastics of silt to very fine sand size. Its provenience is undetermined, but in any case, non-local. The other slightly atypical ceramic is no. 3755 a (third sample in Table 9.4), a thin-walled, orange-decorated bowl. Its clay presents ratios lower than for the group of local ceramics, and clearly in the illite range. The ceramic style is also non-local. Although not very significant, this difference in Si/Al ratio distinguishes this ceramic from the others.

This example shows the use of SEM for ceramic studies besides micro-structure and surface analysis of slip or paint. It characterizes ceramics and helps in the distinction of pastes based on their clay composition, with localized selective microanalyses. This is not possible with bulk analysis techniques such as X-ray fluorescence and neutron activation. It is however, a costly method, not designed for large corpus but excellent to use for fine and detailed analysis of particular ceramics. SEM has been used in archaeology in a wide range of research subjects.

9.7
Determination of Firing Temperature

The three preceding case studies have dealt with problems of provenance and paste characterization, the latter more oriented towards mineral and temper identification. Estimating firing temperature and the firing process is another area of research often tackled in ceramic studies. As we have seen in Chapter 6, the problem of the conversion of minerals in the clay paste to form a ceramic is dependent on time, temperature and the composition (presence or absence of fusing elements). Nevertheless, the objectives of archaeologists or archaeometrists have often been oriented in the direction of determining the maximum temperature in a firing program. Such work will allow one to establish a hierarchy of states of conversion of the ceramic materials in the paste into the final product.

The attempts of hierachization will be valid when the firing program is short, on the order of hours. Here, the maximum temperature attained will be evident since the function of time is not important. However, if the firing program is variable, i.e. more than 12 h, the problem of temperature identification becomes more problematic. It has no sense determining the temperatures of firing in the process of producing porcelain or stoneware-type materials.

There are different ways to address the question of the maximum temperature attained in a firing program, from simple observation of pore and color changes upon refiring a given archaeological sherd to complex physico-chemical methods. A good list of the techniques employed is provided by Rice (2015, see also Rasmussen et al. 2012). All techniques, however, look for the physical and/or chemical transformation of the minerals and paste structure occurring at different temperatures and under different firing conditions (see Chap. 6). These transformations help determine the firing temperature and kiln atmosphere to which a ceramic has been subjected. Among the techniques often used are thermal expansion and Mössbauer spectroscopy, examples of which will be presented here.

The estimation of firing temperature by thermal expansion analysis is based upon the property of fired ceramic bodies to expand when heated up to the original firing temperature, at which point sintering and vitrification resume, inducing shrinkage. This exact pattern may not be observed, showing rapid expansion instead of shrinkage, for example, as gas may be trapped when pores are becoming sealed through vitrification. Similarly, the presence of calcite may produce incoherent expansion curves. Recrystallization, absorbed moisture, and rehydration of minerals in post-depositional context must also be considered when interpreting the graph.

In any case, a strong change in the rate of expansion indicates that the original firing temperature has been reached. This can be corroborated by optical analysis and X-ray diffraction to determine which mineral is present. Firing experiments in reconstructed kilns are another way to assess the rate and length of heating. Thermal expansion analysis is done in the laboratory with a quite simple set of instruments: a dilatometer in a furnace to measure the sample expansion when heat is applied to it and a recording device to register temperature curves in correlation with sample expansion.

The sample is a piece of sherd, carefully cut, of known dimensions.

The study by Kaiser and Lucius (1989) is a good example of this analysis process. The ceramics analyzed are from the late Neolithic period, from three Vinca culture sites in the former Yugoslavia. They are hand-built pots, for the most part fired in a reducing atmosphere and of local production. This study is aimed at better understanding the ceramic production and pyrotechnological developments in the Balkans in relation to incipient copper metallurgy, and hence firing temperature was important to assess. The sherds were placed in a dilatometer (instrument used to measure the amount of expansion of an object due to heating) and heated in an oxidizing atmosphere at the rate of 5°/min up to 800 °C and 1°/min above this temperature. One can see that the temperatures of thermal conversion in this laboratory method are different from the several hours necessary in a normal firing program. The expansion curves are recorded and the resulting profile indicates a severe drop of expansion around 900 °C for a majority of the sherds (70%). The lack of expansion is assumed to show that the ceramic was not fired above this temperature. The recurrent production of ceramics at such high temperature in all three Vinca sites and among all ware categories showed a good control of pyrotechnology, probably influencing the development of metallurgy in the Balkans or perhaps vice versa.

Tite (1969) presents a study which will help underline the care that must be observed when interpreting thermal analysis results. This is an early article on the determination of firing temperature by measurement of thermal expansion. A group of 38 sherds was subjected to thermal expansion measurement. These sherds were from different cultures, sites and time periods, from the Near East and Mediterranean world, from the Roman and post- Roman periods in Europe (England, France), from China, and from Nigeria (a clay figurine). Three subgroups were first formed according to the analysis results. Twenty-eight ceramics showed expansion curves with equivalent firing temperatures higher than 700 °C; five ceramics indicated equivalent temperatures under 700 °C; and four sherds exhibited erratic curves due to the presence of calcite. Preliminary investigation shows variation in the expansion curves in relation to: the presence or absence of a vitrification phase in the sherd, moisture absorption in the sherd after its production and the presence of calcite leading to too high or too low estimates. Better accuracy in estimating firing temperatures is observed for high-fired ware (above 700°C), with an error of only± 20 °C. For ceramics fired under 700 °C, before any vitrification starts, the inaccuracy is much greater.

In order to interpret the analysis results of these ancient ceramics, and in particular of the low-fired group and the ceramics with calcite, a closer look was necessary. X-ray diffraction analysis was performed to determine the mineral composition of the sherds, which may influence the expansion curve. Rehydration or recrystallization may also cause interpretation errors. The presence of primary calcite is an indication of low firing, while secondary calcite (recrystallized) is not, the two being distinguished in thin section by petrography. Another verification involved subjecting the sherds to steam treatment and reconducting thermal analysis to check the degree of moisture absorption and its effect on the measurements.

The result of these analyses when considering ceramic production and pyrotechnology in a world-wide perspective shows firing temperatures ranging from 500 to 1180 °C in most of the culture groups considered. Although Roman ware is usually fired above 900 °C, a few sherds exhibit firing temperature range of 500-700 °C (See Eramo and Maggetti 2013 for a discussion on kilns and firing calcareous clays during Roman times in Switzerland). The four Gallo-Roman sigillate ceramics of this study, are indicated to have all been very well fired (1020-1150 °C), as are the Chinese porcelain and celadon of the Sung and Ching dynasties (960-1190 °C). However, it is known that the celadon wares are the result of temperatures above 1200 °C, as are the Roman wares. It is possible that the presence of an extensive glass/liquid phase may have led to an underestimation of the firing temperature. Given the short period of heating in the experimental protocol, it is not unreasonable to see such differences from actual firing temperatures. However, the method of thermal expansion gives a relative hierarchy which can be very useful in characterizing ceramic firing states.

The post-Roman ceramics from England show firing temperatures that vary between 500 and 950 °C. There is also much variation in the results for the Near East and Mediterranean ceramics, with lower firing temperatures achieved for the Neolithic and some of the Chalcolithic samples and higher temperatures (above 940°C for Bronze Age Ubaid, Mycenaean and Greek Attic wares. Finally, the Nigerian clay figurine must have been fired below 500 °C.

9.8
Mössbauer Spectroscopy

Hess and Perlman (1974) investigated the color of clay and ceramics in relation to the oxidation state of iron in the different phases of the ceramic. Iron is a common coloring element in clays. It combines with oxygen and reacts to firing in varying ways (Chap. 6), with different color results. Understanding why and how a ceramic has a specific color brings us closer to assessing the knowledge of the methods used by ancient potters. Hess and Perlman desired to understand how differences in appearance of dissimilar ceramics were produced with the same clay. The ceramics in question are Philistine and Mycenaen IIICl-style wares from Tell Ashdod, a Bronze Age site in southern coastal Israel. The local clay, of which a sample was taken, is naturally red, of kaolinite-illite composition with an iron oxide component. The determination of the oxidation state of iron was obtained by Mössbauer analysis, (see Chap. 9.3.3.4). The resulting spectra show characteristic patterns indicative of the valence (oxidation state) of iron.

Hess and Perlman (1974) observed unfired and fired clay and three ceramic fragments: one red and one grey-green Mycenaen IIICl ware, and one red Philistine ware. The red color of the Ashdod clay was initially thought to be caused by the presence of hematite (Fe_2O_3) an iron oxide of red color. However, none was observed in the Mössbauer spectra or by X-ray diffraction study, although non-magnetic ferric iron ($Fe3+$) was pre- sent. Once fired in an oxidizing atmosphere, hematite appeared in the

Mössbauer spectra. The problem is that observation of hematite is dependent upon temperature and particle size. If the particles are too small, and measured at room temperature, it will not show its characteristic pattern, but will do so if the sample is cooled or heated, the latter inducing a growth in the size of the hematite crystal. The differences in reds observed in the clays and two ceramics of Tell Ashdod when fired in an oxidizing atmosphere were thus correlated to the particle size of the hematite crystals, yielding a stronger red color when particles are smaller. Thus, when the clay is fired at different temperatures it gives different colors. Finally, the grey-green color of the third fragment studied is thought to be caused by magnetite, as observed from the Mössbauer spectra. The formation process of magnetite, which partly forms from the reduction of Fe_2O_3, suggests an original firing in a reducing environment.

Other clays were also analyzed to further explore the question of color change upon firing. Particularly interesting are two grey calcareous clays from Enkomi, in eastern Cyprus. Their composition is very different from the Israeli clay studied. These grey clays have a high content of iron (higher than for the Israeli red clays), but the resulting color when fired is lighter. This fact is partly explained by the occurrence of hematite within the silicate matrix and not as free oxides. The calcareous composition of the clay dilutes the color to yield a lighter ceramic instead of the dark red ceramics one would normally think would be obtained with this kind of clay. When fired under oxidized conditions, the clays turn light reddish yellow. In a non-oxidizing environment they color light grey. Information on provenance of the clay can also be deduced from the absorption spectra of the grey clays. The clays showed a mixed iron composition of Fe_3+ and $Fe+$ and calcareous compounds ($CaCO_3$ similar to a glauconite-calcite mixture. This similarity in composition suggested a marine origin for the clay used by the modern-day potters of Famagusta, nearby Enkomi, and possibly for some of the ancient ceramics of the region. A good chapter on calcite, carbonates, calcareous and marly clays is found in Albero Santacreu (2015). See also Trindade et al. (2009).

A second study, by Wagner et al. (1991), focuses on assessing the firing conditions of Pre-Columbian Formative ceramics. The Mössbauer analysis is actually part of a larger characterization study including neutron activa- tion analysis and petrography of South American material. The ceramics come from three sites: Montegrande and Batan Grande in northern Peru, and Canapote in northern Colombia; dating from the 3rd to the 1st millennium B. C. For the Mössbauer analysis, 200 mg of ceramic powder were taken. The samples were fired under different conditions (oxidized or reduced), at different temperatures (between 50 and 1200 °C in air, up to 900 °C when reduced). Three major parameters were measured: the quadrupole splitting, the isomer shift and the magnetic hyperfine interaction. The measurements were made at room temperature and at 4.2 °K. When the ware presented color differences in the surface and the core, different measurements were made to assess the firing conditions responsible for the change in color. When paint was present, it was scraped and analyzed independently from the sherd.

The Montegrande sherds usually show a black core, a red outside layer, and a red slip. The Mössbauer spectrum was different for the various layers. The black core contains much grey iron-bearing silicates, that are typically formed in a reducing environment. Incomplete oxidation of carbonaceous matter can also produce a dark core, but the paste composition and the iron state would be different. The red slip applied to the sherds proved to contain the free iron oxide hematite. The complex Mössbauer patterns obtained for Montegrande ceramics were best explained when compared with laboratory-fired mudplaster (a sample of wattle-and-daub taken from an excavated house on the site) in a reducing atmosphere, followed by a late oxidation cycle. According to the authors, the kiln must have been first covered (producing reducing conditions) and then uncovered at the end of the cycle before cooling was completed to produce a surface oxidation, unless oxidation occurred during burial. Other firing processes were also assessed by the Mössbauer analysis, implying an oxidation phase before the reduction, which means that the potter covered his fire after some firing time or with a more complex procedure, implying an oxidation-reduction-oxidation sequence. For the determination of the firing temperature, the temperature data of the laboratory-fired mud plaster was plotted against the Mössbauer parameters. Temperatures of 700 to 1000 °C were found to correspond best to the patterns when a reduction-oxidation sequence was respected. The first part of the firing cycle (the reduction phase) showed a higher temperature than the oxidation phase (ca. 1000 vs. 800 °C). The relative amount of iron components present in the samples was also recorded and served as a group determinant. The sherds were ordered into groups with similar Mössbauer patterns, characterizing the clusters already obtained by NAA (neutron activation analysis).

The Batan Grande site includes a production zone where ceramic kilns and wasters have been found. The kilns are small open structures, partly excavated into the ground. Experimental firing was conducted in the attempt to reproduce the original process (Shimada et al.1994). The experimentally fired ceramics in these kilns served as comparative material for the Mössbauer analysis, as well as laboratory-fired pieces. Local clays were also laboratory-fired to study their physicochemical reactions upon firing and the oxidation state of iron. The wasters could not serve for analysis, being usually discarded for improper firing (i.e. over firing) or defects (severe bloating, cracks) which would not help in determining the correct firing temperature and atmosphere of normal ceramic pieces. For this reason, the researchers used sherds from a nearby modern-day production village. The Mössbauer laboratory results closely matched the results obtained for the experimentally fired ceramic and the clay samples of Batan Grande, with much Fe^{3+}, and little Fe^{2+} in silicates and hematite (Fe_2O_3). The clay sampled in a cut of the site at two different levels is a mixture of illite, smectite, kaolinite, biotite, feldspar and amphibole, with illite being the major component. It is a good example of impure clay with coarse mineral inclusions, and as such, the type of clay most often used in ancient and traditional ceramic productions. The Mössbauer results were compared to the temperature measures of the refired sherds, which indicated that the original firing temperatures were around 600 to 900 °C, with a reduction phase followed by oxidation.

Finally, the Mössbauer analysis of a few Canapote sherds helped in assessing weathering effects during burial in moist soils. The unusual results of refired ceramics prompted this interesting study. The sherds show low calcium content and little or no

Fe^{2+}. An important increase in quadrupole splitting of Fe^{3+} was observed for low temperatures (400-600 and 200-700 °C), which can be explained by two different processes: the oxidation of Fe2+, already present, or the liberation of Fe^{3+} from the clay lattice upon dehydroxilation during refiring in the laboratory. As Fe^{2+} is of low abundance or absent in many sherds the second solution was favoured. The loss of water (and weight) during refiring was an indication that rehydroxilation of the clay in the sherds occurred at some time after the original firing, and also that this original firing was not high enough to destroy completely the crystalline structure of the clay to impede the clay rehydroxilation process. This process was also observed to decrease with depth of burial: the ceramic fragments closer to the soil surface had absorbed more moisture than the sherds further below. Another point was the presence of very fine hematite particles provoking a magnetic hyperfine field distribution. These fine particles can form during refiring from the decomposition of oxyhydroxides, which are formed during burial.

The above examples of Mössbauer analysis show the different use of the technique and its capacity in answering specific archaeological questions related to ceramic production and weathering. Determining which firing process is responsible for the characteristics of the sherds under study or if weathering occurred is not only important for further characterization and provenance studies, but also for understanding the potter's behaviour and his technological knowledge in general.

Bibliography

Acevedo, VJ, López, MA, Freire, E, Halac, EB, Polla, G, and Renoso, M (2012) Estudio de pigmentos en alfareria estilo negro sobre rojo en la Quebrada de Humahuaca, Jujuy Argentina. *Boletin del museo chileno de arte precolombino* 17(2): 39-51.

Acevedo, VJ, López, MA, Freire, E, Halac, EB, Polla, G, and Marte, F (2015) Caracterización arqueométrica de pigmentos color negro de material cerámico de la Quebrada de Humahuaca, Jujuy, Argentina. *Chungara* 47 (2), 229-238.

Cantin, N, and Mayor, A (2018) Ethno-archaeometry in eastern Senegal: The connections between raw materials and finished ceramic products. *J of Archaeol Sc Reports* 21:1181-90.

Cogswell, J, Neff, H, and Galscock, M (1996) The effect of firing temperature on the elemental characterization of pottery. *J of Arch Sci* 23: 283-287

De La Fuente, G, Kristcautzky, N, Toselli, G, and Riveros, A (2005) Petrología cerámica comparativa y análisis composicional de las pinturas por MEB-EDS de estilo Aguada Portezuelo (ca. 600-900 DC) en el valle de Catamarca (Noroeste Argentino). *Estudios Atacameños* 30: 61-78.

Druc, I, and Uribe M (2018) Un volcán en la cerámica. Producción en el altiplano de Isluga, Chile. *Comechingonia* 22(1): 11-36

Gebhard, R, El-Hage, Y, Wagner, FE, and Wagner, U (1988/1989) Early ceramics from Canapote, Colombia, studied by physical methods. *Paleoethnologica Buenos Aires* 5:17-34

Gehres, B and Querré, G (2018) New applications of LA-ICP-MS for sourcing archaeological ceramics: microanalysis of inclusions as fingerprints of their origin. Archaeometry 60(4) 750-763. online issue doi: 10.1111/arcm.12338

Hess, J, and Perlman, I (1974) Mössbauer spectra of iron in ceramics and their relation to pottery colors. *Archaeometry* 16(2):137-152

Kaiser, T, and Lucius W (1989) Thermal expansion measurement and the estimation of prehistoric pottery firing temperatures. In: Bronitsky, G (ed) Pottery technology. Ideas and approaches. Westview Press, England.

Maniatis, Y, Simopoulos, A, and Kostikas, A (1981) Mössbauer study of the effect of calcium content on iron oxide transformations in fired clays. *J Am Ceramic Soc* 64(5): 263-269

Maniatis, Y, Simopoulos, A, and Kostikas, A (1982) The investigation of ancient ceramic technologies by Mössbauer spectroscopy. In: Olin, JS, and Franklin, AD (eds) Archaeological Ceramics. Washington, D.C., Smithsonian, pp 97-108

Newman, ACD (1987) Chemistry of clays and clay minerals. Mineralogical Society, Monograph 6. Longman Scientific and Technical, London

Olsen, SL (ed) (1988) *Scanning electron microscopy in archaeology*. BAR, International Series 452, Oxford, pp 408

Rasmussen, KL, De La Fuente, GA, Bond, AD, Korsholm Mathiesen, K, and Vera, SD (2012) Pottery firing temperatures: a new method for determining the firing temperature of ceramics and burnt clay. *J of Arch Sci* 39: 1705-1716

Rice, PM (2015) *Pottery analysis, A source book, second edition.* University of Chicago Press, Chicago, pp 592

Rye, O (1981) Pottery technology. Manuals on archaeology 4. Taraxacum, Washington DC, 150

Schmitt, A (1993) Apports et limites de la petrographie quantitative: Application au cas des amphores de Lyon. *Revue d'Archéométrie* 17:51- 63

Shimada, I, Elera, CG, Chang, V, Neff, H, Glascock, M, Wagner, U, and Gebhard, R (1994) Hornos y producci6n de cerámica durante el Periodo Formativo en Batan Grande, costa norte del Perú. In: Shimada, I. (ed). *Technología y organización de la producción de cerámica prehispánica en los Andes.* Lima, Pontificia Universidad Católica del Peru, pp 67-119

Shimada, I (ed.) (1998) *Andean Ceramics: Technology, organization and approaches. MASCA Research Papers in Science and Archaeology*, Supplement to Vol. 15, University of Pennsylvania Museum of Archaeology and Anthropology, Philadelphia, pp. 23-61

Stoltman, JB (1989) A quantitative approach to the petrographic analysis of ceramic thin sections. *American Antiquity* 54(1): 147-160

Tite, MS (1969) Determination of the firing temperature of ancient ceramics by measurement of thermal expansion: a reassessment. *Archaeometry* 11:132-143

Verosub, KL and Moskowitz, BM (1988) Magneto-archaeometry: a new technique for ceramic analysis. In: Farquhar, RM, Hancock, RGV, and Pavlish, LA (eds), Proceedings of the 26th International Archaeometry Symposium, Toronto, pp 252- 255

Wagner, U, Gebhard, R, Murad, E, Riederer, J, Shimada, I, Ulbert, C, Wagner, FE, and Wippern, AM (1991) Firing conditions and compositional characteristics of formative ceramics: archaeometric perspective. 56th Annual Meeting of the Society of American Archaeology, New Orleans

10 Some Current Analysis Methods

10.1
Ceramic Analysis: What For and How?

An important point to stress is the purpose of ceramic analysis, i.e. why do we do it? We review here three main objectives of ceramic analysis. These are: (1) classification, (2) the study of pottery technology, (3) provenance studies. These are however only a springboard to explore larger questions dealing with the society where the potters lived and ceramic production took place, group interactions, organization of the craft, for whom, how far, how production evolved, etc. Co-crafting or craft complementarity should also be considered, as ceramic production was co-occurring with other crafts and all this needs to enter the big picture, as the by-products of some crafts, materials or fuels could be used for other productions.

Some analysis methods have been discussed already in earlier chapters, if so we will only review them briefly here. Others will be presented too, but we recommend the reader to search the litterature and Internet when interested by a particular technique as many excellent books, articles, and case studies exist, presenting in detail the advantages, limitations, theoretical background, and use for each. Some references in this regard are given at the end of the chapter.

10.1.1.
Classification

Classification of a ceramic corpus can be done, apart from archaeological criteria, according to compositional, textural and/or granulometric criteria. Such a classification, that is, grouping all ceramics with similar compositional characteristics, is usually undertaken to back up a stylistic typology, as a complement to it, or when no external features can help classify the fragments, as in the case of body sherds. Paste typologies rely on characterization of the inclusions, their grain size and shape, the porosity, and the organization of the grains in the paste. A classification based on body or surface color is less certain and precise. Color variation can be great even within a single sherd. Consider, for example, a red, oxidized, cooking pot with big black spots due to contact with the flames during firing (called fire clouding). In other words, color is determined by different factors, such as the firing atmosphere and temperature, or the type of material. Besides, firing conditions can vary from one batch to another and reflect a production cycle rather than a ceramic type. Paste typologies and ceramic classification can be achieved with all the analysis techniques available, whether optical or chemical. Petrography is a common technique that often paves the way for more complex studies.

10.1.2
The Study of Pottery Technology

The study of pottery technology relates the compositional and textural aspects of the paste to ceramic properties and/or human behaviour. It strives to answer questions like these: Why is a certain paste preferred to others for a certain functional ware? How is permeability achieved? What was the firing temperature? This involves characterizing the paste and interpreting the observations made. It implies identifying the temper and recognizing the way in which the raw materials were prepared (if the materials were crushed, selected, carefully homogenized, etc.), and how the pot was built (i.e. wheel-made or coil-made). Mechanical and physical tests are done with ceramic fragments and laboratory-made tiles to determine the strength, breakage resistance, thermal stress and porosity of different pastes. In pyrotechnology, different methods are used, from petrography to Mössbauer spectroscopy. This study takes into account the presence or absence of minerals that form or are destroyed at certain temperatures and the chemical state of iron (oxidized or reduced) in the different compounds.

10.1.3
Provenance Studies

These are often achieved with chemical analyses, but can also be very successfully conducted with mineralogical studies, e.g. petrography; a very famous example is given by Shepard (1963) in an analysis of Rio Grande glaze paint and, more recently, Mills and Crown (1995), or through micro- analysis with SEM and other microprobes. They are based upon the provenance postulate, formulated by Weigand et al. (1977), that chemical (or mineral) differences exist between sources of raw material and are identifiable by analytical techniques. These differences can be recognized in the objects made from these raw materials. This idea was already present within the concept of geochemical fingerprint as stated by Perlman and Asaro (1969).

This provenance postulate holds true for ceramics, but it is complicated by the complex and composite nature of the material studied. As seen in Chapter 6, it is best to look for a production area, rather than for the exact source of the raw material(s) used. Thus, with few exceptions, sourcing in ceramics is not an objective goal. Remember that clay sources can be highly variable compositionally, numerous, or extinct. Recent studies have shown that when the paste was tempered, the best match is made by comparing the chemical composition of the pot to sampled raw clay and temper mixed in varying proportions. Mineral analysis (petrographic or other) is thus important in determining the temper used or the parent rock and geological origin of the mineral and lithic inclusions in the paste.

10.1.4
Quantitative Studies

Compositional studies, particularly chemical, generate extensive quantitative data. These are easily processed by computer with statistical programs. The data analysis yields groups of ceramic fragments with similar chemical composition, indicating the same provenance. For raw materials, it is postulated that internal compositional variation of a source is less important than intersource variation. If represented by statistical vectorial distances, the variation in the composition of material from the same source should be smaller than the variation between samples from different sources, with composition expressed in concentration of elements. Similarly, objects from the same production center show smaller compositional variations than objects from distinct production loci and are, therefore, clustered together.

Here, we enter the realm of figures and quantitative studies. One very important thing to remember is that statistics are a means of describing reality. The statistical analysis does not constitute reality, but only an approximation of it, more or less accurate according to sample size, distribution parameters and the type of statistics employed. A good example is the representation of ceramic composition in two-dimensional space (a graph with x and y axes), while, in fact, many more dimensions are present (as many as there are chemical variables). This is a practical reduction of complexity that helps us visualize the clustering of ceramics with similar composition. Also, the attribution of one pot to one particular site or composition group is based on statistical probabilities. It is the best fit for a study, with a given number of samples and a particular statistical procedure. Add more subjects to the corpus (more ceramic fragments, more archaeological sites) and the picture might well change. This is normal. Therefore, the use of different analytical and statistical procedures to verify and triangulate data is very important.

10.1.5
Use of Qualitative and Quantitative Studies

Qualitative studies are not to be rejected or considered less important than quantitative ones. For some time, the view prevailed that qualitative data were subjective and therefore not "scientific", and only quantitative analysis could give reliable data, unbiased by the analyst. The influence of the researcher cannot be denied, especially in analysis involving the human eye and mind to determine the nature of the constituents, as with petrography. Grain identification is complex, requiring not only a trained eye, but experience and a feel for the material studied. A petrographer accustomed to studying one kind of ceramic from a particular region may at first be disoriented when looking at another corpus. Post-use and de- positional conditions may alter the material so much that the identification of the inclusions is "obscured" or impossible. It is no wonder that fully automatic computer identification is not possible, as too many variables change simultaneously.

The "subjectivity" of qualitative analysis can be neutralized by cross-examination or verification procedures such as asking a peer to redo the analysis, blind tests, double checking at intervals of days or weeks etc. Identification criteria can be defined, such that they may be used by different persons with the same success. The Munsel color chart is a good example of a means to control subjectivity and unify descriptions. It is a color code for describing soil colors and, by extension, ceramic and other materials. It uses an alphanumeric system to characterize each hue, illustrated by a color patch to which the material studied can be visually compared by means of a hole in the hue table. X-ray diffraction (XRD), scanning electron microscopy (SEM) and other devices equipped with microanalysis programs can identify grains, on the basis of crystallographic characteristics (the distance between layers in the crystalline red are specific to each mineral type) or elemental analysis. However, these analyses are long and often costly, and the information obtained may not be as revealing as the qualitative analysis done by an experienced petrographer. For example, the microanalysis of a grain (i.e. by SEM) will provide its elemental composition, including silicon, iron, calcium, etc., content. The grain identification, however, is not always straightforward. The oxygen component must be recombined with the elements measured in order to obtain the oxide compounds and then chemical formulae known for minerals in geology. For some grains, there is an ambiguity. For ex- ample, some pyroxene and amphibole compositions are the same except for the hydrogen content. Since hydrogen is not analyzed, the calculated oxide contents are the same, hence, the difficulty in mineral species identification by chemistry alone. One must then rely on mineralogical crystallographic properties.

Subjectivity is introduced at each decision-making step. However, decision-making is not restricted to qualitative studies. Quantitative analyses require human intervention, which can influence the results also. It is the researcher who chooses which elements or variables to study, their number, the statistical procedure to follow, the clustering method used, the adequate confidence interval etc. For example, considering some elements and not others in principal component analysis, or the decision to cut the classification dendrogram at the first node or the second, will produce different clusters, with varying archaeological meaning. Many articles have been published on this aspect of statistics. Statistical methods should be used not to make the data say what we want, but to seek structures, com- positional patterns, that can be verified with other procedures. All in all, qualitative and quantitative studies complement each other. In many cases, the knowledge acquired through the qualitative analysis of a corpus will help significantly the interpretation of the quantitative analysis. In provenance studies, for example, chemical analysis plays a major role. The quantitative data generated are statistically treated to yield groups of the same elemental composition, or to ascribe ceramics of unknown origin to a group of ceramics with known provenance. Attribution is made to the group with the highest degree of compositional similarity. Membership may, however, be ascribed by default, because no better choice is available. Petrographic analysis, allowing a (qualitative) identification of the mineral composition, will show in what ways the attribution of membership is erroneous or appropriate.

Let us look at two examples of complementarity between qualitative and quantitative studies.

In the first example, certain ceramics from site X were statistically attributed to site Y, more than 300 km away. Their chemical composition was found to be closer to the composition of the ceramics of site Y. The mineral analysis, however, showed no similarities. The ceramics shared a volcanic composition, leading to a similarity in elemental contents, but the type and shape of the volcanic fragments, identified by petrography, differed greatly, indicating that the provenance could not be site Y. It was necessary to postulate another source, even if not present in the corpus of analysis.

In the second example, some ceramics found at site A were said to come from site B, on the other side of a mountain range. Again, the chemical composition showed similarities that the mineral analysis did not. The petrographic study of the coarse mineral inclusions in the paste of these ceramics suggested a local origin when compared to the local geology. The chemical attribution could have been rejected, but prevailed because of two factors. First, petrography can identify inclusions but not the clay and the very fine non-clay particles. Therefore, it is possible that the similarity observed in the chemical analysis may derive from the matrix and not from the temper (grossly speaking). The coarse inclusions being few, they may not influence the chemical classification as much. Second, a tradition of itinerant potters is known in the region, travelling from one valley to the other. They carry their clay, but can use local temper. The two valleys where sites A and B are found are connected by mountain passes, and intervalley travel is attested in both historic and earlier times. Therefore, it is quite possible that the ceramic attribution is correct, attesting to this tradition of ceramic production by itinerant potters. We see here how different data, quantitative and qualitative, chemical, mineral, geological and ethno- graphic, can combine to yield a better picture of past activities.

10.1.6
Sample Size and Qualitative and Quantitative Studies vs. Time and Cost Invested

The time invested in these studies, qualitative or quantitative, should be measured relative to the research objectives. Quantitative studies call for much time and money. Sometimes, their results are not more telling than qualitative studies and do not justify the effort and the cost involved. This can be the case when the sample size is reduced to a few items and no generalization or comparison can be made. A large corpus, on the contrary, will yield very interesting results, but substantial time and money must be invested in the analysis. Thus, in many projects, a qualitative analysis is undertaken and, based on the results, a smaller corpus is selected for quantitative analysis. This sampling is meant to be representative of the paste compositions recognized in the original population.

As for sample size, some authors suggest taking 5 to 30 samples for each analytical category. In some studies, 10 or 20% of each compositional type is represented. One might say this is really a personal matter: sample size should be adapted to the research objectives, the budget, the schedule, and the type of analysis technique. In terms of costs, routine analysis could be quite inexpensive. In a routine analysis, standards are already calibrated and the analytical and statistical procedures are preset. Analysis is then both rapid and affordable. This approach allows large laboratories specialized in analyzing archaeological material to offer attractive prices. Also, the cost for running one analysis on a single sherd may be as high as that for a full batch of samples, or higher.

The ceramic sampling method for selection (probabilistic sampling vs. pragmatic sampling), the amount (mass) of sample to take, and where to sample in the sherd, are determined by the corpus and the type of ceramic materials available, as well as by the analysis technique used. Ideally, probabilistic sampling insures the reliability of our estimates and interpretation when applied to the larger initial population. Furthermore, if the data distribution is normal, one can use parametric statistics, which offer greater power of generalization of results. In practice, probabilistic sampling is rarely possible because too many individual samples are required. Instead, pragmatic sampling is performed where ceramics are chosen so as to be representative of the different ceramic types or compositional categories observed. The larger the sample, the better the chance of approaching normal distribution (probabilistic sampling). Normal distributions are generally observed in ceramic chemical data, sometimes in granulometry and modal analysis if the corpus is large enough. If not, data standardization or transformation is performed.

Probabilistic sampling offers a high level of internal validity, while the validity of pragmatic sampling is external. For example, if samples are chosen in a subjective (or pragmatic) way so as to be representative of a ceramic style, the analysis results will have less generalizability than if the sampling had been probabilistic, but the significance of the results in relation to a particular research will be objective enhanced. One usually tries to balance internal and external validity, for example by doing a probabilistic sampling out of chosen ceramic categories (with a large enough corpus to do so). In any case, the application of non-parametric statistics, or parametric statistics on transformed data (originally not normally distributed), allows one to achieve generalizability of the results, even if the internal validity of the corpus composition is weak. In the case of trans- formed data, this is truer still if the sample size is large.

As for the amount of sample to analyze, this depends on the coarseness of the ceramic. A well-homogenized paste, fine-grained, will require less powder for the analysis to be representative of the whole ceramic, i.e. 100 to 150 mg. For a coarse ceramic, with large grains, not well kneaded or well homogenized, with inhomogeneous grain distribution, more sampling is needed (i.e. 300 to 500 mg). The sample is ground in a mortar, reduced to a fine powder, and well homogenized. Part or all of the powder is then taken for analysis. There can also be two or more samplings from different places on the ceramic fragment, if it is big enough. The same holds true for thin sections. With a fine and well-homogenized paste, only a small cut will be required to be representative of the whole, whereas for coarse ware, as the grain distribution is inhomogeneous, a longer cut may be necessary to obtain the widest array of grain diversity present in the thin section. The cut is usually perpendicular to the wall, so as to include the inside and out- side of the ceramic. Thus, the surface finish can be studied, as well as the core. In petrography, coarse ceramics are often more informative than a very fine ware, where the mineral and grain size variation is reduced or non-existent. For example, the petrographic analysis of a sigillata ware fragment, with its fine homogenized quartz-rich paste, will not be of much use for provenance studies. In contrast, the study of a Neolithic ceramic production or traditional cooking ware will yield good results in terms of classification by inclusion types and granulometry or for provenance studies.

Geology also plays a role in the success of the analysis, especially in provenance studies. If the corpus consists of sherds from different sites located in the same geological setting (in an alluvial plain, for example), with little or no intersite compositional variation, petrographic analysis will yield poor results. The mineral inclusions of the ceramics from the different sites will look the same. Ceramic productions can, however, be identified, if paste preparation varied from one site to the other or for one type of ware (e.g. cooking ware versus bottles). Chemical analysis, however, might detect slight local variations, allowing for ceramic clustering and site attribution.

All these considerations are part of the research planning done before and during a project. Strategies are often adapted to the context and evolve constantly in accordance with the archaeological observations and the results of the analysis. Even if provenance studies are unsuccessful, the data obtained can yield important information on ceramic production if combined with other analysis techniques, for instance petrography. Information can be obtained on how many production areas or workshops are present based on the different mineral compositions or paste textures identified, how the paste was prepared (with coarse or fine material), and so on. Besides, ceramic thin sections constitute a permanent data bank that can be reanalyzed at will.

10.1.7
Use of Comparative Studies: Sampling Geological Comparative Materials: Compositional Profiles and Petrofacies maps

To know what a region has to offer and what might have been available to ancient potters in terms of clays, sands and deposits, it is important to obtain comparative materials. It is often not sufficient to look at geological maps. Even so it is impossible to be exhaustive, and that hundreds or thousands of years might have changed the landscape, accumulating sediments, it is useful to collect geological samples. This also helps evaluating the choices the potters had or recognize that they choose to use one type of material over another.

Building petrofacies maps, as done by Desert Archaeology researchers (Miksa 1998, Heidke et al 2001, Ownby 2015) yields detailed maps of the mineralogy (types and percentages of rocks, minerals, granulometry) of the sands or sediments in a region or per valley segments (upper, middle, lower) to pinpoint the resource area(s) used for sand temper by the ancient potters.

Clays can be collected over large areas or in the valleys and area around archaeological sites, prepared as clay tiles, fired, and prepared in thin sections for petrographic study. This allows comparing granulometry, angularity and texture of clays to that of the ceramic body. Clay characterization over a wide area is also conducted using chemical analysis with INAA or LA-ICP-MS techniques.

To sample clay or sand, one should look at the different features and factors shaping or influencing the terrain. Looking at geological maps helps see where rock formations change and thus where to collect samples. One should consider the topography, in which direction rain and streams are carrying sediments; collecting before and after river junctions is also important, as well as sampling at different levels of clay beds, road cuts, etc. Asking farmers and local people, when no potters are around, can be useful. Crops will not do well in clayey soils, and farmers will know the quality if the soil. All of this information helps orient material sampling. Of course, going with a potter who knows the area is great, but there is no guarantee that the spots he/she collects material from would have been used in the distant past. Potters usually discard the first layer of soil that might be of lesser clay quality. It is a good idea to do the same. One should collect enough to make test tiles for petrographic analysis, sedimentation tests and chemical analysis. Some people may clean a cut, take shovels of material, at one point or across layers or over a certain longitudinal distance according to the level of details needed. That sample is homogeneized and a smaller quantity is extracted for analysis. This step in a ceramic analysis program is crucial. If possible, conduct experiments adding water to the clay to see how it reacts upon drying, adding sand or another temper according to what the ceramic pastes under study reveal, and of course, firing your samples. These may compare often better with the archaeological ceramics than clay or sands alone. Geological and archaeological samples must then be analyzed with the same method(s) for comparison purposes.

10.2
Physical and Chemical Analysis Methods

The analysis methods discussed below are those used commonly today. Others have been used in the past and certain new methods are beginning to be used in ceramic studies. Our presentation of these methods is designed to indicate the general idea behind the methods more than to give a detailed description of the process involved. Several books and articles are indicated in which one can find more useful descriptions of the methods. These analysis methods inform on the mineral or chemical composition of the samples, and must be chosen according to the objectives of the study and type of sample. In general, it is advised to combine several techniques to reach better understanding of the materials and ceramic production.

Methods Used According to the Objects they Analyze:

Optical or Visual Methods

1. Binocular microscope (or digital USB portable microscope) gives information of the relations of temper grains and clay-rich zones of the paste or ceramic body. These microscopes work in reflective light (with the light shining above the sample) and one can study ceramic cross sections (fresh), paint, surface treatments and wear with great success. However, good mineral identification is difficult with a binocular microscope.

2. Petrographic Microscope study is designed for the identification of the temper materials, mainly their mineralogy, and the characterization of other non-argillaceous inclusions, providing they are big enough to be examined with this type of microscope, as well as the study of the texture and structure of the clay matrix. The method necessitates the use of a thin section, i.e. a thinned slice of the ceramic sherd with a 0.03 mm thickness and the observation is done in transmitted light (the light shines from beneath and through the sample).

3. Scanner from direct or photographic data. This allows for the analysis of a bidimensional image, using computer programs, providing data on granulometry, grain distribution, voids size, quantity and form, working on light and color contrast. The surface analyzed needs to be flat.

Mineral Identification

1. X-ray diffraction (XRD), scanning electron microscope (SEM), and now QEMSCAN are methods used to identify minerals and study the form and structure of crystal aggregates. Information can be derived concerning the crystals in using monomineral samples but always in using aggregates of crystals. Average interplanar distances and the average composition of the atomic planes can be derived from XRD studies. The morphology of aggregates is determined by SEM.

2. Differential thermal analysis (DTA), thermogravimetric analysis (TGA) studies are concerned with the water molecule loss in clays and recrystallization temperatures of clays or other minerals. Should carbonates be present for example, one should detect the loss of CO_2.

3. Transmission electron microscope (TEM) investigates the shapes of individual crystals.

4. Infrared (IR) and Fourrier-Transform Infrared Specstroscopy (FTIR) methods investigate the relations of individual molecules in the crystals and provides information on mineralogical composition and organic compounds. This is very useful for ceramics, pigments, and paints. This scale of observation is within the crystals and multicrystalline aggregates.

5. At the same level or below it are the high-resolution electron microscope (HRTEM) analysis methods. Here the atomic layers or the atoms themselves are visible to the

investigators eye via the electron beam. The scale of observation has been much reduced compared to the other methods.

6. Raman spectroscopy is a (relatively) new method in archaeometry, non destructive, which can be portable and is used a lot for surface analysis (paints, slips, pigments). It allows identifying mineral compounds, oxides, carbon, and compounds of organic origin, among others.

Chemical Analysis

1. Neutron activation (NAA) and

2. X-ray fluorescence (XRF) methods (including portable XRF) give the relative abundance of chemical elements present in the whole sample (bulk analysis), with the advantage to be sensitive to trace and minor elements in addition to some major elements (more so for INAA than XRF).

3. Laser ablation inductively coupled plasma mass spectroscopy (LA-ICP-MS) also provides elemental characterization of trace, minor and major elements, with the advantage of being able to point the laser at a specific area of the sample. It is often used to characterize the clay matrix or certain minerals, rather than the 'whole' sample.

4. Mössbauer analysis determines the oxidation state of iron in the sample, among other characteristics of the environment of the iron (Fe) atoms. It is used in ceramic analysis to assess firing atmospheres and maximum firing temperatures.

Age Determination

Thermoluminescence study is designed to determine the age of a (ceramic) sample. The method excites electrons entrapped outside their normal stabilities. The electrons are put in excited, unstable states by cosmic and other radiation. The method is complicated but has a great potential for dating ceramic materials.

Physical Properties

Density-porosity measurement can be used to give a good idea of the stage of firing attained on a ceramic, as can hardness.

10.3
Visual Methods

10.3.1
Binocular Microscope

A binocular microscope is the stable and slightly sophisticated form of a magnifying glass. It can be very, very useful. The magnification can be reasonably high, allowing one to see details less than a tenth of a millimeter in size. This is sufficient for the larger temper grains and for the details of glaze, slip and other decorative techniques. If a camera is adapted to a reasonably good microscope, one can begin to do quantitative analysis on the non-plastic part of a ceramic paste.

One thing to keep in mind is that the scale of the object should be recorded if any numerical method is used. A simple ruler can be used to estimate the dimension of the field of vision of the microscope and this gives the key to identifying the sizes of grains in the field of vision. Just divide the field of vision into the appropriate units (usually millimeters), and the size of the objects is automatically determined.

Preparation of a sherd for examination can be accomplished by cutting the sherd with a diamond-enforced saw, usually cutting under a stream of water to keep the surface smooth. It is often better to polish the surface after the cut with an abrasive to obtain a still smoother surface which allows a better definition of the grains in the clay matrix. This method is especially effective when the temper grains are of a different and contrasting color to the clay matrix.

10.3.2
Portable Digital Microscope

Description of the use of this new handheld device, as well as examples and types of image analysis possible are given in Chapter 8, section 8.2.3.

10.3.3
Petrographic Microscope

This instrument is more sophisticated, and costly. The investment in the microscope is much higher than for a binocular microscope and the preparation of the sample is more complex, and hence costly. The petrographic microscope requires a thin section, in that the light must shine through the sample in order for the object to be identified; it is a transmission microscope. The thin section of the sherd should be near 0.03 mm thick, the definition of objects in the field of vision is near 0.002 mm. Here, one comes near the clay-sized material and hence a good definition of the non-plastic material in a ceramic can be obtained.

In the same way as with the binocular microscope, one can adapt a photographic or other imaging system to the petrographic microscope in order to do quantitative studies. The advantage of the petrographic microscope is having a clear identification of the temper minerals. The identification of the grains is the key to provenance studies of this material.

The specificity of the petrographic microscope is that the light source passes through the sample. This requires a thin sample (0.03 mm thickness). Thus, one makes a thin section of the sample (pottery sherd) and looks through it. The reason to look through a sample is that the light source interacts with the crystal as it passes through and some wavelengths (colors) move more rapidly than others. This difference in the speed of different colors depends on the chemistry of the mineral phase and upon its crystallography. Hence, different mineral species have different interactions with the light that goes through them and they can be identified by this effect.

The conditions to observe this interaction are a polarization of the light. Normally this is done using a Polaroid sheet. Light beams from the source (a light bulb) are selected which vibrate in only one direction. This is plane polarized light (PPL). The plane polarized light is allowed to pass through the thin section. A second polarizing material is placed in the light beam of rays that leave the sample. This polarizer can be rotated. When it is oriented so that the light beams vibrate at right angles to the first analyzer below the sample, the light is cross-polarized (XPL). If the mineral has no interaction with the light beams, slowing some and speeding others, the grain is black, no light passes through the crossed polarizers. If the mineral does have some interaction with the light spectrum, certain colors are more visible than others. The orientation of the crystal and the interaction give characteristics which allow mineral identification.

In order to correctly identify the minerals present, one should have some grounding in mineralogy and some idea of petrology.

10.3.4
Computer Scanner and Video Systems

The computer scanner is the most easy and precise method of gathering the photographic data from either binocular microscope or petrographic microscopic studies. It can also be used by direct scanning of appropriate samples. One can obtain relatively cheap image analysis systems which allow one to treat and select the information in a very easy and convenient manner. This is "off-the-shelf" software which is tried and proven. Video cameras can be used to the same ends, usually on microscopes.

When the sample has been selected, the image taken and treated to give a homogeneous view of the object, the operation of binarization is necessary. This transforms all shades of grey into either black or white signals. Hence, if the temper grains are lighter than the clay matrix, one selects a grey level which takes all the areas of temper grains and assigns it to a white code. All the rest of the object is black. These operations are shown in Section 8.3.2. With this selection, one can use a particle analysis routine; this is a reasonably easy program to obtain.

The most important point to understand in the process of image analysis is that the initial information is the most important. If the photograph that you use does not have a sufficiently high precision, the end result cannot be satisfactory. For this reason, photographs are much better than television or video camera data. The reason is very simple; a camera has a given definition which is based upon the size of a television screen. Only a certain number of points per linear distance can be taken into a video image. However, a photographic film has thousands more points on it per unit length than any video system, present or in the reasonable future, can have. All those crystals on a film are much more responsive than a video system can ever be. The information stocked on the film takes up little space, and takes little time to be registered. However, on a video or computer system, each point takes up a non-negligible space in the computer; but it is even more difficult to call up these data and transfer and treat them.

This fills up computer discs and memory banks in no time. Therefore, with the best picture to begin with, one can begin to reduce its precision. If you start with the precision of your computer, there is no way to make it better. You can, of course, make it worse; but a photograph has definition so great that it can be reduced, and still most of the details will persist.

Basically, the path to numerical treatment of ceramic data, grain size and shape and abundance, can be treated easily with a scanner, and the results provide the only way to make serious progress in the optical study of ceramics.

10.3.5
Electron Microscopes

These analysis methods are based upon the physics of interaction of a beam of electrons with crystalline (or non-crystalline) matter. Several derivative analysis methods are based upon the same principle. The source of probing energy is a beam of electrons generated much as light was and still is in a light bulb similar to that invented by T. Edison many years ago. A thin wire of a resistant material is linked to a stream of electrical current. The resistance of the material causes it to transform the electrical energy into other forms (light in the case of a light bulb). However, in our new generation machines, the energy sought is the excess of electrons bottled up in the thin resistant wire. These electrons are expelled from the circuit and, in the apparatus, attracted elsewhere by high electrical potentials. The escaped electrons are shaped into a finely focused beam by electro-magnets. This beam is directed upon a target of interest (clays for example) where different physical processes of interaction take place and, depending upon the apparatus, analyzed by different methods.

10.3.5.1
Scanning Electron Microscope, SEM-EDXRF, Scanning Micro-XRF

The SEM uses the flux of secondary and backscattered electrons from the material to form an intensity image of the material bombarded by the electron beam. The resulting cathode tube realization indicates the three-dimensional aspect of the sample. This is very useful in the identification of textures and shapes of mineral grain aggregates. The definition or resolution of the image is on the order of 0.01 μm.

A second use of SEM data is to show the relative electron density of a given flat surface. The number of electrons rejected from a surface by rebound or by electron interaction is proportional to the number of electrons in the constituent atoms of the material. Heavier atoms give off more electrons. Hence, a bit of lead will show up among clays, for example. These methods allow one to investigate textures of ceramics, the relations of clay and temper grains at a small scale.

SEMs are usually coupled to a EDXRF (energy dispersive X-ray fluorescence) detector to conduct chemical analysis for major and minor elements. The latter is more of a quantitative nature, also internal reference systems can allow to calculate relative frequencies of the elements present.

Another 'device' can now be coupled to an SEM, a micro-XRF spectrometer, among the new technologies available for analysis of archaeological materials. It is fully non-destructive, has scanning capacities, enhancing the SEM to sensitivity to trace elements. which can be identified and mapped. It works great for pigment analysis, paints, or core analysis. An exemple of its use in art at the Rijksmuseum for an analysis of a Rembrandt (with great photos of the instrument in place) can be seen at https://www.blue-scientific.com/micro-xrf-art-rijksmuseum/

10.3.5.2
Transmission Electron Microscopes (TEM)

Transmission electron microscopes (TEM) are used to see the shapes of crystals in an essentially two-dimensional plane. The clays are deposited on a carbon-coated copper grid (with holes in it so that the clays can be "seen" by the electron beam). An electron beam is aimed at the sample. The electron beam is absorbed more by the clay than by the carbon film and thus a shadow image of the clay particle is produced. The TEM resolution is greater than that of SEM microscopes by a factor of more than 5 or so. Their principal use is that of identifying the crystal shapes of clays and their dimension. Also, the detection of fluorescence radiation by energy-dispersive techniques (as on electron microprobes) can be used to estimate the composition of individual crystallites. Most clay structures have a hexagonal symmetry which gives either hexagonal forms or those derived from a hexagonal prism. Some crystals are sufficiently elongated to appear as needles.

10.3.5.3
High-Resolution Transmission Electron Microscopes (HRTEM)

The newer generation of electron microscopes give a new dimension, smaller, to the study of clays. Transmission electron microscope micro- photographic methods give resolutions of less than 2 Å (0.00002 mm). This allows one to "see" the levels of atomic planes in the clays with ease and in special cases the atoms themselves. Using energy-dispersion analysis methods, it is relatively easy to determine the composition of a clay mineral when it has a dimension of several unit layers, > 100 Å. It is also possible to determine the homogeneity of an individual mineral crystal on a scale previously undreamed of. The succession of unit layers, the intimate interlayering of the clay structures and their growth patterns, are now easily determined using the high-resolution electron microscope (HRTEM).

The drawback of HRTEM studies is found in the fragile nature of clay minerals, which absorb too much of the electron beam energy and convert it to thermal energy. The time during which one can take a "picture" of the clay mineral atomic layers is very short and there are many poor photos made. Often, several dozens or hundreds of photographs are taken to find one which corresponds to the definitions of focus and sharpness needed to convince clay mineralogist of structural details.

HRTEM is not a routine method of clay analysis but it is a very crucial one to determine the structural relations of a well-defined clay material.

10.4
Mineral Identification by Non-Optical Methods

These methods are indirect, i.e. they depend upon an interpretation of data resulting from interaction of waves or energy with the mineral phases present. They allow one to identify the phases present as individuals or in a mass of objects.

10.4.1
X-Ray Diffraction (XRD)

Most ceramics will have highly transformed clays present. In these cases, the clays are for the most part amorphous and hence give no response to X-ray diffraction, which is based uniquely on a periodic, ordered atomic structure. However, when firing temperatures were low one can still identify clays. XRD is of course a very useful method for clay mineral identification of the raw clays in the area of study. It identifies temper grains with ease, and the identification of certain neoformed minerals can give clues about firing temperatures.

A major problem in X-ray diffraction (XRD) is that most often the clay mineral diffraction maxima are not visible or only faintly so because the particles are very small, and there are not enough atoms present in any one direction (plane) to give a decent intensity of diffracted X-rays. This is aggravated by the fact that the clays are found in essentially a two-dimensional form, the sheet structure. Thus, if the greatest diameter of a clay mineral is 2 μm, the smallest dimension is smaller by a factor of 10 to 20 or more. This reduces the diffraction power of the atomic planes even more in most of the crystallographic directions.

However, a clever mechanism was devised to reinforce the signal coming from the crystal lattice, using a uniaxial powder diffractometer and oriented clay samples. In orienting the clays, making them lie one on the other in the same plane, the diffraction effect is enhanced for the small crystallites. The orientation produces a pseudo-macrocrystal which diffracts approximately as if thousands of the crystallites became a single, large crystal. The resulting diffraction characteristics are thus enhanced in the crystallographic direction parallel to the sheet structure. However, the other diffracting planes in the crystals which are not parallel to the sheet structure are almost totally lost to the diffraction spectrum. The X-ray beam sees only the planes parallel to the orientation direction. As a result, the identification of clay minerals is done almost entirely on the spacing of the crystallographic planes parallel to the sheet structure. This is called the basal spacing of the clays, and it is of fundamental importance to any discussion of clay mineralogy.

X-ray diffraction methods require only several milligrams (0.01g or less) for a good spectrum. A little scraping on the sherd sample is sufficient. The identification of the diffraction peaks is done according to reference tables. Now it is also possible to identify precisely the area to be analyzed on a sample (say only the slip or some specific inclusions) when a camera is linked to the XRD machine. Furthermore, thin sections can be used, providing part of the section be removed from the glass support, for example by immersion in acetone. The thin section is cut with a blade (e.g. 0.5 cm wide x 1 cm long) and the selected part is fixed to a Kepton tube (or other) using glue.

10.4.2
QEMSCAN Quantitative Evaluation of Minerals by Scaning Electron Microscopy

QEMSCAN is a type of hybrid instrument coupling SEM and XRD possibilities allowing one to identify, map and quantify the minerals (aka inorganic phases) present in a sample. The later does not need to be a thin section, but the techique calls for a high polished slab covered with a conductor (e.g. graphite). The machine works using the BSE (back scattering electrons) mode of the SEM with X-ray diffraction emissions. The EDS (energy dispersive spectra) from XRD are collected from thousands of points along the sample in an automated process. The results are analyzed and classified as mineral phases based on an internal database that must be adapted to the corpus studied. The instrument is not easy to use in the sense that it requires a very good knowledge of the database and of the samples to analyze (what minerals are potentially there). When the analyst is an expert, this gives good results. However, as for any technique, data interpretation requires more than analysis performance, and this does not replace the eye of a trained ceramic petrographer to make sense of the human and technological patterns one sees in a thin section. Here again, a combination of skills and techniques are necessary to reach a higher level of understanding of past productions and societies. A good example of the use of this technique is found in Riera-Soto et al. (2019) and Frigolé et al. 2019.

10.4.3
Thermogravimetric Analysis (TGA)

Heating samples at a regular rate (dT/dt) effects weight loss depending upon the retention of the water in or on the clay structures or other phases present in a sample. Destruction of carbonates releasing CO_2 gas is observed in TGA analysis also. The most complex minerals are clays which have at least two, or at times three or four, types of water or hydrogen ions in or on their crystallites. The binding energy of these types of water can be used to identify the minerals present. Several tens to hundreds of milligrams of sample are necessary.

10.4.4
Differential Thermal Analysis (DTA)

This analysis method is based upon the type of thermal reaction, exo- or endothermic, which occurs as a material is heated. Here, not only water loss, or the loss of gasses, is measured, which are endothermic reactions taking in heat for the most part, but also eventual recrystallization and recombinations of the matter present are observed. These are for the most part exothermic reactions, releasing heat. The complexity of the full destabilizations and the problems of kinetics (rate) of these reactions makes interpretations of full DTA diagrams very great, necessitating a long experience in the art of interpretation. Such methods are rarely employed in routine investigations, they are left for the more special applications of mineral species characterization or certain industrial applications such as ceramics.

Thermal differential analysis requires only a small amount of material, several milligrams of sample for a decent determination.

10.4.5
Infrared Spectral Analysis (IR) and Fourrier-Transfrom Infrared Spectroscopy (FTIR)

Infrared spectral analysis (IR) has been used to characterize clays, other minerals and organic materials for some time. The IR method is somewhat less sensitive to the differences in mineral structures than other analysis methods and, as a result, its use as a method of identification, especially in a mixed phase sample, is difficult. However, IR is useful to identify certain features of pure phases. An XRD study is usually necessary before using IR. In IR studies, less than 1 mg are used to obtain a rapid spectrum and as little as 0.1 mg can be used routinely. Much information is available from infrared spectra when organic matter is present. However, silicate and organic compounds give spectra which overlap, and it is usually necessary to get rid of one or the other in order to identify the material present. The problem here is to preserve the characteristics each has when one destroys another, usually by chemical attack.

In these studies, IR spectra are produced by passing a multiwavelength infrared beam through a finely dispersed sample. The normal method uses a sample dispersion (< 1mg) in several hundred milligrams of potassium bromide (KBr). The mixture is pressed to several tons pressure to vitrify the KBr, which then becomes transparent to the infrared radiation. Such dispersions are stable to 600 °C, and thus heat treatment is possible to observe the various states of water or hydrogen ion bonding in the clay or other material in the sample.

The mineral structure absorbs the infrared radiation according to the vibration frequencies of its various crystalline components. These range from the OH (hydroxyl) unit, the CO (carbonate) unit to the SiO_4 and AlO_4 units, then the AlO_6, MgO_6, FeO_6 etc. units and finally to a multimolecular Si_2O_5, or greater units of the silicate networks and C-H-N-O (carbon, hydrogen, nitrogen, oxygen) interactions in organic mole-cules.

The smaller the number of atoms in the vibrating unit, or the lower the mass of an atom, the higher the energy of vibration and the shorter the wavelength absorbed by the vibrator. By convention, most IR spectroscopists (of chemical training) use the inverse of the wavelength, reciprocal centimeters (cm^{-1} to designate the frequency of vibration.

The most useful zones of IR observational spectra for clay minerals are in the OH vibrational region because here one can have access to information not available by other analysis methods. It is of little use to find the chemical composition by IR spectra when there are other, easier methods available which are more precise, electron microprobe for example. However, the bonding energy of the crystalline water (OH units) in the clay structures can be determined with ease using IR methods.

Organic substances are identified in a general way by the C-H and C-N interactions, which do not occur in silicates and which are seen in a distinct region of the IR spectrum. Carbonates give strong bands for C-O interactions occurring in specific regions also. Thus, the different types of material, silicates, carbonates and organic matter can be investigated using the same investigative method.

Infrared analysis requires a very small amount of material, 0.001g or less, and is hence useful when an object should not be modified in its physical aspect.

FTIR or Fourrier-transform infrared spectroscopy is also used in the characterization of ceramics, pigments, and paints, providing information on mineralogical composition and organic compounds. As paint and slip may have organic binders (or be carbon-based), the method can be very useful. The technique applies a Fourier 289rilliant algorithm on the sample signal acquired to get an infrared spectrum. There are now portable FTIR instruments. Bonneau et al 2017 offer a great example of multi-instrument analysis using FTIR, SEM-EDS, and Raman spectroscropy to characterize the paint(s) and help ^{14}C dating parietal paintings. This study reports on protocols to evaluate and prepare the best paint micro-samples to collect in the field causing the least damage to the art work and 289rilliantly shows how the different analysis methods are integrated.

10.4.6
Electron Microprobe (EMP)

The electron microprobe is essentially an X-ray fluorescence spectrometer hooked up to an electron microscope. The idea is simple enough but the physics of the electron-solid interaction and the ensuing escape of the generated X-ray fluorescent radiation have taken quite some time to master, at least thus far. The method is a reliable and reasonably rapid one which can be used to obtain an excellent chemical analysis of a very small quantity of material. The diameter of the analysis spot is near 2 μm and the penetration of the beam which excites radiation that can be detected is about 6 μm in depth.

This analysis method allows one to use a normal petrographic thin section (thickness 30 pm) in order to locate the precise grains of interest. Results are listed in the normal chemical fashion, giving oxide percents or atomic proportions. The method is now routine. The resolution of the method is close to that of the petrographic microscope and therefore the old limitations are still present, i.e. a single clay particle cannot be designated for analysis as in TEM.

Two main types of detection of the electron-excited radiation (X-ray fluorescence) are commonly used, wavelength dispersion (WDS) and energy dispersion (EDS). The first system extracts a portion of the generated secondary X radiation (X-ray fluorescence) into a crystal spectrometer. The system uses the Bragg principle in "reverse" compared to XRD analysis. The white (multichromatic) radiation generated is focused onto a crystal which diffracts it (as would a prism white light) so that the wave-lengths of the characteristic radiation due to the elements in the sample are dispersed in space. A sensitive detector is displaced around the diffracting crystal so that the intensity of the different wavelength radiation is recorded by the machine. Using appropriate corrections, the raw intensities are converted into the initial excited intensities due to the atomic ratios in the target.

Energy dispersion uses the energy of the radiation photons as their method of identification, instead of their wavelength. The conversion of energy into an appropriate signal of an appropriate intensity is done in a crystal similar to a common p-n transistor, a lithium-drifted silicon crystalline material. The p-n gap determines the energy range at which the detector is most efficient. These detectors are more sensitive (efficient) than the wavelength systems, but they give a higher background and hence are less precise than the wavelength methods. For a weak signal, low electron beam current, energy dispersion is more efficient and for a high beam current the wavelength method is better. In the analysis of clays and carbonates, low beam currents are normally better, to avoid destruction of the loosely bound, hydrated, alkali-bearing structure. Therefore, energy dispersion is better in many cases.

Samples are not destroyed by analysis preparation but they must not exceed several centimeters in size in order to be put in the vacuum analysis chamber of the instrument.

10.4.7
Raman Spectroscopy

Raman spectroscopy is another of the modern techniques reaching archaeologists and the ceramic analysis laboratory. This technique is often used in geology and chemistry, and now becomes more frequent for archaeometric studies, in particular for the study of pigments (Fig. 10.1). It is based on the observation of the scattering of light of specific excitation wavelength off a targeted area. The light is shone onto a sample with a laser probe and detectors collect the scattered light giving information on the chemical composition of the sample. Different materials will yield different spectra, which are compared to a reference data base for identification. This is not a quantitative method, but some semi-qualitative data (amount of the material identified) can be obtained from the intensity of the scattered light and area covered under the peaks in the spectra. The sample does not need particular preparation and the technique is non destructive, unless you excite the area too long or with too much energy, which can be damaging.

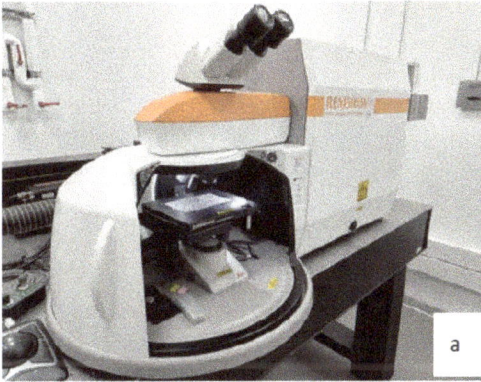

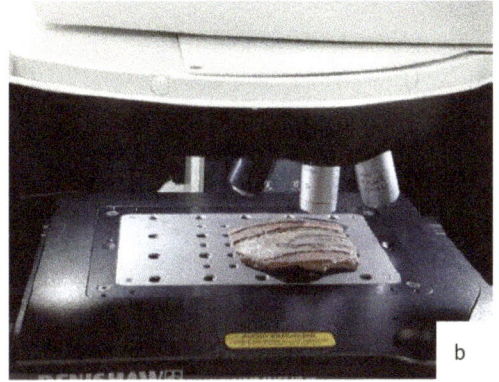

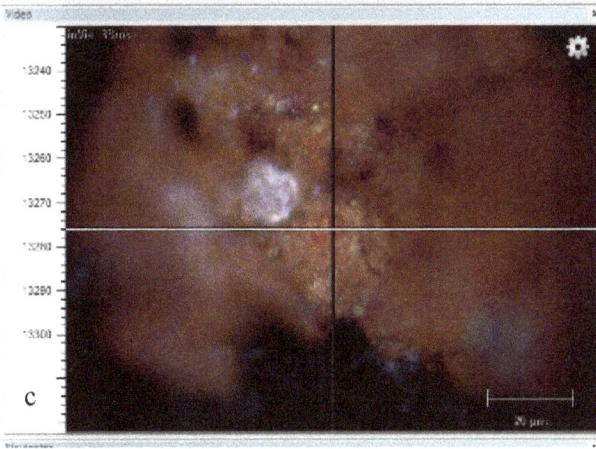

Fig. 10.1. MicroRaman (**a**), used here for the study of pigments on the surface of a ceramic fragment (**b**) The microRaman allows one to focus on a specific area of the sample with the objectives. (**c**) The sample is then subjected to a certain wavelength and amount of time, and a spectrum is acquired (**d**). The later often needs to be 'cleaned' to see the peaks that reveal the presence of certain components, oxides or minerals. Tables exist to identify the peaks.

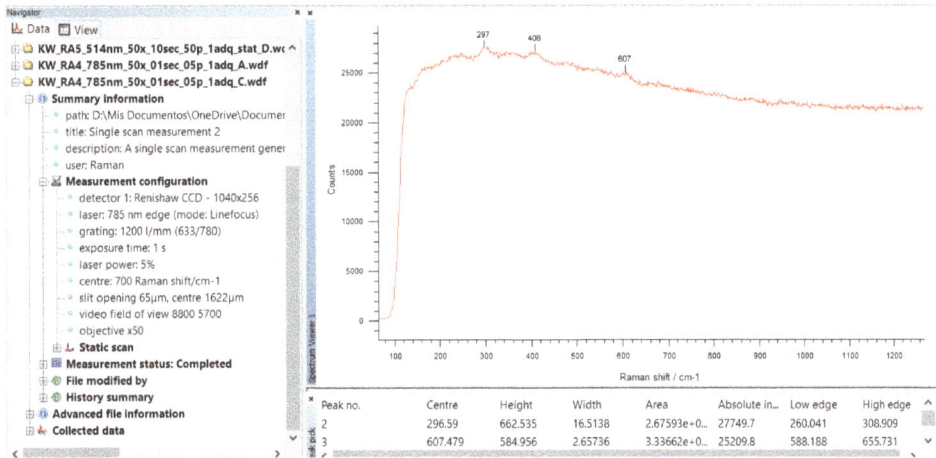

You can analyze samples in a transparent container (e.g. plastic bag, vial, but not glass as it masks the Raman signal). The Raman spectrometer is often coupled to a microscope, which helps target specific areas. When integrated into one device, this is a micro-Raman. You can select areas of a dozen microns in diameter. Like spot analysis with SEM-EDS, various points need to be analyzed to verify the homogeneity of the pigment or surface studied. Portable, or handheld devices exist for fieldwork, analysis in museums, at an archaeological dig or depository.

The technique identifies mineral compounds. It is good for oxides identification, excellent for carbon, pigments and coatings of organic origin. The identification of pure metals do not work with Raman (they do not produce Raman spectra), so this would not work if you wanted to see the composition of the metallic ores in your ceramic or onto its surface.

Excellent studies are now being conducted, often combining different techniques like petrography, microstratigraphy, multispectral analysis, XRD, and SEM-EDS, in addition to or in conjunction with Raman spectroscopy. Most are designed for pigment analysis, slips, paints, and technology of surface treatment. We must also think that the paint layer that we examine is made up of pigment often mixed to a medium or agglutinant. For example, Raman spectroscopy will help identify the red slip on your pot, say hematite (Fe_2O_3) and/or titanium oxide (TiO_2); it could determine that the black design is made with manganese oxides, carbon, iron oxide (e.g. magnetite Fe_3O_4) or combinations thereof, and that the white pigment is hydroxyapatite ($Ca_5(PO_4)_3OH$) mixed with dioxide of Silicium (SiO_2), (see for example Marte et al. 2012, Acevedo et al. 2014, De la Fuente and Martínez 2008, Puente et al. 2019, Vandenabeele et al. 2007).

10.5
Identification of Chemical Elements

10.5.1
Laser Ablation Inductively Coupled Plasma Mass Spectrometry (LA-ICP-MS)

Laser ablation inductively coupled plasma mass spectrometry presents the advantage of being able to conduct elemental analysis of specific parts of a sample, with minimum sample preparation and deterioration as the ablation is nearly invisible to the naked eye. The technique allows to analyze some 57 major, minor and trace elements with a high sensitivity at low cost as compared with INAA, and to collect data on the clay matrix, for example, avoiding the coarser inclusions. The information provided complements the petrographic data, which in many occasions helped interpret the statistical analysis of the chemical data. In other instances, the chemical data shed light on minor mineral variations.

The Elemental Analysis Facility (EAF) at the Field Museum of Natural History in Chicago conducts such analyses following a strict methodology (Dussubieux et al.

2007, Sharratt et al. 2009). Here is a summary of how it works. The instrument used is an Analytik Jena quadrupole mass spectrometer, coupled to a New Wave UP213 laser (which takes care of the ablation of the sample, Fig.10.2). The laser has an ablation diameter of up to 150 microns (0.15 mm). Five samples can be analyzed at a time, along with 3 standards. Great care is taken to avoid contamination of the sample and analytical errors. The first 20 minutes of machine use are not recorded to leave time for stabilization. To address ceramic heterogeneity, 10 spots are ablated per sample. For comparison, for glass, a homogeneous material, only four ablations are performed with a laser diameter of 55 μm.

If the analysis focuses on the clay, which can naturally contain silt and very fine grains, the targeted area is chosen to avoid larger inclusions. A CCD (a charge coupled device image sensor) camera attached to the laser allows for the inspection of the surface of the sample and the selection of an area free of heterogeneity (or temper grains). When it is possible, ablations are performed on a freshly cut surface as far away as possible from the external surface of the artifact as in some cases surface contamination can occur (Golitko et al. 2012). Signal acquisition begins after a 20s pre-ablation period to avoid surface contamination as much as possible. The composition of the ceramic is calculated from the average of the ten measurements carried out on each sample after elimination of any "abnormal" measurements, resulting in RSDs above approximately 20%, that could results from the ablation of temper grains or from any other heterogeneity in the ceramic fabric under the sample surface. The ^{29}Si isotope is used for internal standardization. Concentrations for major elements, including silica, were calculated assuming that the sum of their concentrations in weight percent in glass is equal to 100 percent (Gratuze 1999).

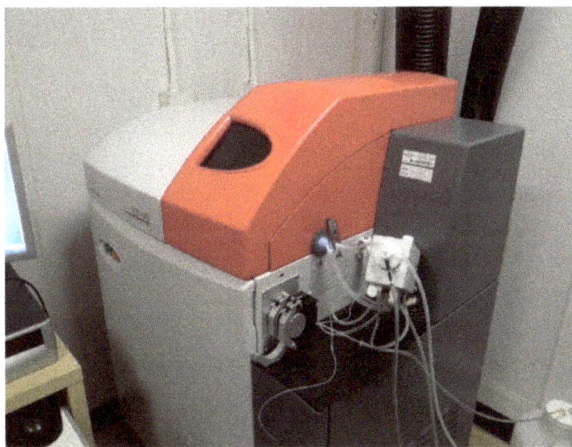

Fig. 10.2. ICP-MS at the Elemental Analysis Facility at the Field Museum in Chicago.

As for other techniques, standard reference materials (SRM) are used to calculate the concentrations of the major, minor and trace elements. Variability in ablation efficiency and signal strength is controlled with the internal standard Si (see also Dussubieux et al. 2007). Attention is paid to the dispersion of each element that may reveal erratic behaviour, due to some analytical problem, low sensitivity (detectability) of the element in the sample or external contamination. An example of LA-ICP-MS analysis of ceramic materials and statistical treatment of the data is found in Druc et al. (2018).

Another very interesting use of LA-ICP-MS is to target specific minerals (say biotite or hornblende) to obtain their chemical composition and compare it to the signature of other similar minerals in other samples and comparative rocks. This allows for provenance studies as particular signatures are linked to specific rock bodies. Gehres and Querré (2018) have conducted a successful study of the ceramics of North-Western France, where they could link the biotites to particular granitic bodies and thus assess ware production and distribution. The laser diameter, however, needs to be small (e.g. 20 µm) which is not the case for all instruments (some laser diameter can reach 100 µm). This is, nevertheless, a very promising method.

10.5.2
Whole Sample Analysis

It is often very interesting to know the composition of the whole sample, without paying attention to what each phase is or what the composition of the phases is. Such methods of chemical investigation are now performed by whole-rock analysis methods which use physical phenomena instead of the classical chemical ones. Chemistry is traditionally involved with reactions between ions and molecules in the aqueous state. The sample is broken into its atomic components by dissolving it in an aqueous acidic solution. In this state, the individual ions or atoms can be selected by their chemical properties. They are reacted with other chemical compounds in order to separate one from the other. This is a method which has been practised for centuries but is now rather rare.

The remains of chemical analysis in solution is done after dissolution of the sample in acidic aqueous solution but the characteristics of the elements are most often determined by ionic interaction with different types of excitation and analysis by physical detectors. Examples are atomic absorption, flame photometry and plasma-induced excitation. Here, the solution is excited by a flame or high temperature means which gives a high energy to the dissolved ions. They emit or absorb energy, which is recorded by spectrometers sensitive to emission in the visible wavelength region.

Another method is to complex ions with another chemical compound which can be excited to give a typical color. The intensity of the color is measured by photometric methods; this is photometry.

These physical methods of identifying the concentrations of ions in solution are considered classical chemistry now, and when one goes to a chemistry laboratory one will most likely get an analysis of one's pot sherd by a combination of these methods. They are good and reliable but not very rapid in execution. Hence they "cost" more, either in time or money. Other physical methods are more commonly used in archaeology these days, they are no less costly or time consuming but they tend to give information on more elements than the wet chemistry physical methods.

10.5.2.1
X-Ray Fluorescence (XRF)

This method is based upon the interaction of X-rays with atoms in a sample. Each atom is excited by a wide range of X-radiation and these atoms reemit X-radiation (X-ray fluorescence) in their turn. The reemission process is proportional to the number of atoms present in a sample. Hence, the intensity of the given, characteristic radiation will give a good idea of the concentration of the chemical species in question. Each atom has a series of X-radiation emissions at specific wavelengths, as do, for example, radio stations (in a well-regulated society, that is). If one can tune in on the given wavelengths of radiation, one can obtain the intensity and hence abundance information.

The X-ray fluorescence radiation is excited by a tube with a cathode of a given metal. Its size is on the order of several tens of centimeters in length. The excitation occurs in a two-stage process. The structure starts out much like a light bulb with a tungsten filament which is heated to boil off electrons and light. In a light bulb, light is the object of the exercise, but in an X-ray tube the electrons are sought. The light emitted is an afterthought. The electrons are directed, by a potential field, to a metal plate which is the target. The metal is pure so that the X-radiation it emits is as little complicated as possible. This target then emits X-radiation, which leaves the tube (made of glass). The excited X-radiation of the target in the tube then strikes the sample which is, in its turn, excited by this energy to give off another set of radiations. This is the fluorescence radiation. Detectors of X-radiation are placed in specific positions around the sample in order to measure the wavelength (element identification) and give an intensity reading (relative abundance measurement).

Detection methods are those described in the section on the electron microprobe (Chap. 10.4.2). Detection limits are on the order of tens to hundreds of parts per million (0.001 to 0.01%) by weight percent in the sample. The material necessary to do this analysis is on the order of a gram or so. The material is ground and melted in a fused bead or left as a powder for analysis. This preparation essentially destroys the object that is used for analysis.

10.5.2.2
Portable X-Ray Fluorescence (pXRF) Analysis

pXRF belongs to the new battery of portable devices available to archaeologists and used for the elemental analysis of ancient materials (Fig. 10.3). While its application for metal characterization has brought its recognition, its efficiency for ceramic characterization is still debated. Much depends upon calibration of the machine for the type of material analyzed, and how the analysis is conducted. It is very attractive to archaeologists for its apparent ease of use, non-destructive character, and ability to register elemental compositions. There is now a great amount of pXRF data to process, but which compositions are we talking about? Unless the sample is free of surface contamination and the methodology is rigorous, one can end up with stray data misleading interpretation. When studying pigments for example, the instrument could pick up some signals from the surface underneath or from the paste if the paint is too thin. Several studies are looking into the matter (in all senses of the term!) and no doubt that better analysis will be possible very soon. For an example of pXRF analysis and protocols see Stovel et al. 2017.

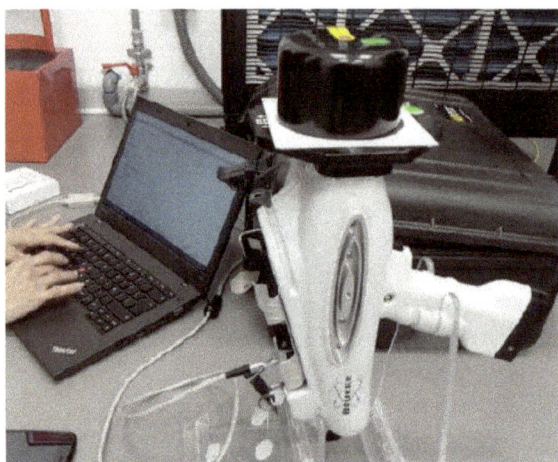

Fig. 10.3. p-XRF equipment operated from a laptop computer.

10.5.2.3
Proton Activated X-Radiation (PIXE)

This method of analysis uses protons to excite X-ray fluorescence radiation instead of X-rays. The system is more complicated in that the tube used to produce the X-rays, instead of being on the order of several tens of centimeters in length, is on the order of several tens of meters in length. This makes for a big machine; it also makes it an expensive machine. The advantage of the PIXE system is that it can excite almost all of the elements possible to a high efficiency in a relatively short period of time, about 20 min or so for a good complete analysis. This means analyses of several parts per million, 0.0001% by weight, for the elements present.

The proton beam can be narrowed to a small spot, less than a micron, and it can be directed on an object outside of the analysis chamber. This is a great advantage for analysis of precious objects which should not be modified.

10.5.2.4
Neutron Activation Analysis (NAA)

Neutron activation (or Instrumental Neutron Analysis - INAA) is a method which uses neutrons to create nuclear reaction in the atoms. These reactions give off high-energy radiation which is analyzed by different types of counters, depending upon the type and energy of the radiation. The reactions occur over different periods of time, some for seconds, some for millions of years. Hence, the counting techniques of radiation intensity must take into consideration several factors: (1) the intensity of the neutron beam, (2) the intensity of radiation for a given period of time for each reaction and (3) the type of radiation measured i.e. its energy. Given a good calculation of these variables, it is then possible to estimate the amount of each element present in a sample.

As one might expect, the machine necessary to prepare a beam of neutrons which will strike a ceramic sample is rather complex. However, the method is very efficient. It can detect elemental presence to less than 1 part per million (< 0.001%). The amount of sample taken from the object and "destroyed", is very small, usually on the order of several milligrams or less. The excited radiation in the sample after neutron bombardment is of different intensity and duration, depending upon the element or atom struck by the neutron beam. The analysis method allows for a large range of elements to be identified, in particular rare and trace elements detected in ppm, as well as minor and some major elements. The trace elements are particularly useful for provenance analysis and to identify raw material sources.

10.5.2.5
Mössbauer Analysis

Mössbauer spectroscopy exploits the property of gamma X-ray emission and absorption by radioactive isotopes that do not recoil; that is, that do not lose the recoil energy when a gamma ray is emitted. The emitted and absorbed energy can later be measured. In the ceramic material, recoil-free elements are few: iron, nickel, aluminium and zinc. Of these elements, iron is the most useful for ceramic technology because of the chemical changes it undergoes that can be linked to specific firing condition and temperature. The iron state is also partly responsible for the color a ceramic takes when fired. With Mössbauer spectroscopy, it is possible to identify the type and amount of iron compounds present, if in a ferric ($Fe3+$) or ferrous ($Fe+$) state. It allows estimation of the temperature and kiln atmosphere, even revealing different oxido-reduction phases. Mössbauer analysis also helps in provenance studies and ceramic classification when clustering sherds with similar Mössbauer signatures.

The method consists in inducing a gamma X-ray emission with a radio-active source (^{57}Co decaying into ^{57}Fe-emitting gamma rays when returning to its ground state) and measuring the amount that are absorbed by the sample, that is by the iron isotopes present. The spectra obtained are characteristic of the chemical state of the iron and its environment. Different parameters are observed, among which are the quadrupole splitting of the peak and the isomer shift or location of the peak. The analysis uses about 200-mg powder samples and requires comparative clay samples to be fired in controlled conditions. The ceramic samples can also be refired in the laboratory to countercheck the results.

10.6
Provenance Based on Zr Age and U-Pb Isotopes

A method used in geology to identify the age of zircons in sedimentary rocks to assess the source of sandstones has rejoined the arsenal of archaeometric techniques used for provenancing ceramics. It is based on the analysis of zircons in detrital quartz or quartz sand found in ceramic bodies. This offers an alternative to optical techniques when trying to source detrital quartz, which is not easy to do.

Zircons are usually tiny and numerous in quartz sands, and very old in geological time, millions of years old. Their age is related to certain geological formations, and it can be calculated by using Pb and U-Pb isotopes ratios. The Zr age-distributions are then compared to those from different possible sand sources. The whole analysis is not necessarily easy to conduct, but very effective. The method calls for isolating the Zr from the ceramic by crushing and separating them from the crystals around, mounting them in epoxy, and polishing the mount for examination under a SEM to determine the spots and Zr large enought (eg. 50 microns or more) to analyze. They are then vaporized into a multi-collector plasma mass spectrometer and the ratios of various isotopes are measured: $^{206}Pb/U^{238}$, $^{206}Pb/Pb^{207}$, $^{206}Pb/Pb^{204}$. Plots of Zr age distributions can then be compared, and the zircons present in a ceramic can be traced to specific valley sections and river tributaries for example. For more see Tochilin et al, 2012 or Kellaway et al 2014. The University of Arizona, in the Department of Geosciences, is one of the few places where this type of analysis is offered to archaeologists.

10.7
Age Determination by Thermoluminescence (TL)

Thermoluminescence is a method currently used to give an approximate age of a pottery sample, or in some cases a good estimation of an age of production. The method is based upon the interaction of high-energy radiation with the components, most often found in a ceramic. The materials most sensitive to this interaction are the minerals quartz and feldspar and glassy materials with high silicate contents, such as melted clay minerals. The materials capture the energy of the radiation and by interaction are left with defects in their structures. The defects are unstable when the sample is heated. The defects liberate energetic particles (less than those which created the defects, of course) in their turn. This energy is in the visible region, and is called luminescence. Hence the term thermo (heated) luminescence (energy in the visible radiation region) is commonly used. The heating process to liberate the defect radiation is a sufficiently low temperature for the heating event which formed the pottery (above 800 °C in most cases) to have effaced any defects formed before the pottery was formed in a furnace. The process of firing essentially sets the thermoluminescent clock to zero.

The energetic radiation which creates defects in the crystalline or glassy structures has several origins. One is cosmic radiation of different energies. This is about 5% or so of the overall amount in most cases. Internal radiation due to disintegration of radio-isotopes of essentially potassium and uranium-thorium (U-Th) in the sample itself account for up to 60% of the radiation. The rest can come from the enclosing sediments, again from disintegration of radioactive isotopes of the elements present. This last category of radiation source assumes an archaeological context. If the sample is above ground, the cosmic radiation will be increazed and the surrounding sediment source will be less.

If the sample has been buried for a long part of its history, a comparison of the radioactivity of the surroundings with that of the sample would be necessary in order to determine the contribution of radiation due to the archaeological environment. Radioactive element content of the sample is also measured. Hence, the source of radiation which interacts with the sample can be determined as a function of time. If one knows the amount of radiation, and its effect on the sample as a function of time, one can estimate the amount of thermoluminescence which will be generated in the sample for each year of deposition in an archaeological context.

The sample is heated to 500 °C in steps to capture all of the thermoluminescent radiation generated by the heating process. Then the sample is subjected to beta radiation in the laboratory in order to determine its susceptibility to this defect-creating mechanism. Comparison of laboratory-induced defects and thermoluminescence is compared to the calculations of radiation induced from the various sources. This determines the efficiency with which the sample captures the radiation energy and trans- forms it into defects which liberate luminescence.

Comparison of the estimated radiation entering the sample from cosmic and other sources plus the internal sources and the efficiency of capture gives a calibration from which one can estimate the age of a sample by the total amount of defects and luminescence liberated by heating. It sounds complicated, but it is not that bad in the hands of experts.

It is obvious that samples rich in quartz and feldspar will be good materials to date. Since potassium feldspar contains large amounts of potassium and hence significant amounts of potassium, which has a radioactive elemental component, these minerals will be useful as an internal thermoluminescent system where the constituent elements in the mineral create the defects necessary to date the pottery.

10.8
Density, Porosity, and Hardness

Density of a ceramic is a good potential measure of its firing degree. The denser a ceramic material, the less air there is in it. The weight of a sherd for a given volume increases as firing stage increases. The higher stage of firing increases the density of the ceramic by melting the clays or grits (sand grains) and by destroying pores in the sample. In this way, density increase means, in general, that the sample loses porosity.

Density can be increased during the early stages of formation of the pot also. If a clay paste is compacted, by pushing it into a mould, for example, the density increases. Pressure eliminates the air pockets in the paste. If the clay paste is formed into flat sheets to be reshaped, it will tend to be, initially, denser than a paste used on a potter's wheel. The act of compression can increase the density of the paste and thus the density of the fired product.

Density and porosity are usually inversely related. Dense, low-porosity products are the result of compaction and/or high firing temperatures for long periods of time.

Methods of determining porosity are not easy to implement. One has to measure the amount of air or water in the sample in order to determine the amount of empty spaces there are in the sample. Density is measured by comparing the weight of a sample in air and when suspended in water. The displacement of water by the sample indicates its volume, and the weight indicates the mass. Density is mass (g) per volume (cm^3). See Chapter 6, section 6.4 for more on porosity.

The classical method of measuring hardness is to determine which material will scratch the sample under investigation. This is derived from a time-tried, and little used, geological-mineralogical test, called the Mohs scale of hardness. An unknown substance is scratched or is used to scratch known materials. This scale is based upon the relative hardness of minerals. Unfortunately, many of these minerals are difficult to procure in the pure state. Further, since one scratches on the other, they are continually being scratched and soon become useless because their surfaces are all scratched and new incisions are difficult to see. However, there are several easily procurable materials which can be substituted for the Mohs minerals. The metals iron and copper are easy to come by. In general, one can use several commonly available materials: a fingernail (easy to come by) a knife (iron or steel, roughly of the same hardness) and copper such as a penny, if you live in the proper country. Aluminium could be used, but in the time of Mohs it seems not to have been a common metal, hence it is not cited.

In any event, an abbreviated scale of hardness could be the following: if your fingernail scratches a ceramic, it is soft. If it scratches your fingernail, it is medium hard. If copper scratches, it is medium hard, but if it scratches copper, the sample is hard. If it is scratched by iron (or steel), it is hard but if it scratches iron, your sample is very hard. These are a very simple set of relative values to determine or class ceramics into general categories of hardness. If one is interested in further precisions, the geologist's Mohs scale is a reference or one can attempt to use the numerical methods devised by materials scientists. This latter analysis is done by allowing a hard point to fall on the sample with different forces. When the point penetrates the sample, the force is noted, and a hardness value is attributed. However, when using ceramics, it is necessary to insure that the ceramic has no large temper grains which are in contrast to the clay-rich ceramic core.

Geologists use the following scale of hardness (Mohs scale) with different minerals as references:

Scale of Hardness

1 Talc	6 Orthoclase (potassic feldspar)
2 Gypsum	7 Quartz
3 Calcite	8 Topaz
4 Fluorite	9 Corundum
5 Apatite	10 Diamond

(fingernail=2, copper coin=3, knife blade=5,
window glass= 5.5, steel file= 6.5)

10.9
Magnetic Analysis

Magnetic analysis uses the magnetic characteristics of ceramics to study provenance and technology, firing temperature, kiln atmosphere and building techniques (Verosub and Moskowitz, 1988). The magnetic properties of the iron oxides and hydroxides like magnetite and hematite are characteristic of their composition and grain size, and can be determined through analysis. The analysis aims at determining the characteristics independent of the geomagnetic field, to yield intrinsic information on the iron compounds. Magnetite and hematite are usually found in the ceramic paste as natural inclusions in the raw material used. Their amount and size are particular to the clay used and the firing conditions. For example, the amount of oxygen available to the ceramic will produce different compounds, hematite being formed in oxidizing atmosphere, while magnetite is formed in reducing conditions. However, impurities may impede or, to the contrary, accelerate the formation of iron compounds, as for calcium-rich pastes, leading to misinterpretation of the magnetic data. Both minerals (magnetite and hematite) are distinguished by their distinct magnetic characteristics, like the Curie temperature, even in very small amounts (down to 1part in 10 for magnetite- Verosub and Moskowitz 1988). The orientation of magnetic particles can also be observed in microscopy, revealing the forming technique used. This orientation causes magnetic susceptibility (anisotropy), which can be measured.

The method measures the degree of magnetization of a mineral, the ease of particle alignment with the magnetic field, and their maximum alignment. The latter is dependent upon temperature, with a decrease in magnetization when temperature increases (Verosub and Moskowitz 1988). Magnetic analysis aims at determining the composition and grain size of the iron compounds to deduce temperature and kiln atmosphere. It is easy and rapid, and non-destructive, unless thermomagnetic analysis is performed, which requires a few hundred milligrams of ceramic.

Bibliography: Ceramic Analysis and Case studies

Attas, M, Fossey, JM, and Yaffe, L (1982) Variations of ceramic compositions with time: a test case using Lakonian *pottery. Archaeometry* 24:181-190

Barlow, JA, and Idziak, P (1989) Selective use of clays at a middle Bronze Age site in Cyprus. *Archaeometry* 31: 66-76

Bertolino, SR, Galván Josa V, and Castellano, G (2011) Ceramic surface paintings and pigments from the Aguada Culture (Argentina): XRD and SEM-EDX archaeometric studies. In: Sarrica, SM (ed), *Paints: Types, Components and Applications, Chemistry Research and Applications*, Nova Science Publishers, New York, pp169-211

Bonneau, A, Staff, RA, Higham, T, and Brock, F (2017) Successfully dating rock art in southern Africa using improved sampling methods and new characterization and pretreatment protocols. *Radiocarbon* 59(3): 659-677

Brodii, NJ, and Steel, L (1996) Cypriot black-on-red ware: towards a characterization. *Archaeometry* 38: 263-278

Bollory, CA, Jacobson, L, Peisach, M, Pineda, CA, and Sampson, CG (1997) Ordination versus clustering of elemental data from PIXE analysis herder-hunter pottery. *J of Arch Sci* 24: 319-327

Castellano, A, D'Innocenzo, A, Pagiliara, C, and Raho, F (1996) Composition and origin of Iapygian pottery from Roca Vecchia, Italy. *Archaeometry* 38:59-65

Goffer, Z (1980) *Archaeological chemistry*, vol 55. John Wiley, NY, pp 376

De la Fuente, G, and Martínez, JM (2008) Estudiando pinturas en cerámicas "Aguada Portezuelo" (*ca.* 600-900AD) del Noroeste de Argentina: Nuevos aportes a través de una aproximación arqueométrica por microespectroscopia de Raman (MSR). *Intersecciones en Antropología* 9: 173-186.

Druc, I, Underhill, A, Wang, F, Luan, F, Lu, Q (2018). A preliminary assessment of the organization of ceramic production at Liangchengzhen, Rizhao, Shandong: perspectives from petrography. *J of Arch Sci Report*s 18: 222-238.

Druc I, Giersz, M, Kalaska, M, Siuda, R, Syczewski, M, Pimentel, R, Chyla, J and Makowski, K (2020) Offerings for Huari ancestors: strategies of ceramic production and distribution at Castillo de Huarmey, Peru. *J of Arch Sci Reports* 30, April. https://doi.org/10.1016/j.jasrep.2020.102229

Druc, I, Inokuchi K, and Dussubieux L (2017) LA-ICP-MS and petrography to assess ceramic interaction networks and production patterns in Kuntur Wasi, Peru. *J of Arch Sci Reports* 12: 151-160.

Dussubieux, L, Golitko, M, Williams, RP, Speakman, RJ (2007) Laser ablation-inductively coupled plasma-mass spectrometry analysis applied to the characterization of Peruvian Wari ceramics. In: Glascock, MD, Speakman, RJ, and Popelka-Filcoff, RS (eds), *Archaeological Chemistry. Analytical techniques and archaeological interpretation*. ACS Publication Series 968. Am Chemical Soc, Washington, DC, pp. 349-363.

Janssens, K, and Van Grieken, R (2004) Non-destructive microanalysis of cultural heritage materials. Elsevier, Amsterdam, London, pp 800

Fergusen JR, Van Keuren, S, and Bender, S (2015) Rapid qualitative compositional analysis of ceramic paints. *J of Arch Sci Reports* 3: 321-327

Frigolé C, Riera-Soto C, Menzies A, Barraza M, and Benítez, A (2019) Estudio de pastas cerámicas del centro-oeste argentino, mendoza argentina: microscopía óptica y QEMSCAN *Boletin de Arqueologia PUCP*, Advances in ceramic and pigment analysis in archaeology, Part 2, 27: 67-85.

Gehres, B, and Querré, G (2018) New applications of LA-ICP-MS for sourcing archaeological ceramics: microanalysis of inclusions as fingerprints of their origin. *Archaeometry* online issue doi: 10.1111/arcm.12338 pp 1-14

Golitko, M, Dudgeon, J, Neff, H, and Terrell, JE (2012) Identification of post-depositional chemical alteration of ceramics from the North Coast of New Guinea (Sandaun Province) by Time of Flight-Laser-Ablation-Inductively Coupled Plasma-Mass Spectrometry (TOF-LA-ICP-MS). *Archaeometry* 54(1): 80-100.

Hughes, MJ, Cowell, MR, and Hook, DR (1991) Neutron activation and plasma emission spectrometric analysis in Archaeology. Occasional paper 82. Department of Scientific Research, British Museum

Hunt, A (ed) (2017) The Oxford handbook of archaeological ceramic analysis. Oxford University Press, Oxford, pp 724

Kellaway, SJ, Craven, S, Pecha, M. Dickinson, WR, Gibbs, M, Ferguson, T, and Glasscock, MD (2014). Sourcing olive jars using U-Pb ages of detrital zircons: a source of 16th century olive jars recovered from the Solomon Islands. *Geoarchaeology* 29: 47-60

Kilikoglou, V, Maniatis, Y, and Grimanis, AP (1988) The effect of purification and firing of clays on trace element provenance studies. *Archaeometry* 30: 37-46

Killick, D, and Hayashida, F (2017). Lung-powered copper smelting on the Pampa de Chaparri, Lambayeque department, Peru. Presented at the 82st Annual Meeting of the Society for American Archaeology, Vancouver, British Columbia.(tDAR id: 430178)

Koh Choo, CK (1995) A scientific study of traditional Korean celadons and their modern developments. *Archaeometry* 37:53- 81

Maniatis, Y, Perdikaksis, V, and Kostsakis, K (1988) Assessment of in-site variability of pottery from Sesklo, Thessaly. *Archaeometry* 30 :264- 274

Marte, F, Acevedo, V, and Mastrangelo, N 2012 Técnicas arqueométricas combinadas aplicadas al análisis de diseños de alfarería "tricolor", de Quebrada de Humahuaca, Jujuy, Argentina. *Boletin del museo chileno de arte precolombino* 17(2), 53-64.

Matson, FR (1985) Glazed brick from Babylon – historical setting and microprobe analyses. In: Kingery, WD (ed) *Ceramics and civilization*, vol II. Technology and style. American Ceramics Soc, Colombus, Ohio, pp 133-156

Miksa, EJ (1998) A model for assigning temper provenance to archaeological ceramics with case studies from the American Southwest. Ph.D. dissertation. Department of Geosciences, University of Arizona, Tucson.

Miksa, EJ, and Heidke, JM (1995) Drawing a line in the sands: Models of ceramic temper provenance. In: Heidke, JM, and Stark, MT (eds), *The Roosevelt Community Development Study: Vol. 2. Ceramic chronology, technology, and economics*. Anthropological Papers 14. Tucson: Center for Desert Archaeology, pp. 133–205

Mills, BK, and Crown, PL (eds) (1995) *Ceramic production in the American Southwest*. University of Arizona Press, Tucson, AZ

Mitri Vzelano, P, Arpa, R, Ferra, E, and Appolonia, L (1990) Roman pottery from Augusta Praetoria (Aosta, Italy): provenance study. *Archaeometry* 32:163-175

Mitri Vzelano, P, Casoli, A, Barra Bagnasco, and Pracco Ancona, MC (1995) Fineware from Locri Epizephiri: a provenance study by coupled plasma emission spectroscopy. *Archaeometry* 37:41-53

Mommsen, H, Beir, T, Diehl, U, and Podzuweit, Ch (1992) Provenance determination of Mycenaenan sherds found in Tell el Amarna by neutron activation analysis. *J of Arch Sci* 19:295-302

Neff, H (ed) (1992) Chemical characterization of ceramic pastes in archaeology. Monographs in world archaeology No 7· Prehistory Press, Madison, Wisconsin, pp 289

Niziolek, LC (2018) Portable X-ray fluorescence analysis of ceramic covered boxes from the 12th/13th-century Java Sea shipwreck: A preliminary investigation. *J of Arch Sc Reports* 21: 679-701.

Olin, JS, Hardbottle, G, and Sayre, EV (1978) Elemental composition of Spanish colonial Majolica ceramics in the identification of provenience. In: Carter, GF (ed) *Advances in chemistry*, vol 171. *Archaeological chemistry* II. American Chemical Society, pp 200-229.

Ownby, M (2015) Ceramic analysis perspectives from afar. In Druc, I (ed), *Ceramic Analysis in the Andes*, Deep University Press, WI, pp. 171-180

Ownby M., Druc, I, Masucci, M (eds) (2017). *Integrative Approaches in Ceramic Petrography.* University of Utah Press, Salt Lake City, pp 233.

Perlman, I, and Asaro, F (1969) Pottery analysis by neutron activation. *Archaeometry* 11:21-52

Pineda, CA, Sampson, CG and Peisach, M (1997) Ordination versus clustering of elemental data form PIXE analysis of herder-hunter pottery: a comparison. *J of Arch Sci* 24:319-327

Pollard, AM, and Hatcher, H (1994) The chemical analysis of oriental ceramic body compositions, part 1.Wares from northern China. *Archaeometry* 36:41- 62

Porat, N, Yellin, J, Heller-Kallai, and Halicz, L (1991) Correlation between petrography, NAA and ICP analyses: application to Early Bronze Egyptian pottery from Canaan. *Geoarchaeology* 2: 133-149

Puente, V, Porto López, JM, Desimone, PM, and Botta, PM (2019) The persistence of the black color in magnetite-based pigments in prehispanic ceramics of the Argentine northwest. *Archaeometry*, 61(5): 1066-1080 DOI:10.1111/arcm.12476.

Quinn, P. (ed.) (2009) *Interpreting Silent Artefacts: Petrographic Approaches to Archaeological Ceramics.* Archaeopress, Oxford, pp. 295

Riera-Soto C, Uribe Rodríguez M, Menzies A, Barraza Bustos M (2019 Avances en petrografía automatizada: cerámicas tempranas de Guatacondo, norte de Chile (900 A.C.-200 D.C.) *Boletin de Arqueologia*, PUCP, Advances in ceramic and pigment analysis in archaeology, Part 1, 26:141-157.

Rice, PM, and Cordell, AS (1986) Weeden Island pottery: style, technology and production. In: Kingery, WD (ed) *Ceramics and civilization*, vol II. Technology and style. American Ceramics Soc, Colombus, Ohio, pp 273-295

Rice, PM (2015) *Pottery analysis. A source book, second edition.* University of Chicago Press, Chicago, pp 592

Sharratt, N, Golitko, M, Williams, RP, and Dussubieux, L (2009) Ceramic production during the Middle Horizon: Wari and Tiwanaku clay procurement in the Moquegua Valley, Peru. *Geoarchaeology* 24(6):792–820.

Shepard, AO (1963) Rio Grande Glaze-paint pottery: test of petrographic analysis. In: Matson, FR (ed) *Ceramics and man.* Aldine, Chicago, pp 62- 87

Shingelton, KL Dell, GH and Harris, TM (1994) Atomic absorption spectrophotometric analysis of ceramic artifacts from a protohistoric site in Oklahoma. *J of Arch Sci* 21: 343-358

Stovel, EM, Cremonte, B, and Echenique, E (2017) Petrography and pXRF at San Pedro de Atacama, Northern Chile. Exploring ancient ceramic production. In Ownby M, Druc I, and Masucci M (eds), *Integrative Approaches in Ceramic Petrography.* University of Utah Press, Salt Lake City, pp. 53-72

Taylor, RJ, and Robinson, VJ (1996) Neutron activation analysis of Roman African red slip ware kilns. *Archaeometry* 38: 231- 243

Tochilin, C, Dickinson, WR, Felgate, MW, Pecha, M, Sheppard, P, Damon, FH, Bickler, S, and Gehrels, GE (2012) Sourcing temper sands in ancient ceramics with U-Pb ages of detrital zircons: A southwest Pacific test case: *J of Arch Sci* 39, 2583-2591

Torres, LM, Arie, AW, and Cowell, MR (1984) Provenance Determination of fine orange Maya ceramic figurines by flame atomic absorption spectometry, preliminary study. In: Lambert, JB (ed) *Archaeological chemistry III,* Advances in Chemistry Series 205. American Chemical Society, Washington, DC, pp 193-213

Vandiver, PB and Koehler, CG (1986) Structure, processing and style of Corinthian transport amphoras. In: Kingery, WD (ed) *Ceramics and civilization*, vol II. Technology and style. American Ceramics Soc, Colombus, Ohio, pp 173-215

Vaughn, K. J., and Neff, H (2004) Tracing the clay source of Nasca polychrome pottery: Results from a preliminary raw material survey. *J of Arch Sci* 31(11): 1577–1586.

Vandenabeele, P, Edwards, HGM, and Moens, I (2007) A decade of Raman spectroscopy in art and archaeology. *Chemical Reviews* 107: 675-686.

Verosub, KL, and Moskowitz, BM (1988) Magneto-archaeometry: new technique for ceramic analysis. In: Farquhar, RM, Hancock, RGV, and Pavlish, LA (eds), *Proceedings of the 26th International Archaeometry Symposium, Toronto*, Archaeometry Laboratory, University of Toronto, Toronto, pp 252-255

Weigand, Ph C, Hardbottle, G, and Sayre, EV (1977) Turquoise sources and source analysis: Mesoamerica and the southwestern USA. In: Earle, TK and Ericson, JE (eds), *Exchange systems in prehistory*. Academic Press, NY, pp 15-34

11   How to Acquire the Knowledge to Do the Job

The preceding chapters presented different aspects of ceramics, the materials used and their properties, the techniques of analysis, and the geologist's and potter's points of view. We now turn to a more practical aspect of the question: how to acquire the knowledge to do the job. Many researchers advocate an opening up of the field. Archaeologists have been encouraged to do the job themselves, or at least to take a geology or archaeometry course to facilitate the dialogue between the analyst and the re- searcher. More and more students are choosing this direction. Although many of these courses are still given outside anthropology departments, a growing number are open to, or especially designed for, students wanting to learn more about materials science and ceramics. Many journals and books are also available to further the researcher's knowledge of the subject.

11.1
Courses in Geology, Chemistry and Physics

Depending on your interests, some courses will be better than others. It is most probable that an interdisciplinary background will be needed. Archaeology is, in any case, an interdisciplinary science. For characterization studies, courses in geology, chemistry and physics are needed. A course in materials characterization, for example, should introduce the different techniques used (SEM, XRD, NAA, XRF, etc.) in terms of the principles of physics on which the analyses are based, how each technique works, why and under what circumstances it is better to use one technique rather than another. Lab activities should also be included, to ensure that the student masters the different microscopes, instruments and analysis programs needed to do ceramic analysis. These courses are often given to engineers in applied science, engineering, or science departments. There are introductory courses (one semester), at the university level, for which the scientific background given at a lower level is sufficient, assuming you are familiar with an electron, a nucleus, and the basic physical forces in action. You do not have to remember how to do integrals, but should be prepared to pull out your science books and keep up your reading and your interest in the subject. To put theory into practice is always convincing. When these basics are mastered, then it becomes easier to learn another program, a new technique. They somehow relate and are based on the same principles.

Once some knowledge of physics is gained, as well as technical skills in manipulating the microscopes and other devices, it is necessary to understand the output, the results of the analysis, or to be able to discuss it with the analyst. Thus, some knowledge of chemistry or geology is also necessary. If your work is oriented towards mineral identification, geology (introduction to geology, to rock-forming minerals, to petrography, etc.) will cover the necessary chemistry as well. These courses explore how minerals and rocks are formed, and crystallography. This applies if you are working in elemental analysis (chemical analysis) as well. Chemistry is approached by looking at how elements combine, materials constitution, oxides, sulphides, phases, etc. Courses in analytical chemistry are thus useful.

For petrographic analysis, a background in geology is, of course, necessary. An introduction to petrography and petrography courses will teach the student how to recognize the different minerals and rock types by using a petrographic microscope. Alongside this, a course on crystallography, with lab practice, could help the student to recognize these rocks in the field. Knowledge of the geology of the different rock types (sedimentary, metamorphic, volcanic, etc.) is very useful. To understand rock formations, it is helpful to know how and why certain types are found in certain places. This is taught in courses on igneous, sedimentary and metamorphic petrography, and these courses combine theory and practice, with many hours spent looking at thin sections. However, the examples studied are always unproblematic thin sections, of unaltered unfired rocks and minerals, with one rock per slide. This is not exactly what an archaeopetrographer would see when looking at his or her ceramic thin sections. Besides, these courses are designed for geologists, not for archaeologists or ceramists. There are very rare exceptions, when the class is given by someone aware of the problems specific to the ceramic material. After having acquired the basics of petrography and mineral identification, the best school is the field, that is, looking at ceramic thin sections under the microscope, if possible with the supervision of a knowledgeable person. Failing the latter, many good books for rock and mineral petrographic identification exist (see the color illustrated petrographic Atlas for the different rock types).

Another area of interest might be exploring ceramic material, testing ceramic properties, and understanding the relationships between function and form. Many of these research approaches require some basic know- ledge of mechanics and physics, often obtained in courses for civil engineers or in another applied science department. The main venue of learning, however, is the lab, and many young anthropologists and archaeo- ceramists have mastered the necessary skills there. Anthropology and archaeology departments usually have a laboratory for physical anthropology, ceramic reconstruction, experiments in stone cutting, etc. The "materials science" orientation is not far from these experimental and field-related aspects of archaeology. Besides, the necessary instruments for these lab tests, porosity studies, and mechanical and thermal stress analyses are relatively easy to acquire and manipulate. They do not call for the environement (and the budget) associated with chemical analysis. Materials studies also benefit from some practical knowledge of ceramic production and technology, obtainable in "hands-on" ceramics courses (in the art department or in community programs).

As well, ceramic ethnography and ethnoarchaeology classes are very helpful. They contextualize ceramic production, show the wide variety of technical options the potter has, and enlarge the analyst's vision beyond what can be seen under the microscope. Acquainting oneself with these subjects, whether by taking courses, reading the many books and articles on the subject, or participating in projects, is a natural step in ceramics studies when one is interested in better understanding ancient and modern ceramic production.

Materials science and ethnography come together through their common goal of understanding ceramics. Ceramics classes (whether ethno- graphic or production-oriented) are often given in anthropology and archaeology departments. These are courses that students in other disciplines (applied sciences, geology) wanting to work with ancient ceramics are encouraged to take. Indeed, it is as important for chemists and geologists to understand the archaeological problematic as for archaeologists to tackle geochemistry while working with ceramics.

To conclude this section, here is a list of useful courses for acquiring the knowledge to do the job. It is not necessary to take them all or to take them concurrently. Much depends on the specialization sought. They may be offered in science, engineering, geology, or anthropology departments. Usually, the interested student gleans his or her courses from all these different departments in a truly interdisciplinary way. From them, the specific content is extracted to construct the specialization pursued. These basic courses are:

Basics of geology and mineralogy
    Mineralogy
    Petrography of rocks (igneous, sedimentary, metamorphic)
    Sedimentology
    Geomorphology

Materials characterization
    Physics of particle interaction
    Structural determinations

Materials properties
    Spectroscopic analyses
    Chemical analyses methods

Data analysis in archaeometry
    Statistics, quantitative methods
    Ceramic technology
    Ceramic ethnography or ceramic ethnoarchaeology

All these should be linked to supervised laboratory practice.

Fortunately, now several institutions offer training in archaeometry (measurement techniques in archaeology), with materials characterization courses for archaeologists and archaeology classes for students with engineering and science background. These programs teach much of what one needs to learn the job. Most of these training programs are recent. They may lead to the obtention of an MS degree or another recognized diploma. These programs are found mainly in England, where the tradition of ceramic analysis is very strong and the United States, as well as in France, Germany, Greece, Chili. Other training centers exist, often linked to one specialization or to an individual who started a laboratory and research center. Many laboratories train a few students, graduate and undergraduate. This is the best training place.

Another venue to learn and expand our knowledge about ceramics and new techniques applied to the analysis of ceramics are the Archaeometry meetings, national and international, Archaeology meetings where sessions are dedicated to different ceramic topics, and occasional meetings and seminars offered by many smaller ceramic groups that now exist. One such group is Ceramic Petrographers in the Americas, and the well-established Ceramic Petrology group in UK.

11.2
Some Journals, Books, and Laboratories Active in the Field of Interest

The following journals, books and laboratories may not be specifically ceramics oriented. They include other materials and research perspectives, such as obsidian provenance, lithic identification, metals, dating, pollen analysis, archaeomagnetism. They are, however, good places to learn about materials related to archaeological problematics. Many discussions and articles, of much importance to the orientation of the research on ceramics, are found in these journals, especially in archaeometry.

Journals
Advances in Archaeological Method and Theory
Archaeometry (three issues per year)
Archaeological and Anthropological Sciences (Springer Verlag)
Archaeological Chemistry (monographs in Advances in Chemistry by the American Chemical Society)
Applied Clay Science (Elsevier)
BAR series (international and British). See the special publications on ceramic issues
Current Anthropology
Clays and Clay Minerals materials (Springer). Journal of the Clay Minerals Society
Geoarchaeology An International Journal
Journal of the American Ceramic Society
Journal of Archaeological Method and Theory
Journal of Archaeological Research

Journal of Archaeological Science and JAS reports

Revue d'Archéométrie (a GMPCA yearly publication – Groupe des méthodes pluridisciplinaires contribuant à l'archéologie)

In Latin America, some countries have started early with archaeometry like Chile and Argentina, others are now more active in the field, and several journals publish archaeometry-related articles, like Chungara, Boletín de Arqueología PUCP, Estudios Atacameños, Comechingonia.

Some Laboratories and Educational centers conducting ceramic characterization or ceramic material studies (among others)

In recent years, several new centers, university laboratories or study groups dedicated to archaeometry and ceramic studies have seen the light. Here are a few of the most well-known.

Archaeological Geophysics Research Group, Bradford University (UK) Department of Archaeology and Prehistory, University of Sheffield (UK)

Archaeometry group at ISKP (Institut für Strahlen- und Kernphysik), Bonn University, Germany

Archaeometry Research Group, Heidelberg, Germany (in particular dating and thermoluminescence)

Archaeometry Laboratory (ERAUB) Dept. de Prehistoria, Historia Antiga I Arqueologia, Barcelona, Catalonia, Spain

ATAM Ancient technologies and archaeological materials, part of the Illinois State Archaeological Survey, USA

Berkeley EDXRF laboratory, Phoebe Hearst Museum of Anthropology, California, USA (specialized in obsidian provenance analysis, but also works with ceramics)

Center de recherches géophysiques, Université de Bordeaux III –CNRS, France (training and research center for graduate students)

Center de céramologie de Lyon, CNRS, France

Elemental Analysis Facility, Field Museum of Natural History, Chicago (LA-ICP-MS, also strong in glass and beads analyses)

Laboratory of Archaeometry, National Center for Scientific Research, Demokritos, Athens, Greece

Laboratoire d'Archéométrie à Rennes (Université de Rennes I), France

Laboratory for Ceramic Studies, Leiden University, The Netherlands (technology, micro-macro analyses, experimental studies)

Materials Science and Engineering TU Delft, The Netherlands (materials characterization in Art and Archaeology)

MIT Massachusetts Institute of Technology, USA, Center for Materials Research in Archaeology and Ethnology (CMRAE) offering BS and PhD in Archaeological Materials (not limited to ceramics, but offers ceramic petrography classes)

Munich Archaeometry Group, Physics Department of the Technical University of Munich, Germany

MURR (Missouri University Research Reactor), in Columbia, USA, where much of the INAA ceramic analysis done in the USA is performed. It also works with the department of geological science at the University of Missouri, USA

Smithsonian Institution, Conservation Analytical Laboratory (CAL), Washington DC, USA

University of Arizona and the Arizona State Museum in Tucson, Arizona,

USA (classes on ceramic petrography (transmitted and reflective analysis and Zircon analysis for archaeological materials (including ceramics and raw materials)

Many of these laboratories and training centers now have a web site on Internet. The Internet is also a good way to learn about new analysis centers for archaeological materials and publication in the field.

11.3
Video Documentaries Related to Topics Discussed in This Book
(I. Druc, free on www.vimeo.com)

Phu Lang, ceramic production in North Vietnam https://vimeo.com/134255181
El paleteado, producción cerámica en Mangallpa, Perú https://vimeo.com/54213959
Paddle-and-anvil, ceramic production in Mangallpa Peru https://vimeo.com/55308616
Producción e itinerancia, entrevista con un alfarero tradicional
        de Ancash, Perú https://vimeo.com/42793251
Atilio Lopez, alfarero tradicional de la sierra argentina https://vimeo.com/97191953
Traditional ceramic production in Tarica, Peru https://vimeo.com/42790326
Arte Maya según Alfredo Gonzáles Castillo (ceramist) https://vimeo.com/35601782
She Kiln (modern ceramic production in Wisconsin) https://vimeo.com/35540095
Ceramic Petrography, an interview with Dr. Jim Stoltman https://vimeo.com/35922518
Murat Ires, Master Potter (Istanbul) https://vimeo.com/35567360
Women potters of Sorkun, Turkey https://vimeo.com/35526032
Cini porcelain production of Kütahya, Turkey https://vimeo.com/35633427
Reduction Firing, The creation of Ottoman Blackware https://vimeo.com/35532313
Andean potters of Conchucos, Peru. https://vimeo.com/35529198

Biographies

Dr. Isabelle Druc is specialized in Andean archaeology, ceramic studies, petrography, archaeometry, petrography, and ethnoarchaeology. She focuses on ancient technologies, traditional productions, communities of practice, and socio-cultural and economic interactions. She is involved in archaeological, ethnoarchaeological and ceramic projects worldwide, has published many articles and books on ceramic analysis and has produced many documentaries and video interviews related to ceramics, traditional arts, and culture. She is affiliated to the University of Wisconsin-Madison and the Field Museum Museum in Chicago.

Dr. Bruce Velde is a very well-known clay mineralogist, with many books and articles to his credit. He has taught and conducted research in the USA and France, where he is established since many years. His interest in ancient ceramics led him to help many students in their studies of ceramics.

Dr. Thi Cuc Phuong Nguyen, Vice Rector, Hanoi University, Vietnam

Dr. Shirley O'Neill, Professor and Dean, Faculty of Education, University of Southern Queensland, Australia

Dr. José-Luis Ortega, Professor, Foreign Language Education, Faculty of Education, University of Granada, Spain

Dr. Surendra Pathak, Head and Professor, Department of Value Education, IASE University of Gandhi Viday Mandir, India

Dr. Charls Pearson, Logic, Semiotics, Philosophy of Science, Peirce Studies, Director of Research, Semiotics Research Institute, China

Dr. Luis Porta Vázquez, Professor, National University of Mar del Plata CONICET, Argentina

Dr. Shen Qi, Assoc Professor, Shanghai Foreign Studies University (SHISU), Shanghai, China

Dr. Timothy Reagan, Professor and Dean, College of Education and Human Development, Univ. of Maine, USA

Dr. Antonia Schleicher, Professor, NARLC Director, NCTOLCTL Exec. Director, ACTFL Board, Indiana University-Bloomington, USA

Dr. Farouk Y. Seif, Exec. Director of the Semiotic Society of America, Center for Creative Change, Antioch University Seattle, Washington, USA

Dr. Gary Shank, Professor, Educational Foundations and Leadership, Duquesne University, Pittsburgh, Pennsylvania, USA

Dr. Kemal Silay, Professor, Flagship Program Director, Department of Central Eurasia, Indiana University-Bloomington, USA

Dr. José Tejada Fernández, Professor, Autonomous University of Barcelona, Spain

Dr. François Victor Tochon, Professor, University of Wisconsin-Madison, Deep Education Institute, President of the International Network for Language Education Policy Studies, USA

Dr. Brooke Williams Deely, Women, Culture and Society Program, Philosophy Department, University of St. Thomas, Houston, USA

Dr. Jianfang Xiao, Associate Professor, School of English and Education, Guangdong University of Foreign Studies, China

Dr. Dan Jiao, Henan University of Technology, Zhengzhou, China

Dr. Danielle Zay, University of Lille 3 Charles De Gaulle, France

Dr. Ronghui Zhao, Director, Institute of Linguistic Studies, Shanghai Foreign Studies University, Shanghai, China

Other referees may be contacted depending the Book Series or the nature and topic of the manuscript proposed.

Contact: publisher@deepeducationpress.org

A LIFE IN SIGNS AND SYMBOLS
BOOK SERIES

SIGNS AND SYMBOLS IN EDUCATION
EDUCATIONAL SEMIOTICS

François Victor Tochon
University of Wisconsin-Madison, USA

In this monograph on Educational Semiotics, Francois Tochon (along with a number of research colleagues) has produced a work that is truly groundbreaking on a number of fronts. First of all, in his concise but brilliant introductory comments, Tochon clearly debunks the potential notion that semiotics might provide yet another methodological tool in the toolkit of educational researchers. Drawing skillfully on the work of Peirce, Deely, Sebeok, Merrell, and others, Tochon shows us just how fundamentally different semiotic research can be when compared to the modes and techniques that have dominated educational research for many decades. That is, he points out how semiotic methods can provide the capability for both students and researchers to look at this basic and fundamental human process in inescapably transformational ways, by acknowledging and accepting that the path to knowledge is, in his words "through the fixation of belief."

In four brilliantly conceived studies, he shows us how semiotic concepts in general, and semiotic mapping in particular, can allow both student teachers and researchers alike insights in these students' development of insights and concepts into the very heart of the teaching and learning process. By tackling both theoretical and practical research considerations, Tochon has provided the rest of us the beginnings of a blueprint that, if adopted, can push educational research out of (in the words of Deely) its entrenchment in the Age of Ideas into the new and exciting frontiers of the Age of Signs. – Gary Shank, Duquesne University.

LANGUAGE EDUCATION POLICY
BOOK SERIES

DISPLACEMENT PLANET EARTH
Plurilingual Education and Identity
for 21st Century Schools

K. Harrison, M. Sadiku, F.V. Tochon (Eds.)

Displacement Planet Earth engages an urgent call for action for displaced families—immigrant and refugee—whose children attend school in the host countries of the U.S., Europe, and Australia. The book develops the basis for a model of cultural and linguistic rights for these diverse students living under migration circumstances. The 19 scholars who contributed to this volume offer an in-depth look at these questions. This volume goes a long way in providing rationales and strategies, urging immediate action. Three sections address the conceptual, the policies and programs, and the narratives of experiences for particular groups, providing a spectrum written by international scholars.

Language Education Policies and teachers' practices can help repair the contextual, psychological ideological and social fabric of human lives and societies impacted by misconceptions based on language ideologies and language status that lead to miscommunication, discrimination, social divisions, violence, war, and human struggle; especially for those displaced.

LANGUAGE EDUCATION POLICY
BOOK SERIES

LANGUAGE EDUCATION POLICY UNLIMITED:
GLOBAL PERSPECTIVES AND LOCAL PRACTICES
François Victor Tochon (Ed.)

This book is a first. Language Education Policy is a new field of study that establishes a cross section between educational policy and language policy studies. It inherits from an abundance of intellectual and methodological traditions while opening new perspectives that focus on the interface between policymaking and its enactment in a classroom or an educational setting. The study of the interface between the macro-policy level of the political stage and the micro-policies of education in practice implies a focus on how policy decisions are translated into regulations that affect the lives of people. 21 authors have contributed to this outstanding volume that situates the stakes in the new field of inquiry with examples in 14 countries. "This essential book shows why language education policy will never work if it is top-down and ignores local contexts and stakeholders. It illustrates the fundamental importance of taking local contexts into consideration and actively engaging and empowering local stakeholders in the development and implementation of all language education policy. A better blueprint for successful language education policy would be hard to find."

DEEP ACTIVISM
BOOK SERIES

Health by All Means.
Women Turning Structural Violence into Peace and Wellbeing
Araceli Alonso and Teresa Langle de Paz

Health by All Means documents the transformation of a community with, for, and by women who experience gender-based structural violence. It can be experienced as a story, a philosophy lesson, a walk with a treasured friend, and, at times, a song, and much dancing. The program was initiated in the context of a university partnership by a faculty member and students at the University of Wisconsin-Madison -including the Department of Gender and Women's Studies, the Global Health Institute and the 4W Women and Wellbeing Initiative. HbAM's success can be attributed to a broad network of actors -including community leaders, students who rotated through and were transformed by it, financial supporters, both individuals and organizations, and from support and recognition by UNESCO.

OUT OF HAVANA
Memoirs of Ordinary Life in Cuba

Araceli Alonso
University of Wisconsin-Madison

Out of Havana provides an uncommon ordinary woman's insight into the last half century of Cuba's tumultuous recent history. More powerfully than an academic study or historical account, it allows us intimately to grasp the enthusiasm, commitment and sense of promise that defined many average Cubans' experience of the 1959 Revolution and the first triumphant decades of the Castro regime. As the story shifts into the final decades of the last century (the 1980s Mariel Boatlift, the so-called "special period in time of peace" [from 1991 to the end of the decade], and the 1994 Balseros or Rafters Crisis), it starts gradually to reveal, with understated yet relentless eloquence, an ultimately insuperable rift between the high-flown official rhetoric of uncompromising struggle and revolutionary sacrifice and the harsh conditions and cruelly absurd situations that the protagonist, along with the majority of Cubans, begin routinely to live out. It is a rare and important document, a unique personal chronicle of an everyday Cuban reality that most Americans continue to know only fragmentarily.

Dr. Araceli Alonso is a 2013 United Nations Award Winner for her activism on women's health and women right. Associate Faculty at the University of Wisconsin-Madison in the Department of Gender and Women's Studies and in the School of Medicine and Public Health, she is the Founder and Director of the award-winning non-profit organization Health by Motorbike

Other books published by Deep Education Press
https://deepeducationpress.org/books.html

- Language Policy or the Politics of Language: Re-imagining the Role of Language in a Neoliberal Society

- Deep Education Across the Disciplines and Beyond: A 21st Century Transdisciplinary Breakthrough

- The Deep Approach to Teaching and Learning World Languages and Cultures: Research on Turkish

- Policy for Peace: Language Education Unlimited

- Science Teachers Who Draw: The Red Is Always There

- Educational Imperialism: Schooling and Indigenous Identity in Borikén, Puerto Rico

- Help Them Learn a Language Deeply Deep Approach to World Languages and Cultures

- Global Language Policies and Local Educational Practices and Cultures

- My Cannibalized Self: An Autoethnography - Biliteracy Development in Japanese Heritage Language Study

- From Transnational Language Policy Transfer To Local Appropriation: The Case of the National Bilingual Program in Medellín, Colombia

- Traditional Potters: From the Andes to Vietnam

- Performing the Art of Language Learning: Deepening the Learning Experience through Theatre and Drama

- Transfer of Learning and the Cultural Matrix: Culture, Beliefs and Learning in Thailand Higher Education

- Family Child Care Relationship-Based Pedagogy: Provider Perspectives on Regulation, Education, and Quality Rating

- Formación y desarrollo de profesionales de la Educación: Un enfoque profundo

Guide for Authors

What our Publishing Team can offer:

➢ An international editorial team, in more than 30 universities around the world.

➢ Dedicated and experienced topic editors who will review and provide feedback on your initial proposal.

➢ A specific format that will speed up the production of your book and its publication.

➢ Higher royalties than most publishers and a discount on batch orders.

➢ Global distribution through Amazon and Barnes & Noble in the U.S., UK, Australia, Europe, Russia, China, South Korea, and many other countries with Expresso Book Machines, printed in minutes on site for in-store pickup.

➢ Fair recognition of your work in your area of specialization.

➢ Quality design. Using the latest technology, our books are produced efficiently, quickly and attractively.

➢ Dissemination through Deep Education campuses.

➢ Book Series: Deep Education; Deep Language Learning; Signs & Symbols in Education; Language Education Policy; Deep Professional Development; Inclusive Education; Deep Early Childhood Education; Deep Activism.

Contact: publisher@deepeducationpress.org

Deep Education Press
https://deepeducationpress.org/index.html

Correspondence for this volume:

Isabelle C. Druc

icdruc@wisc.edu

idruc@mhtc.net

www.ingramcontent.com/pod-product-compliance
Lightning Source LLC
Chambersburg PA
CBHW041016280326
41926CB00094B/4656